MICROBIAL BIOMASS

A Paradigm Shift in Terrestrial Biogeochemistry

MICROBIAL BIOMASS

A Paradigm Shift in Terrestrial Biogeochemistry

editor

Kevin R Tate

Landcare Research, New Zealand

World Scientific

NEW JERSEY • LONDON • SINGAPORE • BEIJING • SHANGHAI • HONG KONG • TAIPEI • CHENNAI • TOKYO

Published by

World Scientific Publishing Europe Ltd.

57 Shelton Street, Covent Garden, London WC2H 9HE

Head office: 5 Toh Tuck Link, Singapore 596224

USA office: 27 Warren Street, Suite 401-402, Hackensack, NJ 07601

Library of Congress Cataloging-in-Publication Data
Names: Tate, Kevin R., author.
Title: Microbial biomass : a paradigm shift in terrestrial biogeochemistry /
Kevin R. Tate (Landcare Research, New Zealand).
Description: New Jersey : World Scientific, 2017.
Identifiers: LCCN 2016034585 | ISBN 9781786341303 (hc : alk. paper)
Subjects: LCSH: Soil microbiology. | Soil microbial ecology.
Classification: LCC QR111 . T25 2017 | DDC 579/.1757--dc23
LC record available at https://lccn.loc.gov/2016034585

British Library Cataloguing-in-Publication Data
A catalogue record for this book is available from the British Library.

Desk Editors: Anthony Alexander/Mary Simpson

Typeset by Stallion Press
Email: enquiries@stallionpress.com

Printed in Singapore

Preface

The year 2016 marks 50 years since Professor David Jenkinson, while working at Rothamsted Research, UK, published the first in a series of papers describing the soil microbial biomass concept and methods for its measurement. Much of the work described in these seminal papers was based on David Powlson's Ph.D. thesis, so I was delighted when he agreed to be the lead author for the first chapter of this book, ably assisted by two co-authors including another of David Jenkinson's protégés, Phil Brookes. This first chapter provides an historical perspective of the soil microbial biomass story and includes a description of David Jenkinson's thinking as he and his colleagues began developing methods for measuring the soil microbial biomass and the major nutrients it contains. Some amusing anecdotes are included in the chapter that show that serious soil research can also have its lighter side. The revolution sparked by this early work has enhanced our understanding of the unifying role the soil microbial biomass plays in terrestrial ecosystem functioning that is often now expressed in terms of ecosystem services, that is, those services that contribute to human well-being. At the heart of this functioning is the mysterious soil microbial world that, before the publication of these first papers, was largely the domain of the soil microbiologist using traditional plate counting and microscopy to study soil microorganisms. The publication of these papers also resulted in an explosion of scientific papers and journals encompassing soil biodiversity, microbial ecology and terrestrial biogeochemistry, aspects that are all featured in the chapters of this book. The earlier impact of this revolution in soil science was reflected in citations in over 200 papers by 1987, and later, as new methods were developed for measuring the microbial biomass and the nutrients it contained, in the Citation Classic (Jenkinson *et al.*, 2004). Incorporation of the microbial biomass as a measureable pool in the simple but robust RothC model continues to be used to simulate and predict changes in soil organic matter in natural and managed agro-ecosystems. David Jenkinson used an early version of this model to make quantitative predictions of future carbon dioxide emissions from soil in response to global warming. This was another first for David as it was published in 1991 in the journal *Nature* about the time world leaders met for the first global climate change meeting in Rio de Janeiro. About 20 years later, and not long before he died, David's concern about climate change prompted him to draft a comprehensive article on this critical issue. Climate change is mentioned in several chapters and features in the final chapter of this book where extreme warming of the Arctic is shown to pose particular challenges to our understanding of the responses of microbial communities.

Some earlier chapters review progress made in our understanding of how the soil microbial biomass unifies and regulates the functioning of terrestrial ecosystems, including the cycling of carbon and nutrients. Other chapters explore the role of the soil microbial biomass in the sustainable management of agro-ecosystems and wetlands (including paddy rice growing). One area of the soil microbial world where experimental aspects exceed existing theory is in explaining its huge biodiversity. Chapter 5 discusses the need for new theoretical models to explain the causes and consequences of soil biodiversity on the structure and functioning of terrestrial ecosystems and their responses to global change.

Two chapters show how some of the mystery of the soil microbial world is being lifted by new tools, including molecular and dynamic isotopic methods. These approaches are revealing new information about the soil's rich functional diversity and the distinct traits of different subgroups and their ecology. Such information has posed new challenges, including how to relate the genomic potentials of specific soil microbial groups to their functional traits and activities. Here there is still an important role for classic culturing techniques. The new insights into the ecology and diversity of guilds within the microbial biomass, when combined with rates of uptake and turnover of different biochemical pools provided by dynamic stable isotope probing, have the potential to enhance the development of new microbially explicit models to improve our understanding of earth system processes. While much still needs to done, these new models are beginning to take us beyond the simpler but still useful models like RothC, in which the microbial biomass was first given expression as the central functional element.

So, publication of this book was made possible with the generous cooperation of a group of the world's leading soil scientists and academics who, like me, are keen to be part of this tribute to David Jenkinson. Every generation produces some outstanding creative thinkers whose influence reaches well beyond their lifetimes and native shores. David Jenkinson is undoubtedly one of these, and his memory will continue to live on in the chapters of this book. I had the privilege of working with him in 1980 and experienced at first hand his intellectual leadership and strong work ethics. David was also a warm friendly man with a strong commitment to helping farmers through his research. This book is our tribute to his intellect and foresights in starting a paradigm shift in biogeochemistry that is leading us to a better understanding of how the soil microbial biomass regulates earth system processes.

This book has been ambitious in concept in that it endeavours to demonstrate how our understanding of soil microbial processes in terrestrial ecosystem functioning has developed over the last 50 years since David Jenkinson first introduced the microbial biomass concept to the world. This enhanced understanding is now crucially important as we grapple with how to feed a burgeoning world population in a warming world. We hope you enjoy reading all the 10 chapters in this book and that you will gain some valuable new insights and pointers to where future research should be directed.

Kevin Tate

Reference

Jenkinson, D. S., Brookes, P. C. and Powlson, D. S. (2004). Measuring soil microbial biomass. *Soil Biol. Biochem.* 36, 5–7.

About the Author

I am a semi-retired scientist whose career began in 1966 with teaching chemistry at Victoria University of Wellington, before becoming a soil scientist in 1968 at the (then) DSIR Soil Bureau in Lower Hutt, New Zealand. Working with long-time mentor/friend and colleague (the late) Dr. Des Ross, we made some of the first measurements of soil microbial biomass outside the UK and then used these measurements to assess the sustainability of various land use practices. Later, I was privileged to work with (the late) Prof. David Jenkinson at Rothamsted Research (UK) and his research group on the microbial biomass in 1980–1981; this collaboration continues with this book, as Chapter 1 is authored by two of Prof Jenkinson's protégés, David Powlson and Phil Brookes. Since 1970, climate change has been an enduring theme through my career, leading to the award of the prestigious Marsden Medal in 2005. In 2011, I was also awarded the inaugural LC Grange Medal for "outstanding contribution to New Zealand soil science". I am a Fellow of the Royal Society of New Zealand, New Zealand Soil Science Society and New Zealand Institute of Chemistry. I continue to work on developing soil-based novel mitigation technologies for reducing methane emissions from dairy farms and, when asked, give talks to community groups on global change.

For relaxation, I enjoy cycling, listening to music and, with my wife Heather, walking, gardening, entertaining visitors and spending time with our children and grandchildren.

List of Contributors

Steve Allison
Department of Ecology and Evolutionary Biology
Department of Earth System Science
University of California, Irvine
USA

Richard D. Bardgett
Faculty of Life Sciences
Michael Smith Building
The University of Manchester
United Kingdom

Philip C. Brookes
Department of Sustainable Soils and Grassland Systems
Rothamsted Research
Harpenden
Herts AL5 2JQ
United Kingdom

Paul L. E. Bodelier
Netherlands Institute of Ecology
Department of Microbial Ecology
Droevendaalsesteeg 10
Wageningen
The Netherlands

Daan Blok
Center for Permafrost
Department of Geosciences and Natural Resource Management
University of Copenhagen
Denmark

Ian D. Bull
Organic Geochemistry Unit
School of Chemistry
University of Bristol
Bristol BS8 1TS
United Kingdom

Lilia C. Carvalhais
Plant-Microbe Interactions Laboratory
School of Agriculture and Food Sciences
The University of Queensland
Queensland 4072
Australia

Alice F. Charteris
Organic Geochemistry Unit
School of Chemistry
University of Bristol
Bristol BS8 1TS
United Kingdom

Bo Elberling
Center for Permafrost
Department of Geosciences and Natural Resource Management
University of Copenhagen
Denmark

Franciska T. de Vries
Faculty of Life Sciences
Michael Smith Building
The University of Manchester
United Kingdom

Richard P. Evershed
Organic Geochemistry Unit
School of Chemistry
University of Bristol
Bristol BS8 1TS
United Kingdom

Merian S. Haugwitz
Center for Permafrost
Department of Geosciences and Natural Resource Management
University of Copenhagen
Denmark

Adrian Ho
Netherlands Institute of Ecology
Department of Microbial Ecology
Droevendaalsesteeg 10
Wageningen
The Netherlands

William R. Horwath
Department of Land, Air and Water Resources
University of California, Davis
USA

Kazuyuki Inubushi
Chiba University
Matsudo
Chiba 271-8510
Japan

Timothy D. J. Knowles
Organic Geochemistry Unit
School of Chemistry
University of Bristol
Bristol BS8 1TS
United Kingdom

Hongwei Liu
Plant-Microbe Interactions Laboratory
School of Agriculture and Food Sciences
The University of Queensland
Queensland 4072
Australia

Hooman Mirzaee
Plant-Microbe Interactions Laboratory
School of Agriculture and Food Sciences
The University of Queensland
Queensland 4072
Australia

Richard Moyle
Plant-Microbe Interactions Laboratory
School of Agriculture and Food Sciences
The University of Queensland
Queensland 4072
Australia

Daniel V. Murphy
School of Earth and Environment and Institute of Agriculture
Faculty of Science
The University of Western Australia, WA6009
Australia

Hirohiko Nagano
Chiba University
Matsudo
Chiba 271-8510
Japan

David S. Powlson
Department of Sustainable Soils and Grassland Systems
Rothamsted Research
Harpenden
Herts AL5 2JQ
United Kingdom

Anders Priemé
Center for Permafrost
Department of Geosciences and Natural Resource Management
University of Copenhagen
Denmark

Vivian Rincon-Florez
Plant-Microbe Interactions Laboratory
School of Agriculture and Food Sciences
The University of Queensland
Queensland 4072
Australia

Peer Schenk
Plant-Microbe Interactions Laboratory
School of Agriculture and Food Sciences
The University of Queensland
Queensland 4072
Australia

Marie Spohn
Department of Soil Ecology
University of Bayreuth
95548 Bayreuth
Germany

Anne K. Steenbergh
Netherlands Institute of Ecology
Department of Microbial Ecology
Droevendaalsesteeg 10
Wageningen
The Netherlands

Elizabeth A. Stockdale
School of Agriculture, Food and Rural Development
Newcastle University
Newcastle-upon-Tyne NEI 7RU
United Kingdom

Wim H. van der Putten
Department of Terrestrial Ecology
Netherlands Institute of Ecology
6700AB
Wageningen
The Netherlands, and
Laboratory of Nematology
Wageningen University
6700ES
Wageningen
The Netherlands

Annelies J. Veraart
Netherlands Institute of Ecology
Department of Microbial Ecology
Droevendaalsesteeg 10
Wageningen
The Netherlands

Jana Voříšková
Center for Permafrost
Department of Geosciences and Natural Resource Management
University of Copenhagen
Denmark

Jianming Xu
College of Environmental and Resource Science
Institute of Soil and Water Resources and Environmental Sciences
Zhejiang University
Hangzhou 310058
China

Contents

Preface		v
About the Author		ix
List of Contributors		xi
Chapter 1	Through the Eye of the Needle — The Story of the Soil Microbial Biomass *David Powlson, Jianming Xu and Philip Brookes*	1
Chapter 2	The Role of the Soil Microbial Biomass in Cycling Nutrients *William R. Horwath*	41
Chapter 3	Managing Soil Microbial Biomass for Sustainable Agro-Ecosystems *Elizabeth A. Stockdale and Daniel V. Murphy*	67
Chapter 4	Microbial Biomass and Functions in Paddy Soil *Kazuyuki Inubushi and Hirohiko Nagano*	103
Chapter 5	Soil Biodiversity and Ecosystem Functioning *Richard D. Bardgett, Franciska T. de Vries and Wim H. van der Putten*	119
Chapter 6	Building Predictive Models for Diverse Microbial Communities in Soil *Steven D. Allison*	141

Chapter 7 Dynamic Compound-Specific Stable Isotope Probing of the Soil Microbial Biomass 167
Richard P. Evershed, Alice F. Charteris, Peter J. Maxfield, Ian D. Bull and Timothy D. J. Knowles

Chapter 8 Emerging Culture-Independent Tools to Enhance Our Understanding of Soil Microbial Ecology 207
Hongwei Liu, Hooman Mirzaee, Vivian Rincon-Florez, Richard Moyle, Marie Spohn, Lilia C. Carvalhais and Peer M. Schenk

Chapter 9 Microbial Ecosystem Functions in Wetlands under Disturbance 227
Anne K. Steenbergh, Annelies J. Veraart, Adrian Ho and Paul L. E. Bodelier

Chapter 10 Arctic Soil Microbial Sensitivity to Seasonal Dynamics and Climate Change 275
Anders Priemé, Daan Blok, Merian Skouw Haugwitz, Jana Voříšková and Bo Elberling

Index 309

1

Through the Eye of the Needle — The Story of the Soil Microbial Biomass

David Powlson*,‡, Jianming Xu† and Philip Brookes*,†

**Department of Sustainable Soils and Grassland Systems, Rothamsted Research, Harpenden, Herts., AL5 2JQ, UK*
†College of Environmental and Resource Science, Institute of Soil and Water Resources and Environmental Sciences, Zhejiang University, Hangzhou 310058, China
‡david.powlson@rothamsted.ac.uk

Birth of the Microbial Biomass Concept

In 1966, David Jenkinson at Rothamsted Research, UK, published a paper that led to a new paradigm in soil science, despite its innocuous sounding title "Studies on the decomposition of plant material in soil. II. Partial sterilisation of soil and the soil microbial biomass." We think that this was the first time that the term *soil microbial biomass* appeared in the scientific literature; it refers to the total mass of all organisms in soil. David Jenkinson later described the microbial biomass as the "eye of the needle through which all organic matter entering the soil must pass". Ten years later, the concept was reinforced and a specific method for measuring the quantity of carbon (C) held in the soil microbial biomass was published in a series of papers by David Jenkinson, David Powlson and others that have been very widely cited (Jenkinson, 1976; Jenkinson and Powlson, 1976a, b; Jenkinson *et al.*, 1976; Powlson and Jenkinson, 1976). The method proposed in 1976 became known as the chloroform ($CHCl_3$) fumigation–incubation method, later abbreviated to FI to distinguish it from the $CHCl_3$ fumigation–extraction method, developed later and termed FE (Vance *et al.*, 1987b). Much of the 1976 research resulted from the Ph.D. work of David Powlson carried out under the supervision of David Jenkinson. Later developments including the development of the FE method for measuring biomass C and methods for measuring N and P in the biomass were largely led by Phil Brookes.

For post-millennial soil researchers, those beginning their work after or around the year 2000, it may be difficult to understand the scientific context of soil microbiology in the 1960s. The genetics revolution had barely started — Watson and Crick had published their discovery of the structure of DNA only 13 years before the Jenkinson (1966) paper. Molecular biological concepts and methods that are now routine were unheard of. The concept of measuring the properties of an entire population, such as the quantity of C held in all living organisms in soil, was not common. At that time, the major focus of soil microbiology was the identification of organisms responsible for specific processes. Processes of interest included ammonium oxidation (nitrification), cellulose decomposition and the degradation of individual pesticides.

By far the major tool available for studying soil microorganisms was culturing under laboratory conditions using agar plates containing different nutrients to give some specificity to the species of organisms likely to grow. Population sizes were assessed by an extension of this approach based on counting the number of colonies developing on plates of nutrient agar from increasingly diluted soil extracts; the method was termed "plate counting".

While many soil microbiologists recognised the limitations of this approach for quantitative studies, little else was available. The concentration of nutrients used for plate counts was usually far greater than likely to be present in soil, thus calling into question the relevance of many findings on population sizes and trends. Despite all this, the approaches used at that time were not without success. For example, in 1952, Selman Waksman at Rutgers University, USA, was awarded the Nobel Prize for medicine in recognition of his discovery of antibiotics produced by organisms in soil. At Rothamsted, UK, Norman Walker discovered a new class of soil bacteria responsible for nitrification in soil: *Nitrospira* to be added to the better known *Nitrobacter* and *Nitrosomonas* species.

By the 1960s, it was becoming clear that the majority of bacteria and fungi in soil did not grow under the laboratory conditions normally used — the term "viable but unculturable" began to be used. This recognition arose due to the so-called direct counting of soil microbes when it was found that the number of organisms observed when soils were examined under the microscope was many times greater than numbers based on plate counts (Skinner *et al.*, 1952). One direct counting technique, not relying on the growth of colonies on nutrient agar, was a method developed by Jones and Mollison (1948) and often called the Jones–Mollison agar film technique. In this method, a known mass of soil is dispersed in a known volume of warm liquid agar and small volumes taken and allowed to solidify as thin films in a haemocytometer slide of known depth. Agar is used not as a source of nutrients but simply as a means of converting a liquid dispersion to a solid film that can be observed under a microscope. The solidified agar film is transferred from the haemocytometer to a microscope slide, stained with an appropriate stain that is reasonably specific to living organisms as opposed to dead organic particles, and the number of stained organisms counted under a microscope. We return to this method later in the chapter.

So far as we can tell, 50 years after the event, David Jenkinson did not initially set out to develop a method for measuring the quantity of organic C held in the soil microbial population. He had entered the world of soil science from a background of a Ph.D. in organic chemistry and initially applied the methods of organic analysis to studying the chemical composition of soil organic matter (SOM). But he rapidly concluded that SOM was such a complex mixture that the chemical analytical methods available at the time were incapable of yielding much useful information, so he turned his attention to the dynamics of SOM rather than its chemical structure. He and others realised that freshly formed SOM, derived from recent inputs of plant material, behaved differently to native SOM, much of which exhibited great stability and resistance to microbial decomposition.

Jenkinson was amongst the first to apply isotopic methods to distinguish between recently derived C and native C in soil. He did this by growing ryegrass in an atmosphere labelled with $^{14}CO_2$ and then adding the ^{14}C-labelled plant material to soil under field conditions for up to 4 years. In an earlier paper Jenkinson (1965) had shown that the labelled grass initially decomposed rapidly — about two-thirds of the added C was evolved as CO_2 within 6 months in the temperate climate of Rothamsted, but thereafter decomposition was far slower with about one-fifth of the labelled C remaining in soil after 4 years. This was a strong indication that SOM could be regarded as existing in pools or fractions that differed greatly in their rates of turnover. We consider that this was one powerful strand of reasoning that led Jenkinson to the concept of the microbial biomass as a discrete and meaningful fraction within soil.

Simultaneously, there was interest in the idea of different rates of turnover of SOM in different fractions from a more practical viewpoint, namely the supply of nitrogen (N) from soil to crops. SOM in agricultural soils typically contains several thousand kg ha^{-1} of N in organic forms, yet only a few percent of this is converted into inorganic forms and thus available for plant uptake each year, typically around 20–200 kg N ha^{-1}. During the 1960s, and still today, there is interest in better predicting the N supply from SOM during a given crop-growing season so that fertiliser applications can be adjusted accordingly. Jenkinson was involved in such work around this time and published a method for predicting N mineralisation based on the extractability of specific forms of organic C from soil using dilute barium hydroxide solution (Jenkinson, 1968).

A third strand of reasoning that certainly had a major influence on his thinking was ongoing research at the time by plant pathologists at Rothamsted and worldwide experimenting with soil fumigants to control soil-born plant fungal diseases and parasitic nematodes. The chemicals used were typically broad-spectrum biocides designed to kill virtually all soil organisms; they included formaldehyde, methyl bromide, trichloro(nitro)methane (known as chloropicrin) and other halogenated hydrocarbons. It was frequently observed in such studies that, even in soils with no apparent pest or disease problem, plant growth was stimulated; in many cases this was attributed to additional mineralisation of N compared to unfumigated soil, though other factors were also thought to be involved as discussed by Jenkinson (1966) and Powlson (1975). The source of this additional N was a matter of interest and, as a key aspect of fumigation was killing of organisms, decomposition of the killed soil microbial cells was considered a likely source. It seemed likely that a recolonising population of microorganisms in the soil would utilise the cells of killed organisms as food with the N in organic forms being mineralised and released as inorganic N. This would also account for the period of enhanced soil respiration observed following soil fumigation, measured as enhanced evolution of CO_2 and uptake of oxygen. This line of reasoning appears to have been the breakthrough in Jenkinson's thinking that led to the idea of a way of measuring the quantity of C held in organisms before being killed by the fumigant. It is interesting that this "eureka moment" came through cross-fertilisation between scientists working in different fields — SOM studies and plant pathology. It is a reminder that all scientists can benefit from hearing about ideas in other fields — something about which Jenkinson was particularly enthusiastic.

Establishing the Chloroform Fumigation–Incubation (FI) Method

^{14}C-labelled C in the soil amounted to only 1.8% of the total soil organic C, yet the CO_2-C evolved from untreated soil when incubated in the laboratory for

Table 1.1 Labelled and unlabelled CO_2 evolved by soil exposed to partial or complete sterilisation treatments, inoculated with fresh soil and incubated for 10 days at 25°C.

Treatment	CO_2-C evolved in 10 days (μg C g^{-1} soil)		Percentage of labelled C in evolved CO_2-C
	^{14}C-labelled	Unlabelled	
None	14.7	146	9.2
Air-drying	23.2	195	10.6
Irradiation (0.25 Mrad)	33.8	238	12.4
CH_3Br vapour	47.5	239	16.6
$CHCl_3$ vapour	49.4	259	16.0
Oven-dried (80°C)	54.6	347	13.6
Oven-dried (100°C)	56.0	493	10.2
Autoclaving (120°C)	58.6	524	10.1

Note: The soil had previously been incubated with labelled plant material for 1 year under field conditions.
Source: Adapted from Jenkinson (1966).

10 days at 25°C was more than five times as enriched (Table 1.1). This indicated that C involved in biological processes leading to CO_2 evolution was far more heavily labelled with ^{14}C than the soil organic C as a whole. All of the treatments tested increased total CO_2 evolution but, importantly, several of them greatly increased the percentage of labelled C in the CO_2 evolved. Fumigation with chloroform or methyl bromide (CH_3Br) almost doubled this proportion to around 16% (Table 1.1). Jenkinson reasoned that the C in cells of microorganisms (the soil microbial biomass) was likely to be the most heavily labelled organic C fraction in the soil, so a treatment giving the greatest increase in ^{14}C labelling of evolved CO_2 must be selectively acting on this fraction. Because $CHCl_3$ was easier to use than CH_3Br and was thought to be less toxic to humans, Jenkinson used $CHCl_3$ fumigation for further work. Although oven-drying and autoclaving increased evolution of labelled CO_2 by a similar amount to fumigation, they also increased evolution of unlabelled CO_2 considerably. Jenkinson interpreted this to mean that these treatments, in addition to killing organisms, also rendered additional non-biomass (and more lightly labelled) forms of organic C decomposable and thus were less selective.

An important conclusion from the results with labelled soil was that $CHCl_3$ had very little effect on the rate of decomposition of non-biomass organic C in soil. Following this conclusion, and several other simplifying assumptions, Jenkinson set out the following expression for estimating the quantity of C in soil microbial biomass (B_C) as follows:

$$B_C = \frac{F_C}{k_C}, \tag{1.1}$$

where F_C is the flush of decomposition following chloroform fumigation defined as (CO_2-C evolved by fumigated soil during a 10-day incubation at 25°C) minus (CO_2-C evolved by untreated soil during the same period) and k_C is the fraction of the biomass C decomposed and evolved as CO_2-C during the incubation.

A preliminary value for k_C was derived by growing a ^{14}C-labelled culture of a single species of soil bacteria, *Nitrosomonas europaea*, adding this to soil and measuring the amount of $^{14}CO_2$ evolved after $CHCl_3$ fumigation and incubation under the standard conditions. It was found that 28% of the bacterial C was evolved (Jenkinson, 1966) giving a k_C value of 0.28, but to avoid an appearance of spurious precision, this was rounded to 0.3.

The Jenkinson (1966) paper also contains descriptions of several experiments designed to test other hypotheses that had been put forward to explain the flush of CO_2 evolution and N mineralisation observed after the so-called partial sterilisation treatments. The results strongly argued against the other hypotheses and supported the idea that the C came from decomposition of organisms killed by the treatments. This paper is elegantly written and warrants re-reading today as an example of a clear but closely argued account of quite complex experiments. Ten years later, David Jenkinson and David Powlson revisited the issues. They examined in detail the impacts of $CHCl_3$ fumigation on soil metabolism (Jenkinson and Powlson, 1976a) and measured the flushes of decomposition caused by fumigation, gamma irradiation, air-drying or autoclaving when applied to a range of soils differing in organic matter content, soil type, pH and management (Figure 1.1). These experiments confirmed the earlier conclusions that chloroform fumigation killed soil organisms, rendering them decomposable and giving rise to the flush of decomposition, but had little effect on the decomposability of non-biomass soil organic C. The flush of N mineralisation following fumigation was also taken as strong indication of an impact on soil organisms followed by their decomposition as organisms are rich in N compared with SOM as a whole: they typically have a C/N ratio of around 5 or less compared with 10 or more in total SOM. Powlson and Jenkinson (1976) found that in a range of soils the ratio (CO_2-C evolved)/(N mineralised) after $CHCl_3$ fumigation was typically about 6 compared to a range of about 12–20 in unfumigated soil.

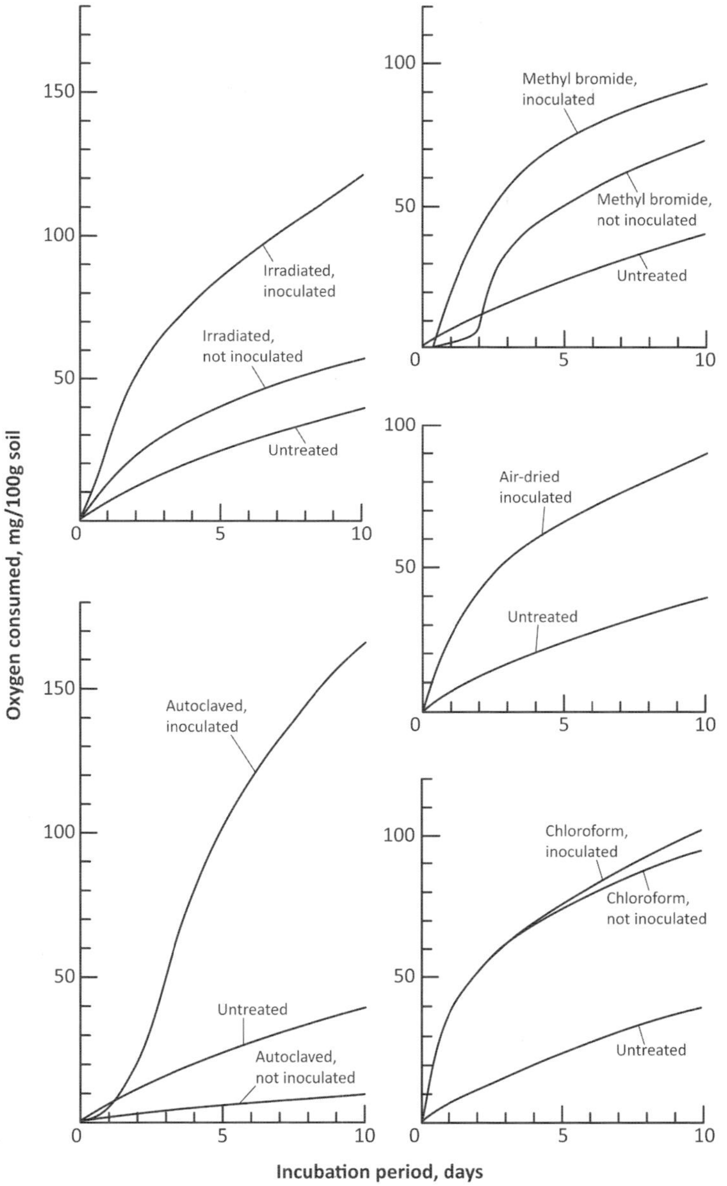

Figure 1.1 Oxygen uptake by soil from the farmyard manure-treated plot of the Broadbalk experiment when incubated for 10 days at 25°C after different partially or completely sterilising treatments, with or without an inoculum of fresh soil. (Adapted from Powlson and Jenkinson, 1976.)

Applying the range of treatments to soils differing in total organic C content demonstrated the complexity of the impacts of the treatments. For example, autoclaving was very effective at killing organisms as shown by the almost complete absence of CO_2 evolution or oxygen uptake if the autoclaved soils were not inoculated with fresh soil (Figure 1.1). It was also clear that this treatment rendered large amounts of non-biomass soil C decomposable in addition to killing organisms. Gamma irradiation at the high dose used, 2.5 Mrad, was thought to be sterilising the soil but considerable CO_2 evolution was observed after treatment, even in uninoculated soil. It was thought that this was either due to the chemical action of free radicals formed during irradiation or to continued activity of enzymes in or released from cells that had been killed in the sense that they were incapable of division but still retained enzymes capable of at least some activity (Jenkinson and Powlson, 1976a). In contrast, air-drying, a treatment to which soils are frequently subjected in the field, had a limited killing effect but had a significant impact on the decomposability of non-biomass organic matter. This is consistent with the observation that the C/N ratio of the organic matter mineralised during the flush following air-drying and rewetting was wider than for the fumigation treatments: generally 8–12, closer to that of SOM as a whole (Powlson and Jenkinson, 1976).

In another of the 1976 series of papers (Jenkinson, 1976), 11 different laboratory-grown organisms were added to soil, fumigated with chloroform and the amount of C evolved as CO_2 measured. Averaged over all the organisms (yeasts, fungi and bacteria), 50% of the added C was evolved giving a value of 0.5 for k_C in Eq. (1.1); this value was recommended in the fifth paper of the series (Jenkinson and Powlson, 1976b) setting out a practical method for measuring microbial biomass C in soil. On the basis of later work, the value of k_C was modified to 0.45 (Jenkinson and Ladd, 1981). In the method proposed in 1976, the CO_2 evolved by fumigated and unfumigated soil is absorbed in sodium hydroxide solution and the quantity determined by titration; in some later developments the quantity of CO_2 evolved is determined by taking gas samples from the incubation vessel and determining CO_2 concentration using gas chromatography. In the detailed studies leading to the final FI method much was learned by monitoring oxygen uptake during incubation, as shown in Figure 1.1, using a manometer connected to the incubation vessel (Jenkinson and Powlson, 1976a; Powlson and Jenkinson, 1976).

In a completely independent test of the FI method, Jenkinson *et al.* (1976) reported, for a set of eight soils, a comparison of microbial biomass C as estimated by this method and as derived from direct microscopy. A modification of the Jones and Mollison (1948) direct counting method was used. Organisms in the agar films were stained using phenolic aniline blue, as originally used by Jones and Mollison: several fluorescent stains were tested but proved unsatisfactory because they did not stain such a wide range of organisms as aniline blue. The main modification to the Jones–Mollison method was to allocate stained organisms to size classes rather than simply count total numbers. Organisms classed as roughly spherical, such as bacteria, were divided into 13 size classes with diameters ranging from 0.3 to 19 μm, and the numbers in each size class counted. Hyphae, regarded as cylinders, were divided into seven diameter classes ranging from 1 to 11 μm, and the lengths in each class measured. In this way a total biovolume for each soil was derived and a biomass C value calculated by making assumptions about dry matter content and C concentration of organisms. This work was incredibly tedious and time-consuming and also subject to a fair degree of subjectivity; three people took part in the work and considerable efforts were necessary to ensure that their results were consistent with each other due to the difficulty of being certain of whether some blue marks were truly stained organisms or artefacts. For seven of the eight soils studied there was a close relationship between biovolume and biomass C as measured by FI. Quite remarkably there was also a close correspondence between actual values for biomass C derived from the two methods (Jenkinson *et al.*, 1976). This numerical correspondence was not taken too seriously in view of assumptions made in converting biovolume to biomass C but the close correlation between the two totally different approaches was taken as strong independent evidence for the validity of the FI method. The conclusion was later strengthened by Vance *et al.* (1987b) using a wider range of soils.

The soil that was an exception to the relationship was acidic (pH 3.9): direct counting of organisms showed a large population but the flush of decomposition following chloroform fumigation, and other treatments, showed a long lag phase and a very small increase in CO_2 evolution compared to untreated soil as shown in Figure 1.2 (Powlson and Jenkinson, 1976). Without fully understanding the reasons, it was concluded that the FI method was not applicable to acid soils (Jenkinson *et al.*, 1976; Jenkinson and Powlson, 1976b). Unfortunately, in subsequent years, it was not uncommon for scientists to ignore this and apply the method to acid soils — then complain that the method did not work!

The FI method immediately attracted controversy. Shields *et al.* (1974) claimed that the value obtained for biomass C was erroneously large, their argument being based on experiments in which ^{14}C-labelled glucose was incubated with soil for 14 days and the flush of ^{14}C-labelled CO_2 after $CHCl_3$ fumigation determined. From their data, they concluded that the calculated size of the newly formed ^{14}C-labelled biomass was unrealistically large, meaning that $CHCl_3$ was

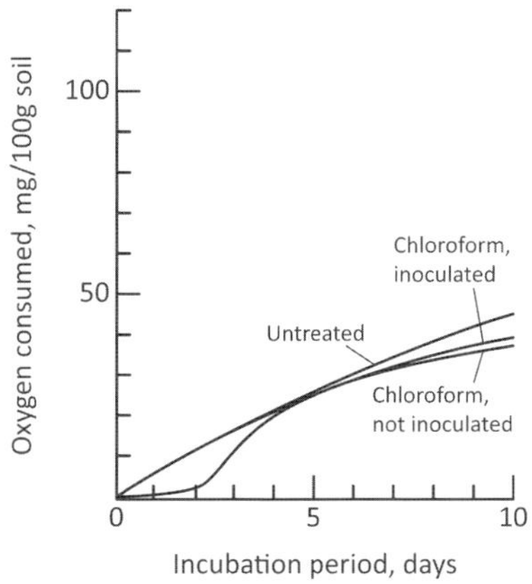

Figure 1.2 Oxygen uptake by an acid soil (pH 3.9) following fumigation with $CHCl_3$ and incubation for 10 days at 25°C with and without an inoculum of fresh soil. (Adapted from Powlson and Jenkinson, 1976.)

rendering non-biomass C fractions decomposable. Jenkinson and Powlson (1976a) conducted several experiments designed to test whether this was true, in part repeating some of the experiments conducted by Shields *et al.* (1974) but also using combinations of $CHCl_3$ fumigation and gamma irradiation. Because of the complexities of the effects of the treatments, the general complexity of SOM and our continuing poor understanding of many aspects of its dynamics, it proved difficult to obtain entirely unambiguous evidence. Similar difficulties arose in arguments over the use of a control in calculating biomass C, as discussed later. However, a crucial issue was that Shields *et al.* (1974) used a k_C value of 0.3, the preliminary value proposed by Jenkinson (1966). In refuting the arguments of Shields *et al.* (1974), Jenkinson and Powlson (1976a) pointed out that using the larger k_C value of 0.5 (Jenkinson, 1976), later revised to 0.45, gave an entirely realistic value for biomass C.

The FI Method: Benefits, Drawbacks and Conundrums Raised

A great advantage of the FI method as proposed and thoroughly tested by Jenkinson (1966) and Jenkinson and Powlson (1976b) was that it made soil microbiology accessible to a far wider range of soil researchers than was previously the case. In fact, Jenkinson once said "now everyone can be a soil microbiologist within 24 hours". Perhaps of broader significance is that the methodology opened a new way of thinking about soil processes that linked chemistry and elemental cycling to the biology driving some of the key processes in soil. In this way, the development of the FI method, and subsequent developments discussed later in this chapter, can truly be regarded as a paradigm shift in soil science. It might be regarded as a bridge between classical soil microbiology, with an emphasis on identifying organisms responsible for specific processes, and modern concepts of biogeochemical cycling. However, the FI method did have some practical disadvantages. Firstly, the method required a 10-day aerobic incubation, often preceded by a pre-incubation to allow effects of disruption due to sampling, sieving and handling of soil to subside — necessary if the CO_2 evolution by the control unfumigated soil is to be truly representative of "untreated" soil. Secondly, it could not be used in acid soils or those containing actively decomposing substrates. Nevertheless, it soon became accepted as a standard method and transformed our understanding of SOM dynamics.

A drawback was the need to purify the $CHCl_3$ to be used for soil fumigation. For many years, $CHCl_3$ purchased from chemical suppliers contained 1% ethanol as a stabiliser to prevent the formation of free radicals. It is essential that all traces of ethanol are removed prior to fumigation otherwise some is retained by the soil and mineralised to CO_2 during the 10-day incubation, giving erroneously inflated biomass C measurements. For many years we used a tedious acid pretreatment of the ethanol-containing $CHCl_3$ followed by washing, distillation and storage over anhydrous Na_2SO_4 at 4°C. However, on storage, even at 4°C, a fraction of the distilled $CHCl_3$ rapidly decomposed to phosgene gas. Not only does this gas have a most unpleasant smell, but it is also extremely toxic, accounting for its use as a chemical weapon in World War I. Fortunately, this tedious purification procedure is no longer necessary as ethanol-free chloroform can be readily purchased containing traces of an alternative stabiliser, usually amylene.

In both the FI and FE methods (described later), it is essential that ethanol-free $CHCl_3$ is used or the ethanol will give inflated biomass measurements: in the FI method $CHCl_3$ will be mineralised to CO_2 and in FE it will be measured as organic C. Table 1.2 illustrates the magnitude of error introduced if $CHCl_3$ containing ethanol is used.

Controversy over the Control in the FI Method

A remarkable feature of the FI method is that, after fumigation and the dissipation of the initial flush of CO_2 due to the mineralisation of the fumigation-killed

Table 1.2 Biomass C ($\mu g\ g^{-1}$ soil) measured following fumigation without and with alcohol-free $CHCl_3$.

Soil number	Pure chloroform	Chloroform + ethanol
1	476 ± 11	884 ± 21
2	1238 ± 38	1658 ± 28
3	291 ± 8	502 ± 10
4	433 ± 16	838 ± 48
5	372 ± 20	677 ± 24

biomass, the rates of evolution of CO_2 in the fumigated and non-fumigated soils are approximately equal for periods of weeks or even months (e.g. Powlson and Jenkinson, 1976a; Kemmitt *et al.*, 2008). The implications of this were largely overlooked during the method development and for a long time after. It implies that the small recolonising microbial population that develops in the fumigated soil after fumigation (initially ca. 10% of that in the corresponding non-fumigated soil) and the large, intact, population in the non-fumigated soil are both mineralising recalcitrant, humified SOM at the same rate. This does seem an unlikely occurrence. Surely the larger population would be expected to mineralise SOM faster than the fumigant-damaged one? Reber (1967) reported that the recolonising population, as well as being smaller, was mainly composed of bacteria, particularly *Pseudomonas* species, which largely utilise labile substrates. Most of the fungal population was eliminated and remained low for many weeks. Other work, using modern molecular methods (Dominguez-Mendoza *et al.*, 2014; Chen *et al.*, 2015), also showed that there are distinctly different microbial community structures in fumigated and non-fumigated soils.

Empirically, it therefore seems highly unlikely that this small recolonising population in the fumigated soil, having limited diversity, could possibly mineralise the large pool of humified and recalcitrant SOM to the same extent as the large, structurally diverse native population. Yet, this is precisely what was proposed in the 1976 papers.

The rates of CO_2-C evolution from the fumigated and non-fumigated soils have repeatedly been shown to be remarkably similar after 4–5 days of aerobic incubation, i.e. once the flush of decomposition of the fumigant-killed biomass is over (e.g. Powlson and Jenkinson, 1976; Wu *et al.*, 1996; Kemmitt *et al.*, 2008). Similarly, Jenkinson (1966), using soil previously incubated with ^{14}C-labelled plant material for several years, found that the ^{14}C enrichments of the CO_2 evolved by fumigated and unfumigated soils were approximately equal after the initial

$CHCl_3$-induced flush of decomposition had subsided. This indicated that the same pool of non-biomass SOM was being mineralised in both cases. However, Voroney and Paul (1984) suggested that the two different microbial populations were incapable of the same rates of SOM mineralisation. They considered that the small recolonising population was incapable of significant organic matter mineralisation so that biomass C (B'_C) was best calculated from

$$B'_C = \frac{F'_C}{k'_C}, \tag{1.2}$$

where B'_C and k'_C are as in Eq. (1.1) but F'_c = CO_2-C evolved from $CHCl_3$-fumigated soil in 0–10 days, i.e. CO_2-C evolved by the unfumigated control soil is not subtracted to derive the value for the flush of decomposition because the quantity of CO_2 derived from non-biomass SOM during the incubation is taken as zero.

Clearly, this is much more than a philosophical debating point. Both Eqs. (1.1) and (1.2) cannot be correct. If a control is not subtracted (Eq. (1.2)), the calculated biomass will be considerably larger than if it is (Eq. (1.1)).

The original data from Jenkinson and Powlson (1976b) is given in Table 1.3. It clearly shows a very consistent increase in calculated biomass C of around 30% if a control is not used. It should be noted that they used the control value obtained over 10–20 days, to overcome initial increases in respiration caused by sampling or plant residues. Nowadays, the standard procedure is to give the soil a conditioning incubation at 25°C and 50% water-holding capacity over soda lime for 5–7 days to trap CO_2 and to permit respiration and microbial activity to stabilise before analysis. With the advent of the FE method described later this procedure is theoretically unnecessary. However, it is still usually adopted by us to minimise possible artefacts arising from soil sampling and disturbance. The approach of Voroney and Paul (1984) would seem to have advantages in that the measurements of biomass without the need to measure a control would be much simplified. In fairness, they argued against this. However, if their approach is used, biomass measurements would be much larger than if the original approach of Jenkinson and Powlson (1976b) is adopted. Indeed, the difference is often much larger than those in Table 1.3, sometimes as much as twice as large (e.g. Schnürer *et al.*, 1985). The modification of Voroney and Paul (1984) was, unfortunately, adopted by many workers, particularly in North America. While some (e.g. Schnürer *et al.*, 1985) wisely published actual CO_2 measurements and biomass C values calculated using both approaches, others did not.

Table 1.3 Biomass C ($\mu g\ g^{-1}$ soil) determined by the FI method calculated with and without a 10–20-day control.

Soil number	Control subtracted (BC1)	Control not subtracted (BC2)	Ratio of BC1/BC2
1	227	356	0.63
2	184	287	0.64
3	524	773	0.68
4	431	637	0.68
5	246	347	0.71
6	242	351	0.71
7	1178	1733	0.68
8	1308	1920	0.69
9	1033	1555	0.66

Source: Data modified from Jenkinson and Powlson (1976b).

After considering all the evidence, we are able to state, with a high degree of certainty, that the original method of Jenkinson and Powlson (1976b) provides the best estimate of biomass C by FI and that of Voroney and Paul (1984) is not valid. Our evidence is as follows:

(1) As stated above, Jenkinson *et al.* (1976) found that there was a remarkably close relationship between biomass C measured by direct microscopy and FI in seven forest, arable and grassland soils having near-neutral pHs. Only in a single acidic soil did this not hold. Vance *et al.* (1987a) found similar results.
(2) The finding that fumigated soil evolves CO_2 at roughly the same rate over long periods as the comparable non-fumigated soil cannot be accounted for by the Voroney and Paul (1984) method. Using their approach means that the magnitude of biomass C is determined by the length of incubation, increasing indefinitely — an improbable event. Surely the respiration rate in the fumigated soil should approach zero after the fumigant-killed biomass has been mineralised? In fact, the observed result is that it returns to almost the same rate as in the non-fumigated soil after 4–5 days. This is very much in line with what would be expected if mineralisation of native SOM proceeded at the same rate in both fumigated and unfumigated soils.
(3) The finding that similar rates of lightly labelled $^{14}CO_2$ were evolved from a long-term ^{14}C-labelled soil with and without fumigation, after the initial flush

was over at 4–5 days (Jenkinson, 1966), strongly suggests that the same pool of pre-existing non-biomass SOM was being mineralised at the same rate in both soils.

(4) Wu *et al.* (1990) heavily labelled a soil with ^{14}C-labelled glucose, so producing a heavily labelled biomass. Following fumigation and incubation, after the initial flush had subsided, fumigated and unfumigated soils both evolved $^{14}CO_2$ at the same rate for many weeks. This should not have occurred according to Voroney and Paul (1984).

(5) The FE method will be discussed in more detail later. It is based on extraction and chemical analysis of biomass components (C, N, P, etc.), unlike the FI method which is based on microbial activity. In the FE method the control is not controversial and biomass C is simply measured from the difference between C extracted from a fumigated minus unfumigated soil. Therefore, apart from the fumigation step, it is a fundamentally different method to FI. Wu *et al.* (1990), using nine soils containing between 0.9 and 4.2% C, reported a linear 1:1 relationship ($y = 0.99x$; $r = 0.99$) between biomass C measured by FI and FE. Clearly, these practically identical biomass measurements, using two very different approaches, could not have been measured from the approach of Voroney and Paul (1984).

(6) Wu *et al.* (1996) incubated a grassland soil with ^{14}C glucose for 27 days and then fumigated it with $CHCl_3$. Biomass C was measured by both FI and FE. Both methods gave extremely similar results. The rates of ^{14}C-labelled C and unlabelled CO_2-C evolved were each near-identical and the CO_2-C evolution curves were approximately parallel. The specific activities of the CO_2-C evolved from both the fumigated and non-fumigated soils were extremely similar after the flush. The specific activity of F_C (measured in the FI method) and E_C (measured by FE) were not significantly different.

These results all concur in suggesting that the original proposal of Jenkinson and Powlson (1976b) is approximately correct and the use of a non-fumigated soil as a control incurs a trivial error in biomass C calculations compared to the approach advocated by Voroney and Paul (1984).

Microbial Biomass C Measurements by the Chloroform Fumigation–Extraction (FE) Method

Following $CHCl_3$ fumigation there is an increase in the C extractable with aqueous solutions from the fumigated compared to the non-fumigated soil. This was first described by Jenkinson and Powlson (1976a) and Powlson and Jenkinson (1976) but, quite remarkably, its significance was not recognised at that time. Probably they were preoccupied with the issues raised in developing the FI method. Brookes *et al.* (1982) published a method for measuring the quantity of phosphorus (P) in the soil microbial biomass based on soil fumigation followed by extraction with 0.5 M $NaHCO_3$. This was the first FE method and stimulated thinking that led towards an extraction method for biomass C.

In 1985 an American Ph.D. student, Eric Vance, joined David Jenkinson's group as a Fulbright Scholar. His aim was to investigate, with Phil Brookes, the use of the FI method in acidic forest soils. Hitherto, the FI method had proved to be invalid in low-pH soils. Eric showed that the reason for the failure of FI under such conditions was that mineralisation of non-biomass SOM was largely inhibited in fumigated acidic soils (Vance *et al.*, 1987a); so it transpired that, in the special case of strongly acidic soils, the contention of Voroney and Paul (1984) regarding the invalidity of the control, as discussed above, was correct. Eric Vance also showed that the extra C released by $CHCl_3$ was closely related to soil adenosine 5′ triphosphate (ATP) content and so could be used as an estimate of biomass C (Vance *et al.*, 1987b). This gave rise to the FE method for biomass C, and the paper describing the method (Vance *et al.*, 1987c) remains the most widely cited paper in the journal *Soil Biology & Biochemistry*. Indeed, some of the original data from Powlson and Jenkinson (1976) was used to calibrate the FE method in Vance *et al.* (1987c). So,

$$B_C = 2.22E_C, \qquad (1.3)$$

where E_C = (organic C extracted after $CHCl_3$ fumigation) minus (organic C extracted from non-fumigated soil).

The factor 2.22 represents the proportion of biomass C extracted by 0.5 M K_2SO_4 solution after $CHCl_3$ fumigation and is analogous to k_C in the FI method; it is based on 45% of the C in the fumigated biomass being extracted after $CHCl_3$ fumigation. The same value is used for biomass N, as discussed later. The factor was not directly derived experimentally. Rather it was obtained from regressions between E_C and F_C (Vance *et al.*, 1987c) and, for biomass N, between F_N and E_N (Brookes *et al.*, 1987). This is why the control issue remains of key importance. If the use of a control in the FI method is wrong, then it follows that the factor 2.22 (equivalent to K_{eC} and K_{eN}) is also wrong and the estimation of biomass C and N by the FE method is flawed. The organic C in the extraction solution (normally 0.5 M K_2SO_4) was originally measured by digestion with potassium dichromate but later a far more convenient method was developed using an automated C analyser (Wu *et al.*, 1990).

The FE method overcame many of the limitations of FI and has now superseded it. Measurements can be made immediately after fumigation rather than having to wait for 10 days as with FI. It may be used in acid soils (Vance *et al.*, 1987a), soils containing actively decomposing substrates under laboratory and field conditions (Ocio and Brookes, 1990; Ocio *et al.*, 1991) and in waterlogged soils including those used for growing paddy rice (Inubushi *et al.*, 1991; Qiu *et al.*, 2015). Neither method can be used in dry soils, at least only after the soil has received a conditioning incubation for 5–7 days, to allow the C released on air-drying to be mineralised (De Nobili *et al.*, 2006). The FE method is ideal for ^{14}C tracer studies involving microbial C dynamics, as the ^{14}C in the soil extracts can be measured directly by scintillation counting.

Biomass N

In the early work developing the original FI method, the flush of N mineralisation following $CHCl_3$ fumigation and incubation was measured and ascribed to N being released from killed biomass. Thus, the flush of N in the FI method was implicitly regarded as an indicator of the quantity of N held in the microbial biomass but it was not until considerably later that an explicit method for measuring biomass N was developed. This only occurred as interest in the FE method for biomass C (Vance *et al.*, 1987c) and P (Brookes *et al.*, 1985) progressed. Biomass N can be measured in the same extracts as biomass C, usually following Kjeldahl digestion, a rather tedious approach. Biomass N is calculated exactly as for biomass C, from

$$B_N = 2.22E_N, \tag{1.4}$$

where E_N = (organic N extracted after $CHCl_3$ fumigation) minus (organic N extracted from unfumigated soil).

Biomass N measurements derived in this way are very closely correlated with biomass C measured by FI and FE (Brookes *et al.*, 1985). Again, biomass N measured by FE overcomes the limitations of biomass C, as discussed above. FE measurements using ^{15}N are quite easily made. The main problem with biomass N is measuring very small concentrations of N in the extraction solutions.

Amato and Ladd (1988) introduced the ninhydrin-N method to measure biomass N. The ninhydrin reagent reacts with ammonium ions and the α-amino N groups in amino acids and proteins to produce an intense purple colour which can easily be measured spectrophotometrically. Ninhydrin-N and biomass N (B_N) measurements are strongly correlated and biomass N can be fairly accurately calculated from

$$B_N = 5.1E_{Nin}, \tag{1.5}$$

where E_{Nin} is (ninhydrin-N extracted from fumigated soil) minus (ninhydrin-N extracted from unfumigated soil).

The ninhydrin method has all the advantages of the other FE methods over FI and is rapid and accurate. It largely superseded the method for biomass N described above that ended with the rather cumbersome measurement of organic N in the extracting solution following FE. However, in studies where ^{15}N labelling of N in the biomass is to be measured, it is necessary to follow the original procedure or something similar as it is essential to measure ^{15}N enrichment in the entire biomass: Ninhydrin-N only represents a small sample of total biomass N.

Biomass P and S

Brookes *et al.* (1982) published a method for measuring biomass P in which the increase in P extractable by 0.5 M $NaHCO_3$ following $CHCl_3$ fumigation is measured. This was the first example of an FE method and laid the foundations for the FE methods for biomass C and N. In the biomass P method a correction is made for released P that is fixed on soil surfaces. Several analogous methods for measuring biomass S have been developed, based on extraction with various reagents, following 24 h $CHCl_3$ fumigation. The extraction reagents include10 mM $CaCl_2$, 100 mM $NaHCO_3$, 16 mM NaH_2PO_4 (e.g. Saggar *et al.*, 1981; Chapman, 1987; Wu *et al.*, 1994).

Why is SOM so Stable? The Regulatory Gate Hypothesis

The FI method was more than a new approach to measure the soil microbial biomass. It also gave us a novel insight into the mechanisms controlling mineralisation of SOM. The global stock of soil organic carbon (SOC) is around 1500 Pg C, twice that in the atmosphere and about three times the amount in global vegetation (Lal, 2004). This C enters soil by photosynthesis and returns to the atmosphere by mineralisation. This exchange is about 10 times that entering the atmosphere from fossil fuels. Direct measurement of soil C stocks is an insensitive means of detecting changes, while the soil processes governing mineralisation and stabilisation of C are poorly understood. The mineralisation of SOC is a fundamental process upon which all life on the planet ultimately depends. It is therefore vital that we have a correct understanding of the key processes involved if we are to better manage this huge C pool. Furthermore, it is remarkable that no one has yet satisfactorily explained the paradox of SOC mineralisation, as outlined above. The paradox is that even if 90% of the soil microorganisms are destroyed by $CHCl_3$ fumigation, following fumigant removal, SOC mineralisation continues at the same rate as in the undamaged soil once the initial decomposition flush is over, usually within about 4 days (Jenkinson and Powlson, 1976a; Wu *et al.*, 1996). This phenomenon has been known for nearly 60 years but has never been satisfactorily explained, even though $CHCl_3$ fumigation forms the basis for all the widely used methods of measuring soil microbial biomass, as discussed above. It is important to note that this mechanism only operates in soils above pH 5.5. Below this pH the C dynamics after fumigation are totally different, decomposition being inhibited by increased Al availability (Vance *et al.*, 1987a).

Humified SOC is thought to comprise, at least in large part, random condensation products of microbial metabolites and has no clearly defined chemical structure. It is highly resistant to microbial degradation, so that it is only mineralised at a very slow rate by the soil microbial biomass. It therefore seems incredible that when the soil microbial biomass is reduced to 10% of its original size by $CHCl_3$ fumigation, it can mineralise the same SOC pool at the same rate as the large undamaged population for many months without any fresh substrate inputs (e.g. Joergensen *et al.*, 1990). It has recently been proposed that the description of SOC given above is incorrect and that it is largely composed of smaller molecules given protection from decomposition through association with minerals and metal oxides (Lehmann and Kleber, 2015) but this remains controversial. However, even if the recently proposed model of SOC is correct, the paradox remains of a small recovery of biomass following fumigation achieving the same rate of SOC decomposition as the far larger population in the unfumigated soil.

It is clear that only an extremely small microbial population survives $CHCl_3$ fumigation as shown by the negligibly small concentration of ATP in fumigated soil before $CHCl_3$ removal (Kemmitt *et al.*, 2008). The population developing during a subsequent incubation must therefore develop from the very small number of surviving organisms, or from a small added inoculum. It is theoretically possible that the community structure of the small recolonising microbial population is the same as that in the intact non-fumigated soil. However, Chen *et al.* (2015) showed that the much diminished soil microbial biomass in the fumigated soil has a much smaller diversity (survivors mainly belong to the classes *Bacilli*, *Eurotiomycetes* and *Sordariomycetes*). Others have also shown vast differences in the community structure of pre- and post-fumigation populations (Reber, 1967; Dominguez-Mendoza *et al.*, 2014). This suggests that the basal (or underlying) rate of SOC mineralisation is unrelated to the size, community structure or activity of the soil microbial biomass. Since SOC is mineralised, although very slowly, by the soil microbial biomass, it seems very strange that a group, or groups, of soil microorganisms have not evolved the ability to mineralise SOC more rapidly, thereby gaining more energy and achieving a competitive advantage over other less efficient microorganisms. If not, why

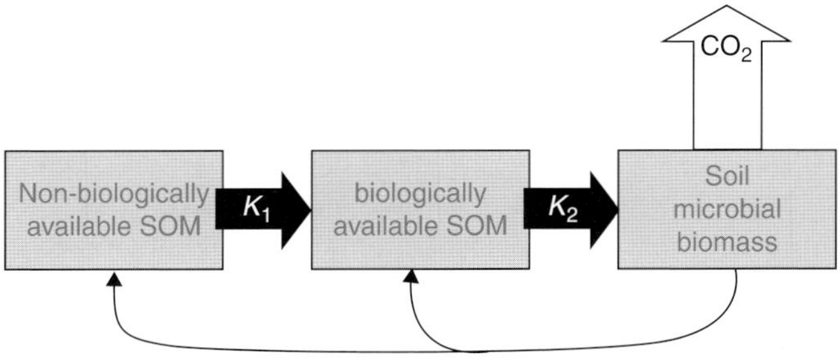

Figure 1.3 Diagrammatic representation of the 'Regulatory Gate Hypothesis'. (Adapted from Kemmitt *et al.*, 2008.)

not? This surely suggests that the mineralisation rate of SOC is not under microbial control. If so, the question is, what is the mechanism?

Kemmitt *et al.*(2008) proposed a new theory to explain this phenomenon. They considered that SOC had to first undergo an abiotic process, or processes, before it could be mineralised by the microbial biomass. They termed this the Regulatory Gate Hypothesis (Figure 1.3).

In Figure 1.3, $\mathbf{K}_1$ is the abiological transformation of non-bioavailable SOC. $\mathbf{K}_2$ is the biological mineralisation of (now) bioavailable SOC. Arrows indicate that the soil microbial biomass may create both non-biologically available and biologically available SOC but is not able to directly influence the rate of $\mathbf{K}_1$. The precise nature of $\mathbf{K}_1$ is not known. It may include processes such as desorption, chemical oxidation, free radical activity, cleavage of phenolic rings (Majcher, 2000) or stabilised exocellular enzymes (thought unlikely). Probably, more than one mechanism operates simultaneously. Kuzyakov *et al.* (2008) suggested that enzymes released from the $CHCl_3$-induced lysis of microbial cells remained active and could mineralise soil organic C for several months. We discount this: exocellular non-stabilised enzymes would be readily utilised as substrate and have been shown to have a short half-life in soil (Burns, 1982; Nannipieri *et al.*, 1983).

We are currently investigating the underlying mechanisms of the Regulatory Gate Hypothesis. In one experiment, we attempted to see whether we could measure abiotic release of mineralisable soil organic C. Six soils were fumigated for 24 hours, the $CHCl_3$ removed, then incubated four times for 4 days to destroy most of the microbial biomass. The soils were then washed carefully with 0.01 M $CaCl_2$ to remove as much $CHCl_3$-released microbial C as possible. They were then very slowly perfused for 63 days with water saturated with $CHCl_3$ (Kemmitt *et al.*, 2008), the $CHCl_3$ being added to inhibit microbial activity. The perfusates

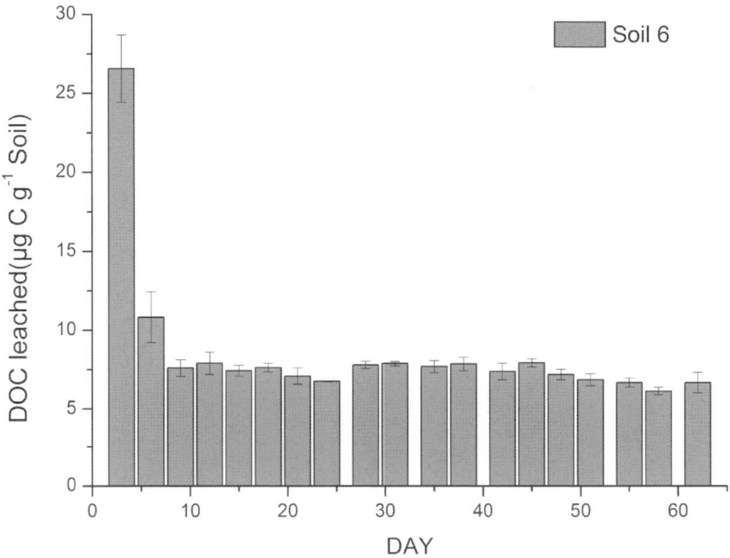

Figure 1.4 Soil organic C leached during a 63-day perfusion of a grassland soil.

were collected at approximately 5-day intervals and analysed for solubilised organic C. In newer work (Yang Feng Chen, unpublished data), the soil in Figure 1.4 was taken from a non-experimental grassland site near Hangzhou, China, that contained 7.02 g kg^{-1} organic C and had a pH of 8.7. After the initial flush of soluble C, the quantity released declined to a low level but the rate of release was remarkably constant for the duration of the experiment. This shows that organic C can be released abiotically into soluble forms at a slow but constant rate, which is in line with the Regulatory Gate Hypothesis. In another experiment (Yan Feng Chen, unpublished data), the perfusates were bulked, freeze-dried and added back to their parent soil. Soil-derived organic C in the freeze-dried extracts was mineralised rapidly, showing that the C released during perfusion was highly decomposable and readily available to the soil population, unlike total soil organic C.

The Regulatory Gate Hypothesis explains why mineralisation of SOC proceeds at the same rate, even if the microbial population size has been decimated, with both activity and diversity drastically changed. It suggests a mechanism whereby SOC is decomposed at the same rate, irrespective of the size of the population in the soil. The results are consistent with the idea that the soil population, whether large or small, damaged or intact, is acting on the same small but constant supply of SOC. This SOC has been transformed abiotically from forms that are not biologically available to forms that are available, and which have been brought into contact with the microbes through diffusion into the soil solution. However, the mechanism(s) responsible for this first and abiotic step remain unidentified. We believe that the other possibility, that the SOC is mainly being mineralised directly by microbial activity, cannot be justified in the light of the above evidence.

Soil Adenosine 5′ Triphosphate Measurements

Cellular adenosine 5′ triphosphate (ATP) has been termed the energy currency of the cell. It can be readily measured in soil, following extraction with a suitable reagent, by the firefly luciferin–luciferase enzyme. The question here is, what is a suitable reagent? A good example of the problem is demonstrated by the long-running debate between the late Professor David Jenkinson and Professor Eldor Paul in the 1980s. Jenkinson and Oades (1979) proposed a method to measure biomass ATP based on ultrasonic dispersion of soil to disrupt microbial cells, followed by extraction of ATP with a reagent containing paraquat (*N,N′*-dimethyl-4,4′-bipyridinium dichloride), phosphate and trichloroacetic acid (TCA). The paraquat (a planar molecule, as is ATP) and phosphate inhibit ATP fixation on the soil colloids, and TCA is a very powerful inhibitor of enzyme activity. The phosphatases and ATPases, once liberated from the ultrasonified cells, would, if not deactivated, almost instantaneously dephosphorylate adenosine 5′ triphosphate (ATP) to adenosine 5′ diphosphate (ADP) and adenosine 5′ monophosphate (AMP) — it was for this reason that Jenkinson and Oades (1979) selected TCA as the extracting reagent. Paul and Johnston (1977) published an alternative ATP method based on extraction with a $NaHCO_3$–$CHCl_3$ reagent, which was later modified by Martens (1985) to also contain P and adenosine (the $NaHCO_3$ reagent). Both this, and the method of Paul and Johnston (1977), consistently measured very much smaller soil ATP concentrations than the TCA reagent. Brookes *et al.* (1983) measured the total soil adenylates ATP, ADP and AMP extracted with both reagents. They found that both reagents extracted similar amounts of total adenylates (i.e. $\sum A_T$ = (ATP + ADP + AMP)). However, ATP comprised 80–84% of A_T with the TCA reagent but only 32–55% with the $NaHCO_3$ reagent: clearly a large discrepancy. Brookes *et al.* (1987) considered that it was due to phosphatases and ATPases retaining activity in the $NaHCO_3$ reagent and denaturing ATP. In contrast, they were immediately and completely deactivated in the TCA reagent. In the TCA method, one set of soils is extracted with a "spike" of ATP to correct for incomplete extraction as some of the ATP released from microbes will be sorbed on soil particles, thus giving a low ATP recovery in the extractant. Martens (2001) considered that the spike approach considerably underestimated the true soil ATP content. He found that when soils were extracted six times, each time with fresh extractant, the sum of the ATP extracted greatly exceeded (by 2–9 times) the ATP measured using the spike approach. In contrast, Contin *et al.* (2002) found no significant difference in the ATP contents of four soils when extracted by the conventional "spike" method and by six separate extractions. Clearly there is an unbridgeable gap between the two sets of data and we still remain uncertain of the reasons. We continue to use the spike and consider it to be approximately correct.

The original TCA method of Jenkinson and Oades (1979) suffered from the disadvantage that paraquat was included to inhibit ATP fixation by soil during ultrasonic extraction. Paraquat, which was used as a herbicide until it was banned in Europe in around 2007, is highly toxic to both plants and animals, including humans. For ATP measurements, the paraquat was purified from Gramoxone, the commercial form in which it was purchased, by alcohol extractions to remove various unwanted components followed by re-extraction with ethanol–charcoal and crystallisation. This is a potentially dangerous operation, as Phil Brookes found when he attempted to hasten the process by boiling the ethanol–paraquat mixture over a Bunsen burner. Shortly after catching fire, a white cloud of ethanol and paraquat vapour poured out of the fume cupboard, out of the door, travelled a few metres down the corridor then dissipated. Unfortunately, the Rothamsted Safety Officer was walking down the corridor, saw what had happened and nearly had a fit. Any fool can follow the instructions in a Methods manual. This fool nearly poisoned a large number of his colleagues and burnt the building down!

Paraquat's dangerous properties always made the TCA method unattractive to many soil scientists. Its banning in Europe meant that a substitute was needed and Redmile-Gordon and Brookes (2011) tested a number of possible alternatives. Imidazole was shown to give the same ATP results as paraquat at concentrations of 1.10 M TCA, 0.25 M P and 0.6 M imidazole (termed the TIP reagent). Imidazole is also toxic, as are many laboratory chemicals, so great care must be taken. However, it is readily available, cheap, avoids the need for purification as with paraquat and no cumbersome conversion factors are required to compare ATP extracted by the two reagents. Apart from the above careless incident, we have experienced no health problems associated with either the TCA or TIP reagent despite frequent use.

The essential features of an ATP extracting reagent are that it must have the property of preventing the almost instant deactivation of ATP to ADP and AMP, as the firefly luciferin–luciferase system only detects ATP. It must also be able to prevent the fixation of ATP to soil via the negatively charged P groups or positively charged adenine groups. Finally, it must not inhibit the luciferin–luciferase enzyme system during the quantitative measurement of ATP. For example, the Paul and Johnston (1977) method for measuring soil ATP, based on extraction with $NaHCO_3$–$CHCl_3$, gave higher recoveries of added ATP but soil ATP concentrations 2–3 times lower than with the TCA reagent. Brookes *et al.* (1987) showed that this reagent failed to inhibit phosphatases and ATPases, which explained their low ATP measurements. They considered that strongly acidic reagents were essential because they inhibit enzyme activity virtually instantaneously and that near-neutral reagents, such as $NaHCO_3$, should be avoided. Other reagents which have been tested include the H_2SO_4–P reagent of Eiland (1983), the H_3PO_4–DMSO–adenosine–urea–Zwittergent reagent (Webster *et al.*, 1984) and the dimethylsulfoxide–Na_3PO_4–NRB reagent (Bai *et al.*, 1998). These, together with the TCA (Jenkinson and Oades, 1979) and TIP reagents (Redmile-Gordon and Brookes, 2011), give comparable results. In waterlogged, that is, paddy soils, ATP can be measured in conjunction with biomass C measurements by FE (Inubushi *et al.*, 1989, 1991, see Chapter 4).

The ATP Concentration of the Soil Microbial Biomass

All living organisms contain ATP, so it must have been selected for its role in cellular energy dynamics very early in the evolution of life. It has an exocellular half-life of less than 1 h (Conklin and McGregor, 1972) and is therefore a very reliable indicator of life in soil. The soil microbial biomass is generally considered as a largely resting population, with a mean cell division time of about 6 months and a much longer turnover time. Jenkinson and Ladd (1981) estimated a turnover time of 2.5 years for the biomass in soil of the Broadbalk continuous wheat experiment at Rothamsted. This long turnover time is a reflection of the very low substrate availability in most soils. Oades and Jenkinson (1979) measured a mean microbial ATP concentration of 10.6 μmol ATP g^{-1} biomass C in a group of Australian and UK soils. Jenkinson (1988) later reported a mean value of 11.2 ± 2.74 μmol ATP g^{-1} biomass C for a much wider range of ATP and biomass C measurements from Australian, Danish, English and New Zealand soils collated from the literature. He laid down strict criteria for including soils in his correlation. These were:

(1) The soils had all received a conditioning incubation of at least several days at 25°C before analysis.
(2) Soil moisture was below 60% water holding capacity.
(3) Soil pH was not below 4.8.

(4) All measurements were done by FI.
(5) Only results obtained with acidic extractants were included.

Since Jenkinson (1988) published the above correlations, there have been some significant changes in biomass methodology. The FE method permits biomass C measurements in acid soils, soils containing actively decomposing substrates and waterlogged soils (not included by Jenkinson, 1988). There is now a much larger literature base than previously. Contin *et al.* (2001) combined all the available literature, including soils below pH 4.8, those containing actively decomposing substrates and soils without a conditioning incubation. The mean biomass ATP concentration was 11.0 μmol ATP g^{-1} biomass C, based on 207 separate analyses, remarkably close to the earlier values.

This analysis demonstrates that, for all practical purposes, the microbial biomass has a constant ATP concentration of around 10–12 μmol ATP g^{-1} biomass C. This is close to the mean biomass ATP concentration reported by Knowles (1977) for a range of prokaryotic and eukaryotic microorganisms undergoing exponential growth *in vitro*. It is because of the observed constant concentration of ATP in biomass that ATP can be used as a proxy for biomass measurements and for testing or calibrating biomass methods, as discussed earlier in relation to FI and FE methods. The data in Figure 1.5(a) is an example of this close relationship.

The relationship is not particularly sensitive to temperature, with only a slow increase in biomass ATP concentrations with increasing temperature. Contin *et al.* (2000) showed that biomass C changed little in grassland and arable soils incubated between 0 and 25°C for 23 days. Soil microbial biomass ATP concentrations did gradually increase with increasing incubation temperature, from about 7 to 10 μmol ATP g^{-1} biomass C. Both values are well within the range of ATP values of exponentially growing microorganisms.

Biomass C can be measured in waterlogged soils by FE and by ATP using either the TCA or TIP extractants (Inubushi *et al.*, 1989, 1991; Qiu *et al.*, 2016). Both biomass C and ATP decline considerably during anaerobic incubation: ATP levels are rapidly restored, but not to the original levels, upon aeration. Biomass C and ATP relationships are shown in a number of aerobically (Figure 1.5(a)) and anaerobically (Figure 1.5(b)) incubated soils, incubated with and without ryegrass as a substrate. The aerobic soils had a mean biomass ATP concentration of 7.9 but in the anaerobic soils it was very much lower, at 2.83 μmol g^{-1} biomass C. Also, in the anaerobic soils there was no observable relationship between biomass C and ATP concentrations. We do not know the reasons for this difference between aerobic and anaerobic soils.

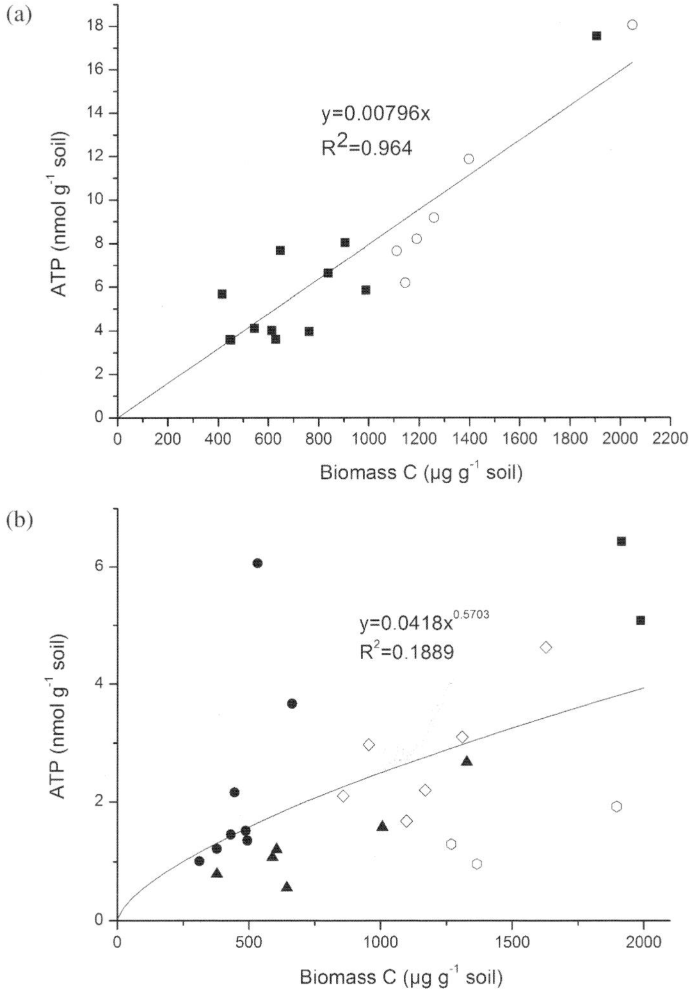

Figure 1.5 Biomass C and ATP measurements in (a) aerobically and (b) anaerobically incubated soils. (Adapted from Qiu *et al.*, 2016.)

The Adenylate Energy Charge of the Soil Microbial Biomass

The adenylate energy charge (AEC) is defined as a linear measure of the metabolic energy stored in the adenylate energy pool (Atkinson, 1977). AEC = [(ATP) + 0.5(ADP)]/[(ATP) + (ADP) + (AMP)]. In actively growing cells it is stabilised between about 0.8 and 0.95, but it can theoretically range from 0 (all AMP) to

1.0 (all ATP). Lower values indicate a population that is stressed, senescent or incapable of cellular biosynthesis (Atkinson, 1977; Karl, 1980). Brookes *et al.* (1983) measured AEC by an enzymatic procedure following extraction with the TCA reagent. The aim was to determine whether the high values for ATP concentration in the soil microbial biomass resulted from (a) a small part of the population having an exceedingly high ATP concentration but with the majority being low in ATP (i.e. low AEC) or (b) a population as a whole that maintained a relatively high and constant ATP concentration (high AEC) independent of whether or not it is in active growth. A grassland soil at a field temperature of 8.4°C at the time of sampling, when extracted moist as soon as possible at 10°C, had an AEC of 0.85. In the same soil, air-dried for 5 days, AEC had decreased to 0.45, and 2.5 hours after rewetting it had increased to 0.76. Very comparable AEC values to those of Brookes *et al.* (1987) were reported by Joergensen and Raubuch (2002) in a moist, sandy soil during an 8-day incubation whether unamended (range 0.81–0.84) or glucose-amended (range 0.76–0.82). Anaerobically incubated soils behave differently: Inubushi *et al.* (1989) showed that AEC decreased rapidly upon anaerobic incubation, falling from a value of 0.76 to values in the range 0.54 to 0.34. Aeration increased AEC, but not to its pre-anaerobic level.

Significance of ATP and AEC Concentrations in the Soil Microbial Biomass

Because the energy inputs to soil from plant material are generally small compared to the size of the population, the soil microbial biomass must be considered as a largely resting, or dormant population, with a very low metabolic rate overall. Blagodatsky *et al.* (2000) and Blagodatskaya and Kuzyakov (2013) estimated that the active pool of soil microorganisms in soil that has not recently received labile substrate addition is very small: generally 1–2% of the total biomass and up to a maximum of 5%. However, the potential is much higher, with organisms able to begin substrate utilisation within a few hours when the active fraction may comprise up to 60% of the total population. The question is: why does the soil microbial biomass maintain an ATP concentration and AEC at levels found in microorganisms in exponential growth, when only a small fraction is undergoing active growth for most of the time? Surely a strategy based on spores or resting cells with low ATP and AEC would be more energy efficient in the normally substrate-poor environment of soil?

De Nobili *et al.* (2001) added small amounts of root extract or glucose (11 or 34 μg C g^{-1} soil), termed "trigger molecules", to three soils. Up to at least seven times more CO_2-C was evolved from the soils than was added in the substrate. The largest response was with the root extract, suggesting that this substrate contained a wider diversity of "trigger molecules" than did glucose. More work is required to characterise this response and to determine the exact molecules that stimulate the biomass and how they are detected. De Nobili *et al.* (2001) hypothesised that the survival strategy of the microbial biomass is based upon the population of "resting cells" using energy to maintain a state of "metabolic alertness" and that it is ready to expend more energy if some cells detect early signals from the substrate, indicating that a more substantial "food event" is to come. The "trigger response" may therefore be an evolutionary response to the fact that although easily available substrate is a rare event in soil, it is never an impossibility. It is hypothesised that a survival strategy based on spores or other resting states may be less advantageous because the process of transforming from spore to a state capable of utilising fresh substrate may be too slow a response. In contrast, investing energy in maintaining both a high ATP concentration and AEC may give an evolutionary advantage.

Some Applications of Microbial Biomass Measurements

Long-lasting damage to soil microbes following soil fumigation for pest control

Although less widely used than in the past, volatile broad-spectrum biocides are still used in some situations for fumigating soils to control soil-borne pests and fungal diseases. Soil fumigation is mainly used for high-value crops in the horticultural industry and especially, but not exclusively, in greenhouses. Methyl bromide was used as a nematocide in various horticultural situations until it was phased out in 2005 under the Montreal Protocol because of its damaging impact on the Earth's ozone layer. As mentioned at the beginning of this chapter, in the 1960s there were many experiments worldwide testing the effectiveness of various fumigants in controlling crop pests and diseases and seeking to elucidate the mechanisms through which they influenced crop yields. Jenkinson and Powlson (1970) used the FI method (though it had not yet been fully developed) to examine the residual impact of past soil fumigation treatments in the field on the microbial biomass. Table 1.4 shows some results, recalculated from this work. In this example, soils had been fumigated with formaldehyde in the field either 6 months or 3 years before they were sampled and analysed in the laboratory.

In the soil that had received three treatments with formaldehyde, the last being 6 months before sampling, biomass C was less than half that in the untreated soil.

Table 1.4 Impact of field fumigation with formaldehyde on the size of the soil microbial biomass as measured using FI, and its rate of recovery.

	Period since last fumigation		
Field treatment with formaldehyde	**Unfumigated control**	**6 months[a]**	**3 years**
Soil microbial biomass C (μg g^{-1} soil)	322	147	242
Microbial biomass as a percentage of that in soil not fumigated in the field (%)	—	47	75

[a] Soil had received a total of three formaldehyde treatments during the previous 2 years.
Source: Adapted from Jenkinson and Powlson (1970).

After 3 years, it had recovered to some extent but was still significantly smaller than in soil that had never been fumigated, showing a slow rate of recovery. Measurements of this type clearly demonstrated the drastic impact of soil fumigation on the soil population, indicating that great caution was required if such biocidal treatments were to be used. Biomass N was affected similarly; at the time there was much interest in the role of N released from killed soil organisms contributing to observed increases in crop growth. If the biomass was as slow to recover as indicated here, less N would be released on subsequent field fumigations compared with the first: this was thought to be one factor contributing to observed declines in crop yield responses to repeated fumigations applied over several years.

Damage to Soil Microbes from Heavy Metal Pollution

All scientists like to believe that their research flows effortlessly from one step to the next, based on their previously carefully planned, and logically based, experiments. If only if it were true! Most of us, if we are really honest, would have to admit that, more than occasionally, we owe a fair deal to luck. The discovery of a negative impact on soil microbes of heavy metals, even at relatively low concentrations, is such an example. One of the less well-known long-term experiments run by Rothamsted Research is the Woburn Market Garden Experiment. It was started in the 1940s to test the value of farmyard manure and sewage sludge compared to inorganic fertiliser on the yields of vegetable crops including carrot and red beet. The produce was often served for lunch in the Rothamsted canteen until a famous Rothamsted scientist decided to analyse the heavy metals in the vegetables. Although what happened next is anecdotal, it comes from a reliable source. Apparently, he rushed into the canteen one lunch time, arms waving and demanded that any vegetables were immediately removed from the menu. He had detected high cadmium (Cd) concentrations in some of them, at levels possibly injurious to human health. Hopefully everyone survived. However, his findings triggered considerable research into the effects of organic fertilisers, especially sewage sludge, which chelates and concentrates heavy metals, and thus influences their dynamics in soil and uptake by crops.

In the 1980s Phil Brookes decided to investigate whether heavy metal availability was increased due to enhanced microbial activity in organically manured soils, including those receiving sewage sludge (now often termed "biosolids"), compared to those given inorganic fertilisers. He did a very simple preliminary experiment which included measuring biomass C in soils from the Woburn Market Garden Experiment that had received inorganic fertiliser, sewage sludge or farmyard manure. At this time, it was believed that heavy metals in soils were at around the (then) European Union permitted concentrations and had no deleterious effects on plants, animals or soil microorganisms. He was amazed to find that soils treated with sewage sludge contained as little as half the biomass C concentrations as those which received farmyard manure or inorganic fertiliser. These were very preliminary results but then with Steve McGrath, an expert on heavy metals in soil and crops, the soils from the entire Woburn experiment were rigorously sampled and analysed for biomass, ATP and heavy metals. The results were conclusive. Heavy metals at quite low concentrations were damaging the soil microbial population (Figure 1.6; Brookes and McGrath, 1984). The biomass concentrations in the metal-contaminated soils were much smaller than in soils given farmyard manure or inorganic fertiliser. Furthermore, in these soils, there was no relationship between biomass C and total soil organic C, in contrast to the usual situation and as shown in the low-metal soils in this experiment (Figure 1.6). As the last applications of metal- contaminated sludge were added around 30 years before this work was done, it was clear that the metals had a long-lasting impact on the soil microbial population.

Shortly after this work was published (Brookes and McGrath, 1984) all hell broke loose. The then UK Prime Minister, Margaret Thatcher, and her Government, were about to privatise the organisations responsible for water supply and sewage treatment. Sewage sludge was normally disposed of by the water treatment plants by providing it free of charge to farmers as an organic fertiliser. Anything that stopped this, even partially, could have serious economic effects and damage water privatisation. The work of Brookes and McGrath (1984) had been noticed by the national and farming press, and this was the last thing the water authorities or the Government wanted prior to privatisation. The work was heavily criticised as irrelevant by the water authority chiefs, as the soil Cd levels in the sewage

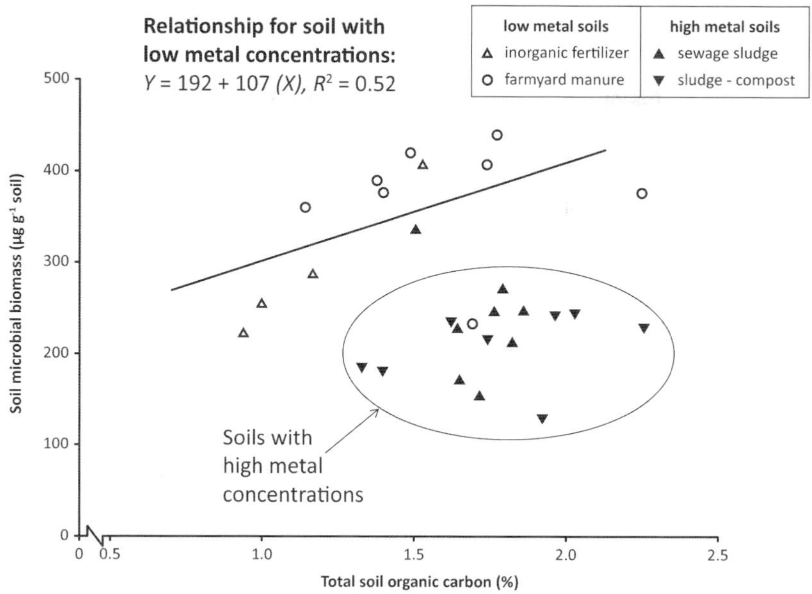

Figure 1.6 Soil microbial biomass and organic C concentrations in soils from the Woburn Market Garden Experiment that had previously received various treatments, leading to either high or low metal concentrations in soil. (Adapted from Brookes and McGrath, 1984.)

sludge-treated soils in the Woburn experiment exceeded European Union permitted limits. This was because it was an old experiment, started long before these limits were set and at a time when sewage sludge was often higher in metal concentrations that was currently the case. Brookes and McGrath clashed with these people at several acrimonious public meetings. However, other work (Chander and Brookes, 1994) showed that in the absence of Cd, zinc–copper interactions could produce similar results, so the effect was not specific to Cd. It was also found that autotrophic N_2 fixation by cyanobacteria was severely inhibited in the metal-contaminated soils (Brookes *et al.*, 1986). This led to other research (McGrath *et al.*, 1988) showing a significant decrease in N_2-fixation by the Rhizobium–clover symbiosis in the metal-contaminated soils, with strong implications for the long-term sustainability of applying metal-contaminated sewage sludge to soil. This body of work led to a review of sewage sludge applications in the UK and a decrease in the permitted metal loadings. All this arose from a few chance biomass C measurements made for entirely different reasons!

Detecting long-term changes in soil organic C content from organic additions

When a soil undergoes a change in management, it can often take a period of years or even decades before a change in total SOC content can be reliably measured (e.g. Poeplau *et al.*, 2011), unless impracticably large numbers of samples are taken (Smith, 2004): this is the case even in replicated field experiments and it is even more difficult to detect changes in soils under practical management in farmers' fields. However, it has been found that the microbial biomass changes more rapidly than SOC as a whole. Thus, although microbial biomass C typically only constitutes 2–4% of total SOC, it can be used as a sensitive indicator of change. This is illustrated in Table 1.5 for impacts of land-use changes for soils in temperate and humid tropical climates.

In the UK temperate climate example, an area of old grassland was ploughed and converted into arable use 22 years before the total SOC and biomass C measurements were made. Total SOC had decreased to 61% of that in the soil that continued under grass but microbial C was only 42% of the grassland soil value. In the area of arable land that was converted from grassland into a bare fallow,

Table 1.5 Total SOC and microbial biomass C in two situations following land-use changes.

Land management	SOC (%)[a]	Soil microbial biomass C (μg g^{-1} soil)[a]
UK, temperate climate		
Old grassland	3.84 (100)	960 (100)
Arable (converted from grassland 22 years previously)	2.34 (61)	400 (42)
Bare fallow (converted from grassland 11 years previously)	1.89 (49)	127 (13)
Nigeria, humid tropical climate		
Secondary forest	1.46 (100)	542 (100)
Arable (cleared from forest 2 years previously)	1.23 (84)	280 (52)

[a] Values in parentheses are SOC or biomass C contents expressed as a percentage of those in grassland soil (UK) or secondary forest (Nigeria).
Source: Adapted from Powlson and Jenkinson (1976).

with virtually no inputs of plant material for 11 years (Barre *et al.*, 2010), SOC declined to 49% of the grassland value but biomass had declined much more to only 13%, clearly showing the sensitivity of biomass C to management changes that influence organic C input. In the tropical example, total SOC in the soil cleared from secondary forest and converted into arable cropping was 1.23% compared to 1.46% in soil still under forest, or 84% of its starting value. Although a change in SOC of this magnitude has major implications for soil properties and fertility, it is not easy to reliably detect given the considerable spatial variability in a situation such as this. However, microbial biomass had decreased to almost half that in the forest soil, a change that was easy to detect and gave a more practical and sensitive means of gauging the direction and rate of change in SOM content following the change in land use.

The modelling study in Figure 1.7 illustrates the movement of total SOC and biomass C to new equilibrium values following a change in management for a

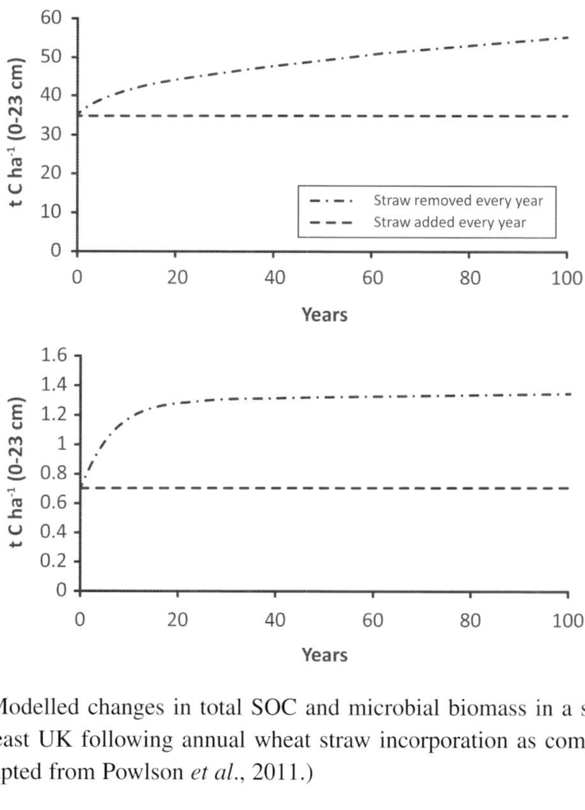

Figure 1.7 Modelled changes in total SOC and microbial biomass in a silty clay loam soil in south-east UK following annual wheat straw incorporation as compared to straw removal. (Adapted from Powlson *et al.*, 2011.)

silty clay loam soil under arable cropping in a temperate climate. It was assumed that total SOC and biomass C were both at equilibrium values when the model run was started and that wheat straw had been removed every year. The RothC model (Coleman and Jenkinson, 1996) was then used to assess the changes resulting from annual incorporation of straw. Figure 1.7(a) shows that the total SOC increased slowly after straw incorporation began and was still increasing after 100 years. In contrast, biomass C (Figure 1.7(b)) increased rapidly during the first 10 years, and by 20 years had virtually reached an equilibrium level with no further increase.

This effect was illustrated by results published in 1987 from two experimental sites in Denmark where spring barley was grown and straw either removed (by burning, which was still allowed at the time) or incorporated every year. At both sites, even after 18 years of continuous straw incorporation, total SOC was only marginally greater than in soil where straw had been burned every year: the small increase in the "incorporated" treatment (Figure 1.8) was not statistically significant. In contrast, microbial biomass C was about 40% greater where straw had been incorporated and easily measurable (Figure 1.8). This was the first time that

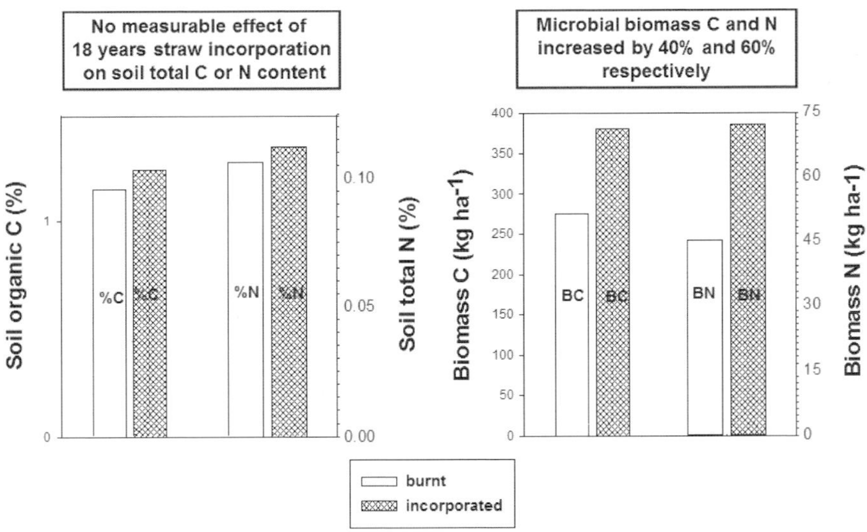

Figure 1.8 The effect of incorporating spring barley straw compared to straw removal by burning on total soil organic C and N and biomass C and N content at a sandy soil site in Denmark after 18 years. The small increases in total soil C or N are not statistically significant but microbial biomass C and N were increased by straw incorporation by 40 and 60%, respectively. (Adapted from Powlson *et al.*, 1987.)

microbial biomass measurements were specifically proposed as an "early warning system" for detecting changes in soil organic C before they are detectable by direct measurements of total SOC. The idea was demonstrated again in a vertisol in a subtropical environment where sorghum residues were either retained or removed in combination with conventional or zero tillage (Saffigna *et al.*, 1989). After 6 years, the treatments that tend to cause an increase in SOM content in the surface soil (residue retentions and zero tillage) caused increases in biomass C and N that were proportionately much greater than those in total SOC or N and much easier to detect.

Concluding Remarks

We now have a suite of methods available to measure the quantities of C, N, P and S held in the cells of living microorganisms in soil, all based on the principle of chloroform fumigation followed by extraction. All of these fumigation–extraction (FE) methods are suitable for measuring biomass in soils containing actively decomposing substrates in both laboratory (Ocio and Brookes, 1989) and field conditions (Ocio *et al.*, 1991), and in acidic soils (Vance *et al.*, 1987c). Biomass measurements can also be made in anaerobic (e.g. paddy) soils following a simple modification (Inubushi *et al.*, 1991; Qiu *et al.*, 2016). We also have the possibility of measuring the turnover rate of C in the biomass following substrate addition (Kouno *et al.*, 2002).

Building on the Jenkinson (1966) paper, we now have the concept of the soil microbial biomass as a meaningful fraction within SOM that can provide considerable insights into soil biology and elemental transformations. The biomass has emerged as a surprisingly large population in view of the small quantities of substrate and energy generally available in soils: in a UK arable soil, the biomass in the plough layer is estimated to have a fresh weight equivalent to about 100 sheep per hectare! But it has a very low metabolic rate, orders of magnitude lower than that of microorganisms growing exponentially under laboratory conditions and with a very slow turnover rate of about 1.2 years. Yet, it has an ATP concentration and AEC comparable to microorganisms growing exponentially *in vitro*. We are still uncertain of the significance of these contradictory characteristics.

Microbial biomass measurements can act as an early indicator of trends in organic matter content resulting from management changes and in some situations a means of detecting damage to soil organisms caused by pollutants. In addition, incubation studies derived from the early work on the chloroform FI method have led to new understanding of the mechanisms influencing the surprising stability of organic matter in soil. Although other approaches are also proving extremely valuable, especially molecular methods and the use of phospholipid fatty acid (PLFA), we owe a huge debt to the insights of David Jenkinson: much of our current understanding of SOM can be traced back to his innovative thinking and experimentation starting with the Jenkinson (1966) paper that was published 50 years ago.

Acknowledgments

We dedicate this chapter to the memory of the late Professor David Jenkinson, who did so much to develop research into the soil microbial biomass. We also thank Yang Feng Chen for supplying unpublished data.

References

Amato, M. and Ladd, J. N. (1988). Assay for microbial biomass based on ninhydrin-reactive nitrogen in extracts of fumigated soils. *Soil Biol. Biochem.* 20, 107–114.

Atkinson, D. (1977). *Cellular Energy Metabolism and its Regulation.* Academic Press, New York.

Bai, Q. Y., Zelles, L., Scheunert, I. and Korte, F. (1998). Determination of adenine nucleotides in soil by ion-paired reverse-phase high-performance liquid chromatography. *J. Microbiol. Meth.* 9, 345–351.

Barre, P., Eglin, T., Christensen, B. T., *et al.* (2010). Quantifying and isolating stable soil organic carbon using long-term bare fallow experiments. *Biogeoscience* 7, 3839–3850.

Blagodatskaya, E. and Kuzyakov, Y. (2013). Active microorganisms in soil: critical review of estimation criteria and approaches. *Soil Biol. Biochem.* 67, 192–211.

Blagodatsky, S. A., Heinemeyer, O. and Richter, J. (2000). Estimating the active and total soil microbial biomass by kinetic respiration analysis. *Biol. Fert. Soils* 32, 73–81.

Brookes, P. C., Landman, A., Pruden, G. and Jenkinson, D. S. (1985). The effects of chloroform fumigation on the release and extractability of soil nitrogen. 2. A rapid direct extraction method for measuring microbial biomass nitrogen in soil. *Soil Biol. Biochem.* 17, 837–842.

Brookes, P. C. and McGrath, S. P. (1984). The effects of metal toxicity on the soil microbial biomass. *J. Soil Sci.* 35, 341–346.

Brookes, P. C., McGrath, S. P. and Heijnen, C. (1986). Metal residues in soils previously treated with sewage-sludge and their effects on growth and nitrogen fixation by blue-green algae. *Soil Biol. Biochem.* 18, 345–354.

Brookes, P. C., Newcombe, A. D. and Jenkinson, D. S. (1987). Adenylate energy charge measurements in soil. *Soil Biol. Biochem.* 19, 211–217.

Brookes, P. C., Powlson, D. S. and Jenkinson, D. S. (1982). Measurement of microbial biomass phosphorus in soil. *Soil Biol. Biochem.* 14, 319–329.

Brookes, P. C., Tate, K. R. and Jenkinson, D. S. (1983). The adenylate charge of the soil microbial biomass. *Soil Biol. Biochem.* 15, 9–16.

Burns, R. G. (1982). Enzyme activity in soil — location and a possible role in microbial ecology. *Soil Biol. Biochem.* 14, 423–427.

Chander, K. and Brookes, P. C. (1994). Effects of Zn, Cu and Ni in sewage-sludge applied singly and in combination on microbial biomass in a sandy loam soil. *Soil Biol. Biochem.* 26, 103–110.

Chapman, S. (1987). Microbial sulphur in some Scottish soils. *Soil Biol. Biochem.* 19, 301–305.

Chen, L., Xu, J., Feng, Y., Wang, J., Yu, Y. and Brookes, P. C. (2015). Responses of soil microeukaryotic communities to short-term fumigation–incubation revealed by MiSeq amplicon sequencing. *Front. Microbiol.* 6, 1149–1153.

Coleman, K. and Jenkinson, D. S. (1996). Roth C-26. 3 — a model for the turnover of carbon in soil. In Powlson, D. S., Smith, P., Smith, J. U. (eds.), *Evaluation of Soil Organic Matter Models Using Existing Long-term Datasets*. Springer, Heidelberg, pp. 237–246.

Conklin, A. R. and McGregor, A. N. (1972). Soil adenosine triphosphate: extraction, recovery and half-life. *Bull. Environ. Contam. Toxicol.* 7, 296–300.

Contin, M., Corcimaru, S., De Nobili, M. and Brookes, P. C. (2000). Temperature changes and the ATP concentration of the soil microbial biomass. *Soil Biol. Biochem.* 32, 1219–1225.

Contin, M., Jenkinson, D. S. and Brookes, P. C. (2002). Measurement of ATP in soil: correcting for incomplete recovery. *Soil Biol. Biochem.* 34, 1381–1383.

Contin, M., Todd, A. and Brookes, P. C. (2001). The ATP concentration in the soil microbial biomass. *Soil Biol. Biochem.* 33, 701–704.

De Nobili, M., Contin, C. and Brookes, P. C. (2006). Microbial biomass dynamics in air-dried soils that have been stored for many decades. *Soil Biol. Biochem.* 38, 1638–1644.

De Nobili, M., Contin, M., Mondini, C. and Brookes, P. C. (2001). Soil microbial biomass is triggered into activity by trace amounts of substrate. *Soil Biol. Biochem.* 33, 1163–1170.

Dominguez-Mendoza, C. A., Bello-Lopez, J. M., Navarro-Noya, Y. E., De Leon-Lorenzana, A. S., Delgado-Balbuena, L., Gomez-Acata, S., *et al.* (2014). Bacterial community structure in fumigated soil. *Soil Biol. Biochem.* 73, 122–129.

Eiland, F. (1983). A simple method for quantitative determination of ATP in soil. *Soil Biol. Biochem.* 15, 665–670.

Inubushi, K., Brookes, P. C. and Jenkinson, D. S. (1989). Adenosine 5′ triphosphate and adenylate energy charge in waterlogged soil. *Soil Biol. Biochem.* 21, 733–739.

Inubushi, K., Brookes, P. C. and Jenkinson, D. S. (1991). Soil microbial biomass C, N and ninhydrin-N in aerobic and anaerobic soils measured by the fumigation–extraction method. *Soil Biol. Biochem.* 23, 737–741.

Jenkinson, D. S. (1965). Studies on the decomposition of plant material in soil. I. Losses of carbon from ^{14}C labelled ryegrass incubated with soil in the field. *J. Soil Sci.* 16, 104–115.

Jenkinson, D. S. (1966). Studies on the decomposition of plant material in soil. II. Partial sterilization of soil and the soil microbial biomass. *J. Soil Sci.* 17, 280–302.

Jenkinson, D. S. (1968). Chemical tests for potentially available nitrogen in soil. *J. Sci. Food Agric.* 19, 160–168.

Jenkinson, D. S. (1976). Effects of biocidal treatments on metabolism in soil. IV. The decomposition of fumigated organisms in soil. *Soil Biol. Biochem.* 8, 203–208.

Jenkinson, D. S. (1988). The determination of microbial biomass C and N in soil. In Wilson, E. R. (ed.), *Advances in Nitrogen Cycling in Agricultural Systems.* CAB International, Wallingford, pp. 368–386.

Jenkinson, D. S. and Ladd, J. N. (1981). Microbial biomass in soil: measurement and turnover. In Paul, E. A. and Ladd, J. N. (eds.), *Soil Biochemistry*, Vol. 5. Marcel Dekker, New York, pp. 415–471.

Jenkinson, D. S. and Oades, J. M. (1979). A method for measuring adenosine triphosphate in soil. *Soil Biol. Biochem.* 11, 193–199.

Jenkinson, D. S. and Powlson, D. S. (1970). Residual effects of soil fumigation on soil respiration and mineralization of nitrogen. *Soil Biol. Biochem.* 2, 99–108.

Jenkinson, D. S. and Powlson, D. S. (1976a). Effects of biocidal treatments on metabolism in soil. I. Fumigation with chloroform. *Soil Biol. Biochem.* 8, 167–177.

Jenkinson, D. S. and Powlson, D. S. (1976b). Effects of biocidal treatments on metabolism in soil. V. A method to measure the soil microbial biomass. *Soil Biol. Biochem.* 8, 209–213.

Jenkinson, D. S., Powlson, D. and Wedderburn, R. W. M. (1976). Effects of biocidal treatments on metabolism in soil. III. The relationship between soil biovolume, measured by optical microscopy, and the flush of decomposition caused by fumigation. *Soil Biol. Biochem.* 8, 189–202.

Joergensen, R. G., Brookes, P. C. and Jenkinson, D. S. (1990). Survival of the soil microbial biomass at elevated temperatures. *Soil Biol. Biochem.* 22, 1129–1136.

Joergensen, R. G. and Raubuch, M. (2002). Adenylate energy charge of a glucose treated soil without adding a nitrogen source. *Soil Biol. Biochem.* 34, 1317–1324.

Jones, P. C. and Mollison, J. E. (1948). A technique for the quantitative estimation of soil microbes. *J. Gen. Microbiol.* 2, 54–69.

Karl, D. M. (1980). Cellular nucleotide measurements and applications in microbiological ecology. *Microb. Rev.* 44, 739–796.

Kemmitt, S. J., Lanyon, C. V., Waite, I. S., Wen, Q., Addiscott, T. M., Bird, N. R. A., O'Donnell, T. and Brookes, P. C. (2008). Mineralisation of native soil organic matter is not regulated by the size, activity or composition of the soil microbial biomass. A new perspective. *Soil Biol. Biochem.* 40, 61–73.

Knowles, C. J. (1977). Microbial metabolic regulation by adenine nucleotide pools. *Symp. Soc. Gen. Microbiol.* 27, 241–283.

Kouno, K., Wu, J. and Brookes, P. C. (2002). Turnover of biomass C and P in soil following incorporation of glucose or ryegrass. *Soil Biol. Biochem.* 34, 617–622.

Kuzyakov, Y., Bladowskya, E. S. and Blagodasky, S. (2008). Comments on the paper by Kemmitt, *et al.* (2008). Mineralization of native soil organic matter is not regulated by the size, activity or composition of the soil microbial biomass — a new perspective.

[*Soil Biology and Biochemistry*, 40, 61–73]. The biology of the regulatory gate. *Soil Biol. Biochem.* 41, 435–439.

Lal, R. (2004). Soil carbon sequestration impacts on global climate change and food security. *Science*, 304, 1623–1627.

Lehmann, J. and Kleber, M. (2015). The contentious nature of soil organic matter. *Nature* 528, 60–68.

Majcher, E. H. (2000). Evaluation of CO_2 during birnessite-induced oxidation of C-14 labelled catechol. *Soil Sci. Soc. Am. J.* 64, 157–163.

Martens, R. (1985). Estimation of the adenylate energy charge in unamended and amended grassland soils. *Soil Biol. Biochem.* 17, 361–366.

Martens, R. (2001). Estimation of ATP in soil: extraction methods and calculation of extraction efficiencies. *Soil Biol. Biochem.* 33, 973–982.

McGrath, S. P., Brookes, P. C. and Giller, K. E. (1988). Effects of potentially toxic metals in soil derived from past applications of sewage sludge on nitrogen fixation by *Trifolium repens* L. *Soil Biol. Biochem.* 20, 415–424.

Nannipieri, P., Muccini, L. and Ciardi, C. (1983). Microbial biomass and enzyme efficiencies: production and persistence. *Soil Biol. Biochem.* 15, 679–685.

Oades, J. N. and Jenkinson, D. S. (1979). Adenosine triphosphate content of the soil microbial biomass. *Soil Biol. Biochem.* 11, 201–204.

Ocio, J. A. and Brookes, P. C. (1990). An evaluation of methods for measuring the microbial biomass in soils following recent additions of wheat straw and the characterization of the biomass that develops. *Soil Biol. Biochem.* 22, 685–694.

Ocio, J. A., Brookes, P. C. and Jenkinson, D. S. (1991). Field incorporation of straw and its effects on soil microbial biomass and soil inorganic N. *Soil Biol. Biochem.* 23, 171–176.

Paul, E. A. and Johnston, R. L. (1977). Microscopic counting and adenosine 5′ triphosphate measurements in determining microbial growth in soils. *Appl. Environ. Microbiol.* 34, 263–269.

Poeplau, C., Don, A., Vesterdal, L., *et al.* (2011). Temporal dynamics of soil organic carbon after land-use change in the temperate zone — carbon response functions as a model approach. *Glob. Change Biol.* 17, 2415–2427.

Powlson, D. S. (1975). Effects of biocidal treatments on soil organisms. In Walker, N. (ed.), *Soil Microbiology*. Butterworths, London, pp. 193–224.

Powlson, D. S., Brookes, P. C. and Christensen, B. T. (1987). Measurement of soil microbial biomass provides an early indication of changes in total soil organic matter due to straw incorporation. *Soil Biol. Biochem.* 19, 159–164.

Powlson, D. S., Glendining, M. J., Coleman, K. and Whitmore, A. P. (2011). Implications for soil properties of removing cereal straw: results from long-term studies. *Agron. J.* 103, 279–287.

Powlson, D. S. and Jenkinson, D. S. (1976). The effects of biocidal treatments on metabolism in soil. II. Gamma irradiation, air-drying and fumigation. *Soil Biol. Biochem.* 8, 179–188.

Qiu, G., Chen, Y., Yu, L., Xu, J. and Brookes, P. C. (2016). The microbial ATP concentration in aerobic and anaerobic Chinese soils. *Soil Biol. Biochem.* 92, 38–40.

Reber, H. (1967). Untersuchungenuber die Weiderbesiedlungeineschemischentseuchten Boden. *Z. Pflanzenk Planzen.* 74, 427–438.

Redmile-Gordon, M. and Brookes, P. C. (2011). Evaluation of substitutes for paraquat in soil microbial ATP determinations using the trichloroacetic acid based reagent of Jenkinson and Oades (1979). *Soil Biol. Biochem.* 43, 1098–1100.

Saffigna, P. G., Powlson, D. S., Brookes, P. C. and Thomas, G. A. (1989). Influence of sorghum residues and tillage on soil organic matter and soil microbial biomass in an Australian Vertisol. *Soil Biol. Biochem.* 21, 759–765.

Sagger, S., Bettany, J. R. and Stewart, J. W. B. (1981). Measurement of biomass S in soil. *Soil Biol. Biochem.* 13, 493–498.

Schnürer, J., Clarholm, M. and Rosswall, T. (1985). Microbial biomass and activity in an agricultural soil with different organic matter contents. *Soil Biol. Biochem.* 17, 611–618.

Shields, J. A., Paul, E. A. and Lowe, W. E. (1974). Factors influencing the stability of labelled microbial materials. *Soil Biol. Biochem.* 6, 31–37.

Skinner, F. A., Jones, P. C. T. and Mollison, J. E. (1952). A comparison of a direct and plate-counting technique for the quantitative estimation of soil microorganisms. *J. Gen. Microbiol.* 6, 261–271.

Smith, P. (2004). How long before a change in soil organic carbon can be detected? *Glob. Change Biol.* 10, 1878–1883.

Vance, E. D., Brookes, P. C. and Jenkinson, D. S. (1987a). Microbial biomass measurements in forest soils: determination of k_c values and tests of hypotheses to explain the failure of the chloroform fumigation–incubation method in acid soils. *Soil Biol. Biochem.* 19, 689–696.

Vance, E. D., Brookes, P. C. and Jenkinson, D. S. (1987b). Microbial biomass measurements in forest soils: the use of the chloroform fumigation–incubation method in strongly acid soils. *Soil Biol. Biochem.* 19, 697–702.

Vance, E. D., Brookes, P. C. and Jenkinson, D. S. (1987c). An extraction method for measuring soil microbial biomass C. *Soil Biol. Biochem.* 19, 703–707.

Voroney, R. P. and Paul, E. A. (1984). Determination of k_C and k_N *in situ* for calibration of the chloroform-incubation method. *Soil Biol. Biochem.* 16, 9–14.

Webster, J. J., Hampton, G. J. and Leach, F. R. (1984). ATP in soil: a new extraction procedure. *Soil Biol. Biochem.* 16, 335–342.

Wu, J., Brookes, P. C. and Jenkinson, D. S. (1996). Evidence for the use of a control in the fumigation–incubation method for measuring microbial biomass carbon in soil. *Soil Biol. Biochem.* 28, 511–518.

Wu, J., Joergensen, R. G., Pommerening, B., Chaussod, R. and Brookes, P. C. (1990). Measurement of soil microbial biomass C by fumigation–extraction — an automated procedure. *Soil Biol. Biochem.* 22, 1167–1169.

Wu, J., O'Donnell, A. G., He, Z. L. and Syers, J. K. (1994). Fumigation–extraction method for the measurement of soil microbial biomass-S. *Soil Biol. Biochem.* 26, 117–125.

2

The Role of the Soil Microbial Biomass in Cycling Nutrients

William R. Horwath

Department of Land, Air and Water Resources,
1110 Plant & Environmental Science Bldg.
University of California Davis
One Shields Ave.
Davis, CA 95616-8627, USA
wrhorwath@ucdavis.edu

Introduction

Soil microbial biomass is an important mediator of nutrient cycling in soils. The annual production of massive amounts of plant detrital material is largely decomposed by the soil microbial biomass and faunal components of soil. The decomposer activity of the soil microbial biomass is an important ecosystem trait that assures the recycling and stabilisation of essential nutrients that maintain net primary production (NPP). Despite the progression of genomic techniques to characterise DNA and RNA for taxonomic and functionality associated with decomposition processes and nutrient cycling, soil scientists and biogeochemical modellers are still relegated to conceptually aggregating the microbial biomass into a black box that performs these functions. The minimalist approach has been highly successful in quantifying soil carbon (C) sequestration, decomposition and nutrient cycling. In this chapter, the basic components of C and nutrient cycling are presented. It covers the evolution of nutrient cycles, the soil habitat, nutrient content of the soil microbial biomass and soil, decomposition of plant inputs to soil, nitrogen (N) mineralisation and uptake pathways, priming effect, influence of climate change on nutrient cycling and future research needs.

Evolution of Nutrient Cycles in the Terrestrial Environment

Microorganisms were the first life forms to evolve on the earth over 3.5 billion years ago. These early microorganisms were chemoautotrophs (obtaining energy from the oxidation of electron donors such as hydrogen sulphide and ferrous iron) and photoautotrophs (use of photon capture to acquire energy). Both were capable of converting carbon dioxide (CO_2) into organic compounds for biomass production. The evolution of photosynthetic cyanobacteria in the earth's early oceans began depleting the high levels of CO_2 and introduced molecular oxygen (O_2) into the atmosphere. These microorganisms required nutrients such as phosphorus and molybdenum to sustain NPP and relied on the nascent physiochemical processes, such as rock weathering, to provide a source of these essential nutrients. With photosynthesis providing an abundant C energy source, these microorganisms evolved the process of biological N fixation to increase N availability and introduced organic and reactive inorganic N to the oceans. On land, one of the first organisms was the lichen, a symbiosis between green alga (Trebouxia) or cyanobacteria (Nostoc) and fungi, primarily the Ascomycota that introduced photosynthesis and biological N fixation into the terrestrial environment about 450 million years ago. Lichens produced organic acids that accelerated rock weathering to increase the supply of essential nutrients and introduced organic matter and organic N into the terrestrial environment. This allowed for the evolution of advanced vascular plant life, such as Cooksonia during the late Silurian and early Devonian. The evolution of higher plant life accelerated the weathering of rocks through root exudates and decomposition of organic matter, providing for the increasing need for nutrients to support the burgeoning increase in NPP. The proliferation of vascular plants led to the vast deposition of organic matter to soils, the source of fossil fuels, such as coal, that helped propel humanity into the industrial age. It is hypothesised that the inability of early heterotrophic bacteria to degrade lignin, an essential component of the secondary cell wall of vascular plants, was partly responsible for the burial of large amounts of lignaceous plant material during the Carboniferous and Permian Periods (Robinson, 1990). As lignin-degrading organisms evolved later in the Paleozoic Period, the decomposition of dead plant materials was likely to have accelerated and led to an increase in nutrient cycling to support ever expanding higher plant life in the form of forests.

Today, the soil microbial biomass has evolved into an ecosystem trait that regulates energy transfers and nutrient cycling. These functions are critical for maintaining ecosystem productivity or NPP at all levels of the food web (Figure 2.1). Net primary producers are responsible for the majority of annual C inputs into soil. There is a consortium of micro- and macro-organisms or fauna capable of degrading and utilising specific molecular structures and components of plant litter, exudates and leachates. The soil microbial biomass is a key player throughout the soil food web, being both free living and essential symbiotic components of the digestive tracts of micro- and macro-fauna. The gut-inhabiting

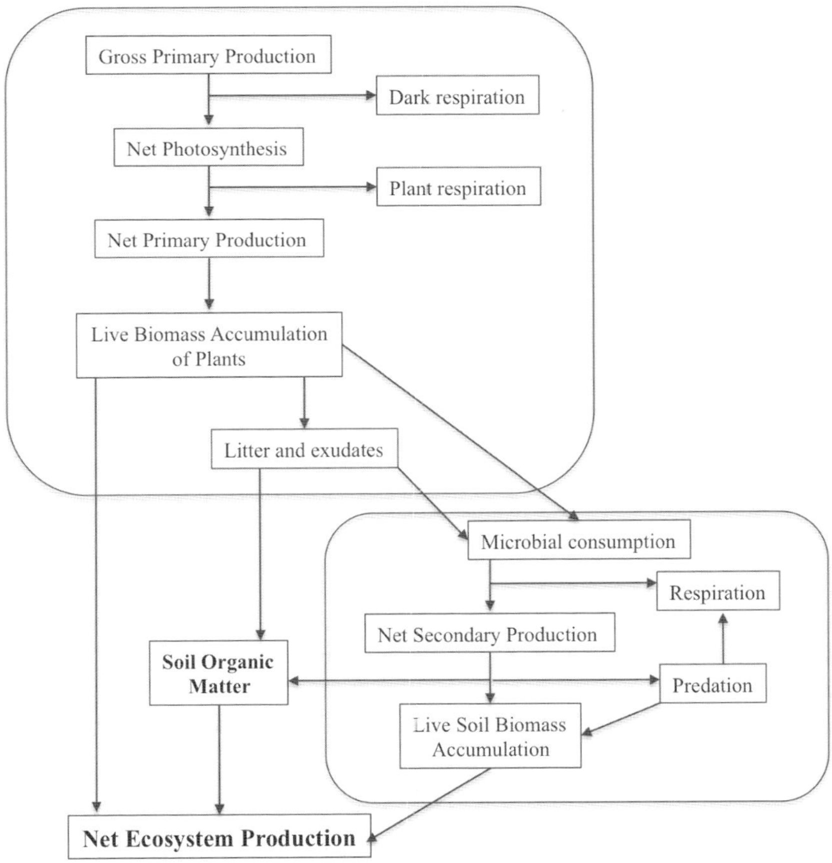

Figure 2.1 The plant (primary producers) and microbial biomass (secondary producers) components of photosynthesis and decomposer activity that interact to produce SOM. The sum of all activities is termed NEP.

Table 2.1 Soil, environmental and organismal factors affecting the distribution and activity of the soil microbial biomass.

Soil factor	**Environmental factor**	**Organismal factor**
Mineralogy	Rainfall	Plant cover
Parent material	Temperature	Net primary production
Soil age	Rain shadow	Vegetation history
Topography	Exposure (north vs. south)	Animal (grazing)
Soil pH	Elevation	Human influence
Water holding capacity	Landscape depression	Animal (burrowing)
Water infiltration	Water springs	
Particle size (sand, silt and clay)	History of burning	
Bulk density (g/cc)	Cultivation	
SOM content	Riparian influence	
Fertility	Pollution	
Erosion		

bacteria in termites are an example of a symbiosis that degrade secondary cell-wall polysaccharides such as cellulose. The turnover of NPP provides the energy source and substrates for soil microorganisms to build new biomass, termed net secondary production (NSP). The decomposer activity of the soil biomass spans days to decades and is dependent and regulated by temperature and moisture. Only a small fraction of NPP and NSP is transformed into soil organic matter (SOM), another ecosystem trait that imparts changes to soil properties, such as water-holding capacity and storage of essential nutrients both in its structure and through ion exchange. The residual C remaining after decomposition is net ecosystem production (NEP). The NEP is dynamic and tends to be an ecosystem characteristic influenced primarily by climate and secondarily by edaphic factors such as soil mineralogy (Rasmussen *et al.*, 2005). Other soil, environmental and organismal factors affecting NEP are shown in Table 2.1.

Soil Microbial Biomass Habitat

The soil microbial biomass is located in a complex matrix of soil minerals and organic matter. A gram of soil can contain more than 10^9 bacteria and hundreds of metres of fungal hyphae (Paul and Clark, 1996). The mass of microbial and faunal components per hectare is measured in tonnes (Table 2.2). The soil

Table 2.2 The mass of soil biomass components in soil from Killham (1994).

Soil biomass component	Tonnes per hectare
Bacteria	1–2
Fungi	2–5
Actinomycetes	1–2
Protozoa	Up to 0.5
Nematodes	Up to 0.2
Earthworms	0–2.5
Other fauna (collembola, mites, arthropods, etc.)	Up to 0.5

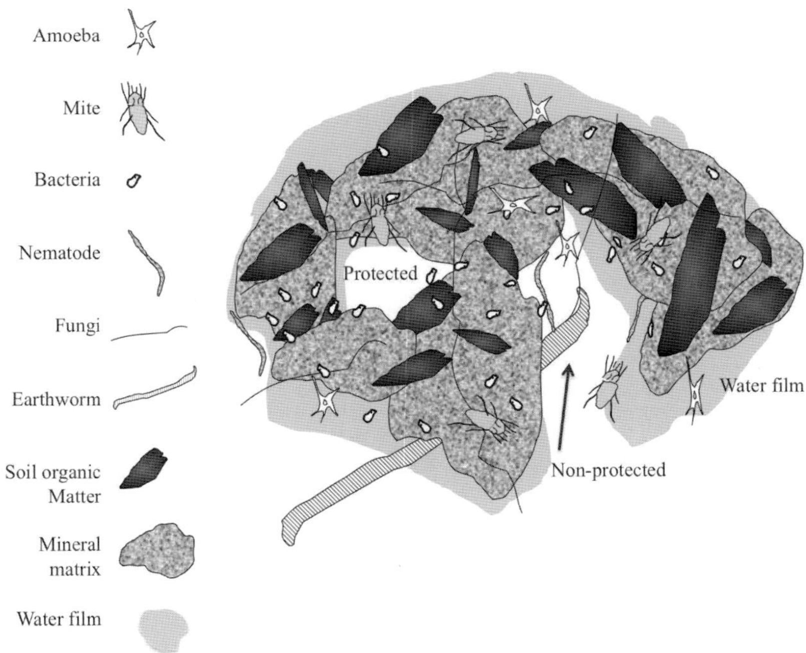

Figure 2.2 The soil ecosystem composed of minerals and SOM that interact to form aggregate structures that produce niches composed of different habitats and substrates. The protected space is isolated from O_2 exchange and substrates. The unprotected or open continuous pores have ample gas exchange and substrate flows.

microbial biomass inhabits numerous soil niches composed of different habitats and substrates (Figure 2.2). In most soils, the soil microbial biomass comprises about 5% of total soil C and about 1% of total soil N (Smith and Paul, 1990). Microbial biomass and activity is highest in the surface soil, about 10 cm in depth

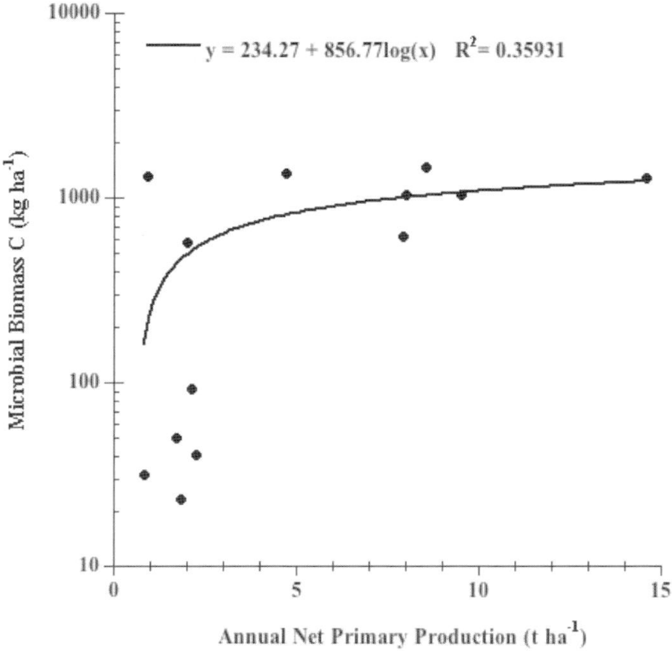

Figure 2.3 The relationship of NPP and soil microbial biomass C in late-successional ecosystems for the 10 cm soil depth. (Adapted from Zak *et al.*, 1994.)

in undisturbed systems. In agricultural systems, ploughing can homogenize the microbial biomass into the 20–30-cm soil depth. Both root growth and above-ground litter deposition enrich the shallow soil with C inputs. Figure 2.3 shows the relationship between soil microbial biomass C and annual NPP in late-successional ecosystems. The microbial biomass C begins to asymptote in size (700–1000 kg C) at about 2 t ha^{-1} of annual NPP. This suggests that other factors such as climate or nutrient limitation are the primary factors affecting the size of the microbial biomass. As NPP increases, there is likely to be increased competition for essential nutrients between the vegetation and the microbial biomass. In addition, as NEP accumulates in the form of SOM, available nutrients are tied up in its structure and immobilised. The stoichiometry of stable SOM is approximately 10% N and 1% each of P and S regardless of biome vegetation or soil type (Horwath, 2014). The soil microbial biomass declines rapidly with soil depth, relying on the movement of dissolved organic C and N and nutrients to support growth and persistence.

Soil microorganisms inhabiting smaller areas, such as non-continuous capillary pores, are protected from predation. They require adaptive strategies to cope with limited C inputs and nutrients. Those organisms inhabiting larger unprotected areas such as soil macropores benefit from continuous substrate and nutrient inputs but are subjected to increased grazing pressure by macrofauna, such as protozoa and nematodes. The concept of *r* and *K* selection describes the distribution of microorganisms based on ecological niche theory of substrate availability and growth (Killham, 1994). A soil microorganism adapted to bountiful energy and nutrient sources is designated as *r*-selected. Microorganisms existing under low energy and nutrient-deprived or oligotrophic conditions are termed *K*-selected. The distribution of the soil biomass can be related to specific soil size fractions (Bolton *et al.*, 1993). The greatest concentration of microbial biomass C is found in the largest and smallest soil particle sizes (Figure 2.4). The large particle sizes, >50 μm, are likely to be associated with larger pores. The small fractions (<2 μm) are likely to be distributed across both protected and unprotected areas. Though interesting, using operationally defined size fractions does

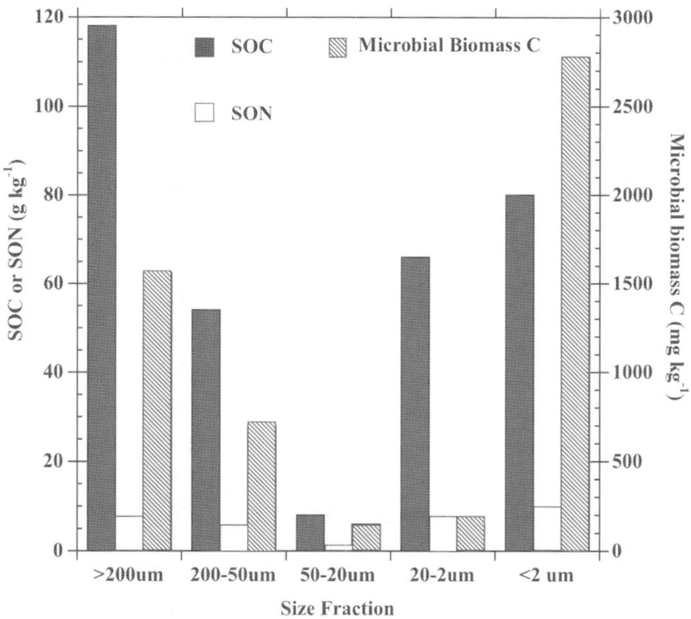

Figure 2.4 The distribution of soil microbial C in different C and N containing soil size fractions. (Adapted from Bolton *et al.*, 1993.)

not account for how the particles are arranged in soil and therefore the spatial distribution of the soil biomass cannot be easily interpreted. The physicochemical selection pressures lead to diversity in the composition and structure of the microbial community. The concept has been used to predict microbial activity in soil biogeochemical models.

The C, N and P in Plants, Microbial Biomass and SOM

Among plants, microorganisms and soils, there are defined stoichiometric ratios of C:N:P. The C inputs in ecosystems are primarily from NPP, which is constrained by the availability of N and P. The C:N:P ratios in SOM typically average 100:10:1 or about 55% C, 4–5% N and 0.5% P (Horwath, 2014). Both plants and microorganisms contain about 45–50% C. Plant tissue typically contains 1–3% N, whereas microorganisms contain 10–15% N. Plant tissue contains about 0.2% P, whereas microorganisms contain 1–1.5% P. The strict stoichiometry of C, N and P in organisms was observed in marine phytoplankton (Redfield, 1958). The elemental relationship referred to as the "Redfield Ratio" provides insights into the biogeochemical constraints of nutrient availability on NPP and NSP. Cleveland and Liptzin (2007) suggested that a similar relationship was found in soil microbial biomass with an average C:N:P ratio of 60:7:1. The relationship was consistent across different biomes. In contrast, Hartman and Richardson (2013), though finding similar stoichiometric relationship globally, found that this relationship was weakly correlated to microbial respiration, an important measurement of activity associated with nutrient cycling.

The relationship of soil C to microbial biomass is also constrained (Figure 2.5). A strong correlation exists from about 0 to 50 g soil C kg^{-1}, but the response asymptotes at about 1 g microbial biomass C kg^{-1} soil at higher levels of soil C. Similarly, the relationship of soil N to soil microbial N is strong until the soil N content reaches 2 g N kg^{-1} (Figure 2.6) or soil C content of 10 g kg^{-1} (Figure 2.7), then asymptotes at about 100 mg N kg^{-1} soil. A weaker relationship exists between soil P and soil microbial P (Figure 2.8). Hartman and Richardson (2013) suggest that N and P limit microbial activity and size. Nitrogen is needed to synthesise proteins, while P availability could limit the rate of protein synthesis by limiting the rate of production and turnover of ribosomes. As suggested earlier, increases in soil C are associated with increased NPP that could lead to competition for essential nutrients. In highly productive ecosystems, perennial plants can store essential nutrients from the soil pool and increases in SOM can immobilise nutrients. This could in part explain the apparent limits to microbial biomass C, N and P as soil C content increases.

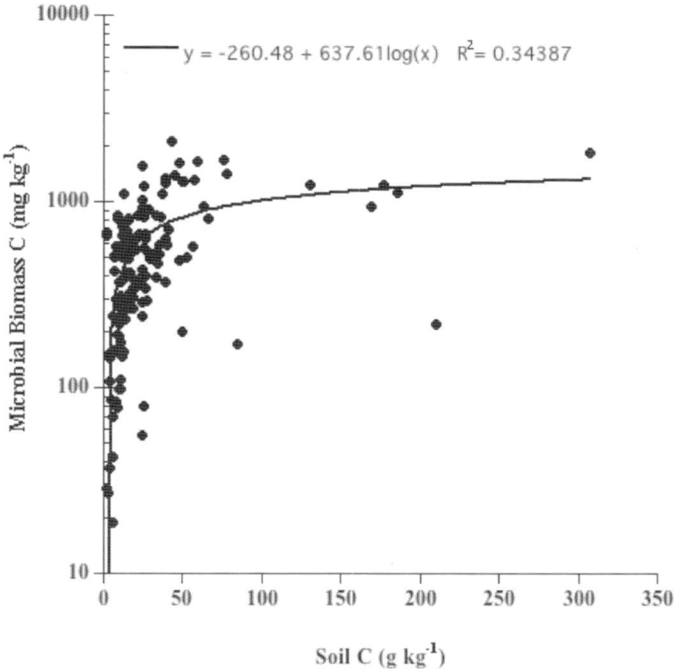

Figure 2.5 The relationship between soil microbial biomass C and soil C. (Adapted from Lynch and Panting, 1980; Brookes *et al.*, 1984; Doran, 1987; Powlson and Brookes, 1987; Kaiser *et al.*, 1992; Bolton *et al.*, 1993; Witter *et al.*, 1993; Zak *et al.*, 1994; Dalal, 1998; Witter and Kanal, 1998; Yao *et al.*, 2000; Barbhuiya *et al.*, 2004; Devi and Yadava, 2006; Cleveland and Liptzin, 2007; Iyyemperumal *et al.*, 2007.)

Decomposition of Plant Inputs

The most important function of the soil microbial biomass is decomposition of plant, microbial and SOM. During decomposition, the microbial biomass acquires (immobilises) and releases (mineralises) nutrients from plant litter. Without this vital activity, dead plant biomass or detritus would accumulate and limit available nutrients for NPP and NSP. The mineralisation of nutrients from the detritus is dependent on its quality. Detrital quality is primarily related to its N content. In both plant and microbial tissue, proteins contain the majority of the total N. Up to 70% of plant proteins are located in the photosynthetic enzyme ribulose bisphosphatase in leaves. Microorganisms contain up to 55% protein. From a bulk plant litter perspective, the cut-off for mineralisation during decomposition is below a C:N ratio of 25:1. The lignin content of litter can also

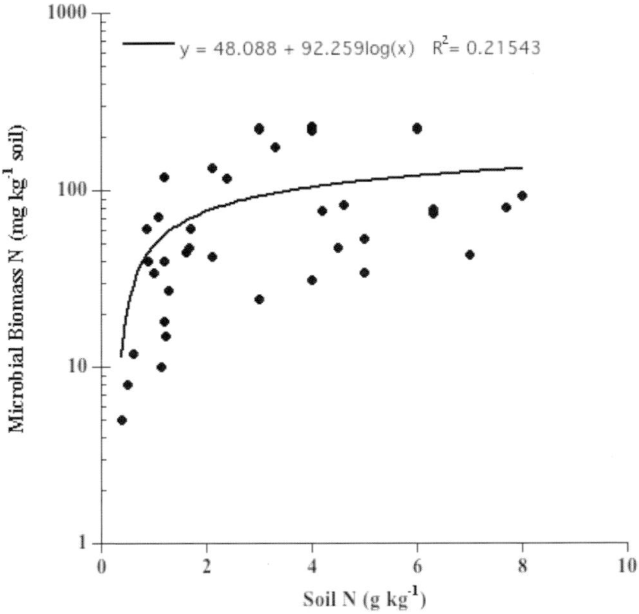

Figure 2.6 The relationship between soil microbial biomass N and soil N. See Figure 2.5 for data sources.

predict N mineralisation with increasing lignin content, leading to less N mineralisation (Jensen *et al.*, 2005).

The decomposition of the various constituents of both plant and microbial biomass is accomplished by groups of microorganisms that specialise using enzymes to degrade specific substrates. The various constituents of plants and microorganisms are biochemically diverse, requiring a diverse set of enzymes to degrade litter and expose the various energy and N-rich components. The first stage in decomposition involves degrading simple compounds of plant litter, such as simple lipids, sugars and starch. Lipids are main components of cell walls and are likely to be degraded quickly during the decomposition process. They represent a diverse class of compounds spanning fatty acids to complex sterols, phospholipids, cutins and suberins. High organic matter soils contain the most lipids. Clay soils contain more lipids than sandy soils. Microorganisms use a diverse set of lipases to degrade lipids. Representative microorganisms include *Pseudomonas aeruginosa* and *Burkholderia* sp. The extent of lipid degradation depends on their chemical complexity with long-chain aliphatic fatty acids and phospholipids degraded relatively quickly, while complex cutin and suberin are resistant to decomposition.

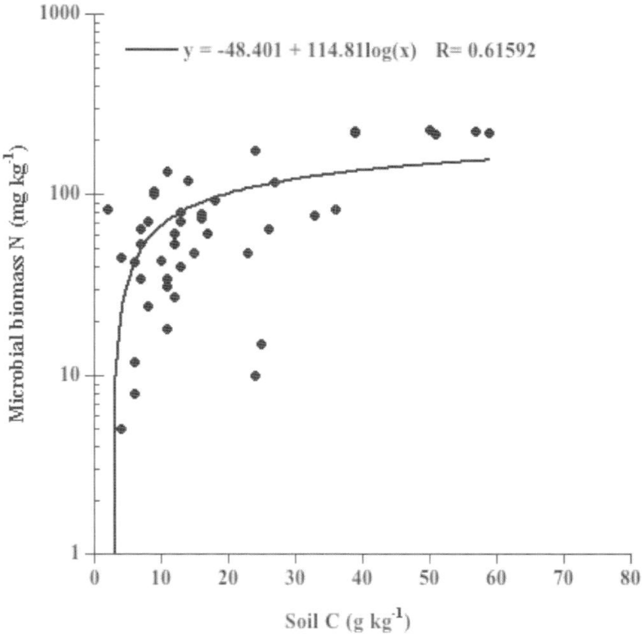

Figure 2.7 The relationship between soil microbial biomass N and soil C. See Figure 2.5 for data sources.

Starch is a simple polysaccharide used as an energy-storage compound in plants. The enzyme amylase degrades starch readily. Many bacteria and fungi produce amylases to convert insoluble starch into soluble, easily metabolised glucose. Representative bacteria include *Bacillus* (Firmicutes), and the Proteobacteria *Pseudomonas*, *Escherichia coli* and *Serratia marscens*. Fungi include *Trichoderma* and *Aspergillus* (both in the phylum Ascomycota) and *Rhizopus* (Zygomycota).

Once the simple components of litters become limiting, a subset of the microbial biomass initiates the decomposition of more complex cell structures. The plant secondary cell wall contains both complex polysaccharides like hemicellulose, pectins and cellulose, and complex hydrocarbons in the form of lignin. On an annual basis, hemicellulose, pectins and cellulose represent the largest biosynthetically produced polymers on earth. Hemicelluloses and pectin are synthesised from 5-C sugars or glycans. Hemicelluloses are reduced to simple sugars by several enzymes known collectively as pectinases. The degradation of pectin and hemicellulose occurs simultaneously. The degradation of hemicellulose/pectin is accomplished by pectinolytic soil bacteria that include *Erwinia* and

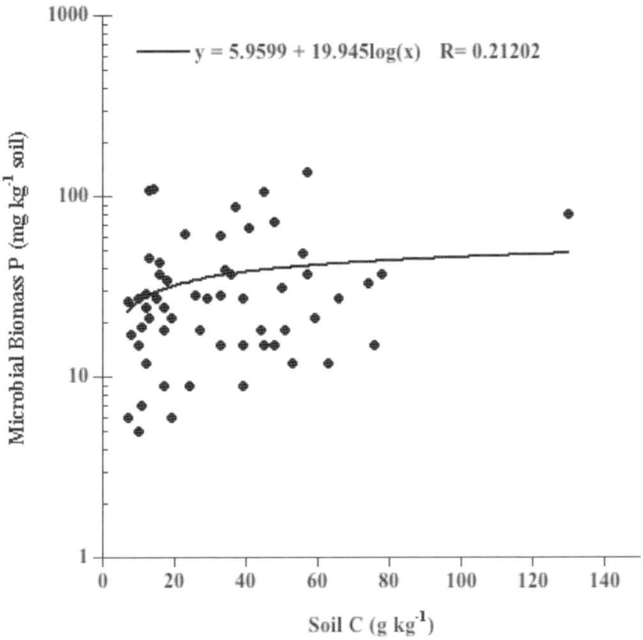

Figure 2.8 The relationship between soil microbial biomass P and soil C. See Figure 2.5 for data sources.

Pseudomonas, both Proteobacteria, *Arthobacter* (Actinobacteria) and *Bacillus* (Firmicutes). Common pectinolytic fungi *Aspergillus*, *Sclerotina*, *Penicillum* and *Fusarium* and *Rhyzopus* (Zygomycota).

Cellulose is an important component of the plant cell wall and exists as microfibrils. They are decomposed by a group of enzymes called cellulases. A wide range of fungi can partially degrade cellulose with representatives that include *Trichoderma*, *Acremonium celluloyticus*, *Penicillium*, *Fusarium* and *Agaricu.* Bacteria are also capable of degrading cellulose but to a limited extent. Common soil bacteria include *Cellulomas* (Actinobacteria), *Cellovibrio* and *Pseudomonas* (Proteobacteria) and *Bacillus* (Firmicutes). Anaerobic bacteria include *Acetobacter* (Proteobacteria), *Bacteriodes* (Bacteroidetes), *Clostridium* and *Ruminococcus (Firmicutes)* and *Fibrobacter* (Fibrobacteres).

Lignin is a unique hydrocarbon that allowed terrestrial plants to evolve a complex vascular tissue surrounded by a secondary cell wall. The basic structure of lignin is based on the phenyl propanoid unit, consisting of an aromatic ring and a 3-C side chain. The random assembly of hydrocarbons produces a hydrophobic non-specific structure that is difficult for enzymes to attack. Fungal lignin degraders represented by Imperfecti (Deuteromycota) and Ascomycetes (Ascomycota) perform incomplete degradation, preferring the polysaccharides exposed as the secondary wall structure collapses. Brown and white rot fungi (from the Phylum Basidiomycota) are more efficient degraders. They degrade lignin through demethylation to expose polysaccharides. Contrary to many laboratory and field studies, lignin degradation proceeds rapidly *in situ* exposing energy-rich polysaccharides and nutrients for microbial growth (Horwath and Elliott, 1995). Wedin *et al.* (1995) showed that lignin does not accumulate in soils using a natural abundance ^{13}C approach to verify the ability of the soil biomass to degrade lignin to increase litter and nutrient turnover.

During NSP, the turnover of the soil microbial biomass or NSP also provides substrates for microbial growth and activity (Throckmorton *et al.*, 2012). A major component of the fungal cell wall is acetylglucosamine, which contains N and is likely to be a source of mineralisable N. Bacterial cell walls are also composed of *N*-acetylglucosamine and *N*-acetylmuramic acid chains. The cross-linkage peptidoglycan (murein) structures are composed of repeating units of l-alanine, d-alanine d-glutamic acid and either lysine or diaminopimelic acid. Representative microorganisms that degrade chitin include *Streptomyces*, *Pseudomonas*, *Bacillus* and *Clostridium*.

Soil Microbial Biomass and Nutrient Cycling

Microorganisms are thought to be more competitive for nutrients due to their constant activity and larger surface area compared to plant roots. This perception is fostered by the fact that all root systems contain a sizeable community of microorganisms, both free-living and symbiotic, defined as the rhizosphere (Tinker, 1984). Plant species can affect nutrient cycling in a variety of ways through litter quality, symbiotic specificity, patterns of nutrient uptake, root exudates, storage of nutrients, herbivory and a host of other factors such as competition for water (Hobbie, 1992). The functioning of ecosystems is highly related to nutrient cycling. Evidence suggests that below-ground biodiversity is a strong regulating factor that shapes above-ground biodiversity (Tilman *et al.*, 2006). However, microorganisms, both free-living and symbiotic, are dependent on plant C inputs and therefore plants create a positive feedback on both soil biomass size and activity. This was noted early in Figure 2.3, where NPP was shown to be a strong determinant of microbial biomass size. Microbial biomass size, however, is not likely to be related to microbial diversity. There is still a notable lack of information on the spatial and temporal distribution patterns of soil biodiversity that is needed to predict interactions of plant communities with the physicochemical properties of soil (Bardgett and van der Putten, 2014).

The turnover of C, N and P through the microbial biomass is important to understand ecosystem-level nutrient cycling. The turnover of microbial C is a measure of the soil biomass activity and can be related to N and P mineralisation and availability (Hartman and Richardson, 2013). The soil biomass serves both as a sink (immobilisation) and source (mineralisation) of nutrients playing a key role in influencing the availability of nutrients for NPP and NSP. Plants also influence the cycling of nutrients by competing directly for nutrients through root and mycorrhizae-mediated uptake. For example, in forest systems, ectomycorrhizal mycelium can comprise one-third of the total soil microbial biomass (Högberg and Högberg, 2002).

Nitrogen mineralisation

Nitrogen isotope pool dilution studies along with methods to determine the N content of the soil microbial biomass have been extensively reported in the soils literature. A widespread conclusion of these studies is that gross N-mineralisation is often positively correlated with microbial biomass size and soil C and N concentrations (Booth *et al.*, 2005). As expected, there is a negative relationship between soil C-to-N ratio and N mineralisation. Overall, N mineralisation is highly controlled by substrate availability, which is related to NPP, and to quality factors such as C-to-N or lignin-to-N ratio of litter inputs. Though simplistic, these narrow ecosystem attributes have been useful in predicting N availability and soil C accumulation in biogeochemical models (Parton *et al.*, 1988). The C-to-N ratio of new litter inputs most accurately describes N availability initially, but this relationship diminishes rapidly for multi-year observations. This is especially true in agricultural soils where both soil N and microbial biomass size is often reduced. However, since more than 50–60% of the litter is decomposed in the first year, the use of the litter C-to-N ratio remains important for describing N cycling (Qiao *et al.*, 2014).

The common view on the outcome of N mineralisation in soils is that plants use only inorganic N left over from the activity of the soil microbial biomass. This is the conceptual basis of most biogeochemical models that describe the N cycling relationship between decomposition of SOM, plant inputs and plant uptake (Figure 2.9). This characterisation of decomposition is now understood to include depolymerisation of the most abundant N pool in plant litter and soil that

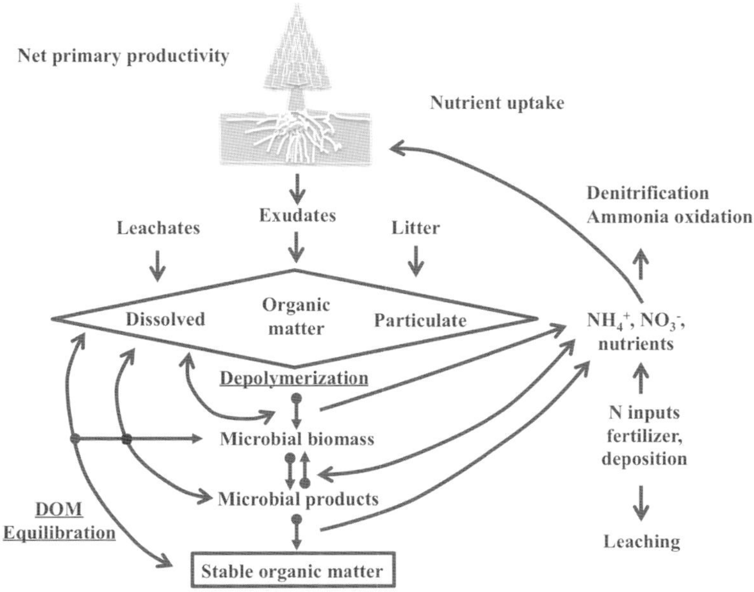

Figure 2.9 Conceptual model describing N cycling relationships between decomposition of SOM and plant inputs and plant uptake. The role of depolymerisation and DOM equilibrium is underlined.

is composed of proteins (Schimel and Bennett, 2004). The depolymerisation is performed by enzymes such as proteases that reduce proteins to soluble peptides and amino acids. Both plants and microorganisms can use these soluble N sources, but demonstration of plant uptake in significant amounts is lacking. The uptake of peptides by plants is less probable and likely requires conversion to amino acids. Näsholm *et al.* (1998), in a review, confirm that plants, especially those with mycorrhizal-infected roots, have the ability to take up amino acids. However, they concluded that there is insufficient evidence that the uptake of organic N by plants supplies significant quantities of the N required. Moran-Zuloaga *et al.* (2015), using position-specific ^{15}N-labelling of alanine, concluded that the majority of N taken up by roots is deaminated outside roots. This suggests that most of the organic N in the rhizosphere is converted into inorganic N by the microbial community before root uptake.

The importance of the direct utilisation of organic N uptake by the soil microbial biomass is highly debated. The uptake of organic N for growth and maintenance would provide a competitive advantage to the microbial biomass by avoiding competition with plants. However, gross N mineralisation studies using ^{15}N pool dilution assume all litter and soil N is mineralised to ammonium before immobilisation or nitrification by the microbial biomass (Kirkham and Bartholomew, 1954). Barraclough (1997) proposed a modification to the Kirkham and Bartholomew (1954) set of equations on mineralisation–immobilisation turnover (MIT) by accounting for the C-to-N ratio of the biomass and substrate to describe the mineralisation of more than one amino acid using the following equation:

$$P = \left(1 - S * \frac{E}{B}\right),$$

where P is the proportion of N mineralised from the substrate, S the C-to-N ratio of the substrate, B the C-to-N ratio of the uptake of N by the soil biomass and E the ratio of assimilated C to the substrate C. During decomposition of two amino acids, the value of E/B predicts the proportion of N mineralised from each of the amino acids.

To account for the possibility of direct uptake of amino acids by the soil biomass, Manzoni and Porporato (2007) proposed a mathematical model that simulates the simultaneous operation of the direct and MIT routes. Based on the model, the fraction (n) of the organic N taken up via the direct route is calculated with the following equation:

$$n = 1 - \left(\frac{\mathrm{C}}{\mathrm{N}}\right)_{\mathrm{s}} \frac{\mathrm{Min}_{\mathrm{gross}}}{\mathrm{DEC}},$$

where (C/N) is the C-to-N ratio of the substrate, $\mathrm{Min}_{\mathrm{gross}}$ the gross N mineralisation rate and DEC the C flux from the decomposing substrate. DEC equals the respiration rate divided by the respired C fraction. The fraction of direct uptake of organic N by the soil microbial biomass was 55–62% under varying C and N availability (Geisseler *et al.*, 2009). Interestingly, protease activity is not correlated to gross N mineralisation and was more dependent on the form of N with increased protein leading to higher expression of enzyme activity (Allison and Vitousek, 2005). This suggests that under ample available N supply, protease activity may be induced as a response to the need for C (Geisseler and Horwath, 2008). The results show that when NH_4^+ availability is high, the utilisation of organic N sources is low. In contrast, when N availability is low, enzymes such as proteases are activated and the direct uptake of N by the microbial biomass is favoured. These processes are collectively called N regulation (Geisseler *et al.*, 2009). The regulation of N cycling is important in influencing the availability of N and key to the outcome of competition for N between soil microbial biomass and plants. Plants can use mycorrhizal relationships to influence this competition, however, as stated earlier, evidence on the significance of the direct uptake of organic N by this symbiosis is lacking (Näsholm *et al.*, 1998).

Most studies concerning N mineralisation have used substrate inputs such as plant litters and inorganic N to study N cycling in soil. An underappreciated source of available N in soils is the N associated with stable SOM. The mineralisation of stable organic matter as a source of N is difficult to differentiate from that associated with labile soil N such as recent plant litter inputs and particulate organic matter, both free and occluded within aggregates. The mineralisation rate of SOM is measured in decades compared to turnover times of years for plant litters and other labile soil C pools. It is therefore assumed that stable SOM contributes less to N cycling than labile pools (Horwath, 2014). However, mineralisation is not likely to be the only pathway where stable N is converted into labile N sources.

SOM sources for carbon and nitrogen cycling

Kemmitt *et al.* (2008) posited that in the absence of fresh inputs of litter and exudates, dissolved organic matter in soil solution originates from stable SOM. They described the process as an equilibrium where the microbial biomass consumes dissolved organic matter (DOM) and, in turn, more would equilibrate into soil solution (Figure 2.9). This process is linked to microbial activity indirectly, where the actual source of DOM is not produced from decomposition activities directly. It is, however, influenced by microbial activity, through its consumption of DOM where it creates a concentration gradient allowing more DOM to equilibrate from stable organic matter to replenish the soil solution. Toosi *et al.* (2012) confirmed this equilibrium process in sterile soils after adding ^{13}C-labelled DOM and observed that only 80% of the added DOM was recovered and 20% originated from the stable organic pool. This occurred after repeated applications, with the total amount of C added in DOM recovered but consisting of 20% unlabelled DOM from the SOM, showing an active exchange process.

The process of equilibrium from stable organic matter into soil solution exhibits a zero-order rate function due to the large amount of stable soil C compared to inputs from other sources such as root exudates and litter leachates. This soil characteristic is critical for maintaining microbial activity. The constant input from all sources to the DOM pool is required to maintain stable nutrient cycling rates required to sustain NPP and NSP. For this reason, soils with large amounts of organic matter maintain a higher level of NPP, creating a positive feedback where the stable SOM is constantly replenished.

Priming effect

Soil priming is a process describing the effect of recent inputs associated with root exudates and litter, such as crop residues, on the microbial biomass utilisation of other soil C sources. The term was first used by Bingeman *et al.* (1953) about the effects of fresh organic material on the increased decomposition of SOM. Other early studies observed the effect of added ^{15}N-labelled N on the increased mineralisation of soil N (Jansson, 1958). In soil microbial studies, the addition of glucose increased the mineralisation of soil C (van Veen *et al.*, 1985). Jenkinson *et al.* (1985) used the term "added N interaction" to describe the effect that N addition to soil has on soil organic N. The priming effect can be positive or negative. In positive priming, fresh organic inputs increased the mineralisation of soil C and/or N. Negative priming occurs when the background rate of soil C or N mineralisation declines in the presence of organic inputs. The influence of organic inputs on the decomposition of other soil C or N pools will influence nutrient cycling rates by both affecting microbial activity and the amounts of nutrients available for NPP.

In past studies, soil priming gained attention due to its effect on soil C stability on the fate of soil N (Kuzyakov *et al.*, 2000). The priming effect has been attributed to laboratory incubations where soil disturbance, such as sieving, was thought to influence microbial activity. Kuzyakov (2010) defined the priming effect as an interaction between living microbial biomass and dead organic matter and is now shown to occur in any soil. It is a sequence of processes occurring in the rhizosphere and detritus sphere and induced by the continuous input of fresh organic materials. In the rhizosphere, the priming effect linked to microbial activity is influenced by the nature of root exudates, availability of nutrients and water relationships (deficit to abundance) related to evapotranspiration (Cheng, 2009). The source of SOM being mineralised can range from labile, such as free particulate to stable organic matter. However, the source of C and N needs further investigation as many studies on priming effects add a labile ^{13}C- or ^{15}N-labelled input and cannot differentiate the source (labile to stable) of soil C or N being affected (Moreno-Cornejo *et al.*, 2015). Though the priming effect has been demonstrated in numerous studies, its influence on nutrient cycling is not well established. Priming is a ubiquitous characteristic affecting microbial activity and the amounts of nutrients available for primary producers.

Turnover of NPP Globally

The turnover of NPP represents the first step in recycling nutrients in an ecosystem. The total global terrestrial above-ground plant biomass is estimated at 686 Pg C (Table 2.3). Total global below-ground organic C is approximately

Table 2.3 Mean soil C contents, C inputs, terrestrial C stocks and turnover for the major earth biomes.

Biome	Mean soil C (kg m^{-2})	C inputs (kg m^{-2} $year^{-1}$)	Terrestrial C stock (Pg C)			K ($year^{-1}$)	Turnover (year)
			Plant	Soil (0–3 m)	Total		
Tropical forests	12	2.03	340	692	1032	0.187	5.3
Temperate forests	8.7	0.85	139	262	401	0.162	6.2
Boreal forests	16.4	0.5	57	150	207	0.105	9.5
Tropical savannas	5.4	0.48	79	345	424	0.063	15.9
Temperate grasslands	13.3	0.3	23	172	195	0.039	25.4
Mediterranean shrub land	7.6	0.46	17	124	141	0.06	16.7
Deserts	3.4	0.08	10	208	218	0.099	10.1
Tundra	19.6	0.1	2	400	402	0.019	52.3
Croplands	7.9	0.48	4	248	252	0.159	6.3
Wetlands	72.3	0.17	15	450	465	0.001	945.4

Adapted from Amundson (2001), Jobbágy and Jackson (2000) and IPCC (2007).

3051 Pg in the upper 3 m of soil (Jobbágy and Jackson, 2000). The accumulation of soil C in the soil surface is from above-ground litter accumulation and root inputs, particularly fine root turnover. Inorganic soil C or soil carbonates are estimated to be up to 748 Pg to a depth of 100 cm but are rarely included in global C estimates (Batjes, 1996). Forests tend to have the highest NPP ranging from 0.5 to 2.03 kg biomass m^{-2} $year^{-1}$. Forests also tend to have the highest turnover rates (<10 years) of NPP resulting in less soil C in relation to C inputs. The high turnover rate is required to cycle nutrients rapidly and sustain NPP. Climatic conditions of higher rainfall and temperature compared to grass systems maintain a high activity of the microbial biomass in forests. Approximately half the soil C turns over very slowly, with mean residence times often of thousands of years (Paul *et al.*, 2001). Therefore, using total soil C to calculate ecosystem soil C turnover is misleading. Approximately 20% of the soil C, including primarily the light or particulate organic C fraction, turns over completely in less than 3 years (Horwath *et al.*, 2002). This represents the majority of NPP and a significant amount of total nutrients recycled, particularly available N and P.

The highest accumulation of C is not always associated with high NPP. Wetlands with low redox potential that retards turnover rates have high soil C (33 kg m^{-2}) but relatively low C inputs (0.17 kg m^{-2} $year^{-1}$). In cold tundra and boreal forest soils, turnover of NPP is limited by temperature, causing C and N to accumulate as litter and particulate organic C. Systems where NPP accumulates due to lack of decomposition can become nutrient-limited. These systems can benefit from disturbance, such as drought in wetlands and fire in forests, to accelerate nutrient cycling.

Climate Change Impact on Soils

Climate change has two pronounced effects that can affect NPP and NSP. Rising atmospheric CO_2 levels should lead to higher NPP as suggested by numerous growth chamber studies over the last three decades. Associated increases in air and soil temperature have also been shown to have positive effects on NPP, but this largely depends on moisture availability.

Ecosystem-level experiments that manipulate atmospheric CO_2 and soil temperature have provided valuable insights into the impacts of ongoing climate change. Carbon dioxide manipulation experiments such as free-air CO_2 enrichment (FACE) and open top chambers have revealed that terrestrial plants, particularly tree species, respond initially to elevated CO_2 but soon acclimate to previous growth rates (Ainsworth and Long, 2005; Huang *et al.*, 2007). The results confirm early growth chamber experiments showing light-saturated C fixation, diurnal assimilation, growth and above- and below-ground production are likely to be limited by soil resources, mainly nutrients and water. Though stomatal conductance decreases under elevated CO_2 conditions, it is apparently insufficient to maintain increases in growth alone (Ainsworth and Long, 2005). Specific reasons for the lack of a CO_2 fertilisation effect besides changes in soil resource availability and environment (increased temperature/changes in moisture status) have been difficult to discern and are likely to result from a negative interaction of combined factors. One important factor that could have a negative outcome on nutrient cycling is that elevated CO_2 often leads to C-enriched biomass or higher C-to-N ratio litters that could negatively influence decomposition rates (Cotrufo *et al.*, 1994). Silva and Horwath (2013) reviewed 66 studies on tree growth in the last century and concluded only six showed increases in NPP. The review confirms that other factors, such as progressive nutrient limitation or drought stress, are likely to offset any positive potential gains from CO_2 fertilisation. The results from elevated CO_2 tree experiments are important since forest ecosystems have the highest global NPP, which may be impacted negatively by elevated CO_2 and result in a reduced terrestrial C sink.

Changes in NPP as climate changes, specifically warming, could lead to increased decomposition rates and lower soil C storage potential. Davidson and Janssens (2006) estimated that the C stored in the litter layer above the mineral soil will decrease by 15% by 2100. The loss of the soil litter layer will most likely lead to the same magnitude of C depletion in mineral soil horizons (Jenkinson *et al.*, 1991). The change in the structure and physiological response of microbial communities to elevated CO_2 and temperature are also likely to influence decomposition rates (Zogg *et al.*, 1997; Allison *et al.*, 2010). Increases in soil temperature will increase microbial activity through the Q_{10} effect, although the effect diminishes with increased warming compared to increases starting at a lower temperature (Kirschbaum, 1995). Fontaine *et al.* (2004) showed that increased soil C inputs can lead to decreases in soil C as a result of nutrient limitation, in turn leading to higher decomposition rates or a positive priming effect. In a meta-analysis of elevated CO_2 effects on plant growth and nutrient cycling, de Graaf *et al.* (2006) showed that the microbial C-to-N ratio increased by 3.8%, microbial N and C by 5.8 and 7.1%, respectively, and soil respiration by 17.1%. They concluded that soil C increases only occurred when N inputs increased.

Changes in the priming effect caused by increased microbial activity and changes in their physiological response has been hypothesised to change potential C sinks in soil to sources under elevated CO_2 and warming (Carney *et al.*, 2007). Soil warming experiments generally show increased microbial activity and N mineralisation, but the results are highly variable depending on landscape and biome (Rustad *et al.*, 2001). These factors will likely lead to progressive nutrient limitation through immobilisation of nutrients in the soil biomass and potentially in aggrading litter-derived poor-quality particulate organic C pools. Regional N deposition from anthropogenic sources could reverse some of the negative aspects of elevated CO_2. However, the negative consequences of N deposition including increased greenhouse gases, soil nutrient depletion from acidification, eutrophication, etc. will likely have a greater negative environmental impact.

Future Research Needs

The nutrient cycling activities of the soil microbial biomass is an important ecosystem trait that influences NPP and NSP. In their search for energy and nutrients to grow and persist, the soil microbial biomass relies on plant litter and soil nutrients. In their search for energy, nutrients may be made available for plant growth. For this reason, the soil microbial biomass is considered the source and sink of essential plant nutrients. Future research directions should reveal the different microbial taxa involved in processing the complex inputs from NPP and NSP, so that the understanding of factors affecting N cycling can move beyond the black box. Specifically, more research is needed on the direct uptake of organic N by microorganisms. Currently most biogeochemical models assume that inorganic N is the primary N source for both plants and the microbial biomass, as suggested in the set of equations developed for MIT six decades ago by Kirkham and Bartholomew (1955). The soil "priming effect" is currently a hot topic in soil science but factors explaining the phenomenon are still in the hypothesis stage. The priming effect is known to be influenced by temperature, moisture and substrate quality, all factors expected to change with elevated CO_2 and warming. The outcome of no or minimal CO_2 fertilisation effects on NPP in elevated CO_2 experiments was not anticipated and suggests we cannot rely on terrestrial systems as a major C sink to combat climate change without a better understanding of the controls on soil resources. The wide diversity and function of the soil microbial biomass is currently poorly understood. New genome-level techniques will provide insights into taxa and functionality, something currently not well understood or linked, but needed to cope with and mitigate the effects of climate change on the microbial biomass and soils.

References

Ainsworth, E. A. and Long, S. P. (2005). What have we learned from 15 years of free-air CO_2 enrichment (FACE)? A meta-analytic review of the responses of photosynthesis, canopy. *New Phytol.* 165, 351–371.

Allison, S. D. and Vitousek, P. M. (2005). Responses of extracellular enzymes to simple and complex nutrient inputs. *Soil Biol. Biochem.* 37, 937–944.

Allison, S. D., Wallenstein, M. D. and Bradford, M. A. (2010). Soil-carbon response to warming dependent on microbial physiology. *Nat. Geo.* 3, 336–340.

Amundson, R. (2001). The carbon budget in soils. *Ann. Rev. Earth Planet. Sci.* 2, 535–562.

Barbhuiya, A. R., Arunachalam, A., Pandy, H. N., Arunachalam, K. A., Khan, M. L. and Nath, P. C. (2004). Dynamics of soil microbial biomass C, N, and P in disturbed and undisturbed stands of a tropical wet-evergreen forest. *Eur. J. Soil Biol.* 40, 113–121.

Bardgett, R. D. and van der Putten, W. H. (2014). Below-ground biodiversity and ecosystem functioning. *Nature* 515, 505–511.

Barraclough, D. (1997). The direct or MIT route for nitrogen immobilization: a ^{15}N mirror image study with leucine and glycine. *Soil Biol. Biochem.* 29, 101–108.

Batjes, N. H. (1996). Total carbon and nitrogen in the soils of the world. *Eur. J. Soil Sci.* 47, 151–163.

Bingeman, C. W., Varner, J. E. and Martin, W. P. (1953). The effect of the addition of organic materials on the decomposition of an organic soil. *Soil Sci. Soc. Am. Proc.* 17, 34–38.

Bolton Jr., H., Smith, J. L. and Link, S. O. (1993). Soil microbial biomass and activity of a disturbed and undisturbed shrub-steppe ecosystem. *Soil Biol. Biochem.* 25, 545–552.

Booth, M. S., Stark, J. M. and Rastetter, E. (2005). Controls on nitrogen cycling in terrestrial ecosystems: a synthetic analysis of literature data. *Ecol. Monogr.* 75, 139–157.

Brookes, P. C., Powlson, D. S. and Jenkinson, D. S. (1984). Phosphorus in the soil microbial biomass. *Soil Biol. Biochem.* 16, 169–175.

Carney, K. M., Hungate, B. A., Drake, B. G. and Megonigal, J. P. (2007). Altered soil microbial community at elevated CO_2 leads to loss of soil carbon. *Proc. Natl. Acad. Sci.* 104, 4990–4995.

Cheng, W. (2009). Rhizosphere priming effect: its functional relationships with microbial turnover, evapotranspiration, and C–N budgets. *Soil Biol. Biochem.* 41, 1795–1801.

Cleveland, C. C. and Liptzin, D. (2007). C:N:P stoichiometry in soil: is there a "Redfield ratio" for the microbial biomass? *Biogeochemistry* 85, 235–252.

Cotrufo, M. F., Ineson, P. and Rowland, A. P. (1994). Decomposition of tree leaf litters grown under elevated CO_2: effect of litter quality. *Plant Soil* 163, 121–130.

Dalal, R. C. (1998). Soil microbial biomass—what do the numbers really mean? *Aust. J. Expt. Agric.* 38, 649–665.

Davidson, E. A. and Janssens, I. A. (2006). Temperature sensitivity of soil carbon decomposition and feedbacks to climate change. *Nature* 440, 165–173.

De Graaf, M. A., van Groenigen, K. J., Six, J., Hungate, B. and van Kessel, C. (2006). Interactions between plant growth and soil nutrients cycling under elevated CO_2: a meta-analysis. *Global Change Biol.* 12, 2077–2091.

Devi, N. B. and Yadava, P. S. (2006). Seasonal dynamics in soil microbial biomass C, N and P in a mixed-oak forest ecosystem of Manipur, North-east India. *Appl. Soil Ecol.* 31, 220–227.

Doran, J. W. (1987). Microbial biomass and mineralizable nitrogen distributions in no-tillage and plowed soils. *Biol. Fert. Soils* 5, 68–75.

Fontaine, S., Bardoux, G., Benest, D., Verdier, B., Mariotti, A. and Abbadie, L. (2004). Mechanisms of the priming effect in a savannah soil amended with cellulose. *Soil Sci. Soc. Am. J.* 68, 125–131.

Geisseler, D. and Horwath, W. R. (2008). Regulation of extracellular protease activity in soil in response to different sources and concentrations of nitrogen and carbon. *Soil Biol. Biochem.* 40, 3040–3048.

Geisseler, D., Horwath, W. R. and Doane, T. A. (2009). Significance of organic nitrogen uptake from plant residues by soil microorganisms as affected by carbon and nitrogen availability. *Soil Biol. Biochem.* 41, 1281–1288.

Hartman, W. H. and Richardson, C. J. (2013). Differential nutrient limitation of soil microbial biomass and metabolic quotients (qCO_2): is there biological stoichiometry of soil microbes? *PLoS ONE* 8, 1–14.

Hobbie, S. E. (1992). Effects of plant species on nutrient cycling. *Tree* 7, 336–339.

Högberg, M. N. and Högberg, P. (2002). Extramatrical ectomycorrhizal mycelium contributes one-third of microbial biomass and produces, together with associated roots, half the dissolved organic carbon in a forest soil. *New Phytol.* 154, 791–795.

Horwath, W. R. (2014). *Carbon Cycling: The Dynamics and Formation of Organic Matter. Soil Microbiology, Ecology and Biochemistry*. Elsevier, London, pp. 339–382.

Horwath, W. R., Devêvre, O. C., Doane, T. A., Kramer, A. W. and van Kessel, C. (2002). Soil C sequestration management effects on N cycling and availability. In Kimble, J. M., Lal, R. and Follett, R. F. (eds.), *Agricultural Practices and Policies for Carbon Sequestration in Soil*. Lewis Publishers, Boca Raton, FL, pp. 155–164.

Horwath, W. R. and Elliott, L. F. (1995). Ryegrass straw component decomposition during mesophilic and thermophilic incubations. *Biol. Fert. Soils* 21, 227–232.

Huang, J. G., Bergeron, Y., Denneler, B., Berninger, F. and Tardif, J. (2007). Response of forest trees to increased atmospheric CO_2. *Crit. Rev. Plant Sci.* 26, 265–283.

IPCC. (2007). In Solomon, S., Qin, D., Manning, M., Chen, Z., Marquis, M., Averyt, K. B., Tignor, M. and Miller, H. L. (eds.), *Climate Change 2007: The Physical Science Basis*. Contribution of Working Group I to the Fourth Assessment Report of the Intergovernmental Panel on Climate Change. Cambridge University Press, Cambridge, United Kingdom and New York, USA.

Iyyemperumal , K., Israel, D. W. and Shi, W. (2007). Soil microbial biomass, activity and potential nitrogen mineralization in a pasture: impact of stock camping activity. *Soil Biol. Biochem.* 39, 149–157.

Jansson, S. L. (1958). Tracer studies on nitrogen transformations in soil with special attention to mineralisation immobilisation relationship. *Ann. R. Agric. Coll. Sweden* 24, 101–361.

Jenkinson, D. S., Adams, D. E. and Wild, A. (1991). Model estimates of CO_2 emissions from soil in response to global warming. *Nature* 351, 304–306.

Jenkinson, D. S., Fox, R. H. and Rayner, J. H. (1985). Interactions between fertilizer nitrogen and soil nitrogen—the so-called 'priming' effect. *J. Soil Sci.* 36, 425–444.

Jensen, L. S., Salo, T., Palmason, F., Breland, T. A., Henriksen, T. M., Stenberg, B. *et al.* (2005). Influence of biochemical quality on C and N mineralization from a variety of plant materials in soil. *Plant Soil* 273, 307–326.

Jobbágy, E. G. and Jackson, R. B. (2000). The vertical distribution of soil organic carbon and its relation to climate and vegetation. *Ecol. Appl.* 10, 423–436.

Kaiser, E. A., Mueller, T., Joergensen, R. G., Insam, H. and Heinemyer, O. (1992). Evaluation of methods to estimate the soil microbial biomass and the relationship with soil texture and organic matter. *Soil Biol. Biochem.* 24, 675–683.

Kemmitt, S. J., Lanyon, C. V., Wen, W. Q., Addiscott, T. M., Bird, N. R. A., O'Donnell, A. G. and Brookes, P. C. (2008). Mineralization of native soil organic matter is not regulated by the size, activity or composition of the soil microbial biomass e a new perspective. *Soil Biol. Biochem.* 40, 61–73.

Killham, K. (1994). *Soil Ecology*. Cambridge University Press, New York.

Kirkham, D. and Bartholomew, W. V. (1954). Equations for following nutrient transformations in soil, utilizing tracer data. *Soil Sci. Soc. Proc.*, 18, 33–34.

Kirkham, M. D. and Bartholomew, W. V. (1955). Equations for following nutrient transformations in soil utilizing tracer data. II. *Soil Sci. Soc. Am. Proc.* 19, 189–192.

Kirschbaum, M. U. F. (1995). The temperature dependence of soil organic matter decomposition, and the effect of global warming on soil organic C storage. *Soil Biol. Biochem.* 27, 753–760.

Kuzyakov, Y. (2010). Priming effects: interactions between living and dead organic matter. *Soil Biol. Biochem.* 42, 1363–1371.

Kuzyakov, Y., Friedel, J. K. and Stahr, K. (2000). Review of mechanisms and quantification of priming effects. *Soil Biol. Biochem.* 32, 1485–1498.

Lynch, J. M. and Panting, L. M. (1980). Variations in the size of the soil biomass. *Soil Biol. Biochem.* 12, 547–550.

Manzoni, S. and Porporato, A. (2007). A theoretical analysis of nonlinearities and feedbacks in soil carbon and nitrogen cycles. *Soil Biol. Biochem.* 39, 1542–1556.

Moran-Zuloaga, D., Dippold, M., Glaser, B. and Kuzyakov, Y. (2015). Organic nitrogen uptake by plants: reevaluation by position-specific labeling of amino acids. *Biogeochemistry* 125, 359–374.

Moreno-Cornejo, J., Zornoza, R., Doane, T. A., Faz, Á. and Horwath, W. R. (2015). Influence of cropping system management and crop residue addition on soil carbon turnover through the microbial biomass. *Biol. Fert. Soils* 51, 839–845.

Näsholm, T., Ekblad, A., Nordin, A., Giesler, R., Högberg, M. and Högberg, P. (1998). Boreal forest plants take up organic nitrogen. *Nature* 392, 914–916.

Parton, W. J., Stewart, J. W. B. and Cole, C. V. (1988). Dynamics of C, N, P, and S in grassland soils: a model. *Biogeochemistry* 5, 109–131.

Paul, E. A. and Clark, F. E. (1996). *Soil Microbiology and Biochemistry*. Academic Press, New York.

Paul, E. A., Collins, H. P. and Leavitt, S. W. (2001). Dynamics of resistant soil carbon of midwestern agricultural soils measured by naturally occurring ^{14}C abundance. *Geoderma* 104, 239–256.

Powlson, D. S. and Brookes, P. C. (1987). Measurement of the soil microbial biomass provides an early indication of changes in total soil organic matter due to straw incorporation. *Soil Biol. Biochem.* 19, 159–164.

Qiao, Y., Miao, S., Silva, L. C. R. and Horwath, W. R. (2014). Understory species regulate litter decomposition and accumulation of C and N in forest soils: a long-term dual-isotope experiment. *For. Ecol. Manage.* 329, 318–327.

Rasmussen, C., Southard, R. J. and Horwath, W. R. (2005). Modelling energy inputs to predict pedogenic environments using regional environmental databases. *Soil Sci. Soc. Am. J.* 69, 1266–1274.

Redfield, A. (1958). The biological control of chemical factors in the environment. *Am. Sci.* 46, 205–221.

Robinson, J. M. (1990). Lignin, land plants and fungi: biological evolution affecting Phanerozoic oxygen balance. *Geology* 15, 607–610.

Rustad, L. E., Campbell, J. T., Marion, G. M., Norby, R. J., Mitchell, M. J., Hartley, A. E., *et al.* (2001). A meta-analysis of the response of soil respiration, net nitrogen mineralization, and above-ground plant growth to experimental ecosystem warming. *Oecologia* 126, 543–562.

Schimel, J. P. and Bennett, J. (2004). Nitrogen mineralization: challenges of a changing paradigm. *Ecology* 85, 591–602.

Silva, L. C. R. and Horwath, W. R. (2013). Explaining global increases in water use efficiency: why have we overestimated responses to rising atmospheric CO_2 in natural forest ecosystems? *PLoS ONE* 8, e53089. doi:10.1371/journal.pone.0053089.

Smith, J. L. and Paul, E. A. (1990). In Bollag, J. M. and Stotzky, G. (eds.), *Soil Biochemistry*, vol. 6. Marcel Dekker, New York, pp. 357–396.

Throckmorton, H. M., Bird, J. A., Dane, L., Firestone, M. K. and Horwath, W. R. (2012). The source of microbial C has little impact on soil organic matter Stabilization in forest ecosystems. *Ecol. Lett.* 15, 1257–1265.

Tilman, D., Reich, P. B. and Knops, J. M. H. (2006). Biodiversity and ecosystem stability in a decade long grassland experiment. *Nature* 441, 629–632.

Tinker, P. B. (1984). The role of microorganisms in mediating and facilitating the uptake of plant nutrients from soil. *Plant Soil* 76, 77–91.

Toosi, E. R., Doane, T. A. and Horwath, W. R. (2012). Abiotic solubilization of soil organic matter, a less-seen aspect of dissolved organic matter production. *Soil Biol. Biochem.* 45, 12–21.

van Veen, J. A., Ladd, J. N. and Amoto, M. (1985). Turnover of carbon and nitrogen through the microbial biomass and a sandy loam and a clay soil incubated with [^{14}C(U)]glucose and [^{15}N]$(NH_4)_2SO_4$ under different moisture regimes. *Soil Biol. Biochem.* 17, 747–756.

Vitousek, P. M., Aber, J. D., Howarth, J. W., Likens, G. E., Matson, P. A., Schindler, *et al.* (1997). Technical report: human alteration of the global nitrogen cycle: sources and consequences. *Ecol. Appl.* 7, 737–750.

Wedin, D. A., Tieszen, L. L., Dewey, B. and Pastor, J. (1995). Carbon isotope dynamics during grass decomposition and soil organic matter formation. *Ecology* 76, 1383–1392.

Witter, E. and Kanal, A. (1998). Characteristics of the soil microbial biomass in soils from a long-term field experiment with different levels of C input. *Appl. Soil Ecol.* 10, 37–49.

Witter, E., Martensson, A. M. and Garcia, F. V. (1993). Size of the soil microbial biomass in a long-term field experiment as affected by different N-fertilizers and organic manures. *Soil Biol. Biochem.* 25, 659–669.

Yao, H., He, Z., Wilson, M. J. and Campbell, C. D. (2000). Microbial biomass and community structure in a sequence of soils with increasing fertility and change in land use. *Microb. Ecol.* 40, 223–237.

Zak, D. R., Tiullman, D., Parmenter, R. R., Rice, C. W., Fisher, F. M., Vose, J., *et al.* (1994). Plant production and soil microorganisms in late-successional ecosystems: a continental-scale study. *Ecology* 75, 2333–2347.

Zogg, G. P., Zak, D. R., Ringelberg, D. B., MacDonald, N. W., Pregitzer, K. S. and White, D. C. (1997). Compositional and functional shifts in microbial communities due to soil warming. *Soil Sci. Soc. Am. J.* 61, 475–481.

3

Managing Soil Microbial Biomass for Sustainable Agro-Ecosystems

Elizabeth A. Stockdale* and Daniel V. Murphy[†,‡]

**School of Agriculture, Food and Rural Development, Newcastle University, Newcastle-upon-Tyne NE1 7RU, UK*
[†]School of Earth and Environment and Institute of Agriculture, Faculty of Science, The University of Western Australia, WA 6009, Australia
[‡]daniel.murphy@uwa.edu.au

Introduction

Soil is central in delivering many of the environmental goods and services on which society depends. Such critical soil functions include food and fibre production, storage, filtering and transformation of water, minerals, pollutants and carbon (C) and nutrient cycling (Table 3.1). The value of these goods and services is significant, and human welfare can be directly linked to the efficient and effective functioning of soil. Our ability to deal successfully with global challenges including feeding the increasing world population, decreasing the reliance on synthetic fertilisers and managing greenhouse gas emissions is closely linked to a better understanding of the functioning of soil organisms and how they respond to external factors (climate and management). The biotic interactions between above- and below-ground ecosystems and the feedbacks between them play a crucial role in regulating ecosystem functioning at local, regional and global scales (Bardgett and Wardle, 2010). Deliberative manipulation of above-ground ecosystems within agro-ecosystems has been central to the growth and development of human civilisation (Diamond, 2005). Such changes have largely

Table 3.1 Multi-functional roles of soil, together with examples of the microbial processes associated with these functions, considering the roles of soil within sustainable agricultural systems.

Function	**Examples of microbial processes involved**
Supporting economically viable crop yields of good quality	
Supports and anchors plant growth	Stabilisation of soil aggregates through fungal networks and bacterial exudates
Suppresses crop disease	Infection of roots by mycorrhizal fungi
Provides and buffers physicochemical environment for plant growth	Decomposition of organic materials, humus formation
Controls water availability for plant growth	Plant–mycorrhizal associations
Controls nutrient availability for plant growth	Mineralisation, dinitrogen fixation, phosphorus solubilisation
Minimising negative environmental impacts of agriculture	
Provides resilience to environmental change and management impacts	Horizontal gene transfer
Buffers water flows to reduce off-site impacts	Synthesis of hydrophobic compounds, formation of biofilms
Buffers nutrient losses to reduce off-site impacts	Nitrification
Buffers impacts of pollution	Degradation of pesticides and other xenobiotics
Source/sink of greenhouse gases	Denitrification, methane oxidation
Self-maintenance of above- and below-ground ecosystems	Carbon cycling, immobilisation

Source: Adapted from Gonzalez-Quiñones *et al.* (2011).

had negative impacts on soil functioning, and there has consequently been increased interest in protecting and enhancing the ability of soils to deliver multiple functions — the soil quality paradigm (Doran and Parkin, 1994). The below-ground processes, which contribute to the delivery of these anthropocentrically defined functions, result from the emergent properties and ecosystem processes arising from the interactions of soil organisms and plant roots with physical, chemical and biological factors in the soil.

The function of a huge diversity of soil organisms and their interactions (predator–prey and extensive omnivory) are the driving force behind many soil functions (Coleman *et al.*, 1984; Stockdale and Brookes, 2006; Bardgett and Wardle, 2010). Jenkinson (1977) coined the phrase "eye of the needle" through which all nutrients must pass to describe the key role of the soil microbial community, which is dwarfed in terms of mass by the amounts of dead organic matter in soil but nonetheless drives the decomposition and turnover of the entire organic matter pool. Decomposition is widely recognised as a keystone process supporting multiple ecosystem functions:

- Cycling of organic matter and C.
- Structure formation and stabilisation.
- Nutrient supply (especially nitrogen (N), phosphorus (P) and sulphur (S)).
- Greenhouse gas balance and atmospheric quality.
- Degradation of added chemicals.
- Disease suppression and crop protection.
- Water cycle — regulation of drainage and run-off volumes and water quality.

Organic matter, soil organisms and habitat factors all need to be taken into account in considering how the soil ecosystem functions (Wardle and Giller, 1996; Young and Ritz, 1998). For each ecosystem function, the ways in which the amount, activity and diversity of soil biota contribute to the delivery of each function can be described (Stockdale *et al.*, 2006). Such approaches show that simply cataloguing the parts of the below-ground ecosystem is not enough to understand function; data on the connectivity between the parts within their temporal and spatial context is also required (Raes and Bork, 2008). In the soil ecosystem, spatial and temporal heterogeneity in both the physicochemical factors defining the soil habitat space and external drivers, such as climate and management impacts, are very high (Ettema and Wardle, 2002). Changes in any of these driving factors can influence, and thus alter, interactions between soil organisms and hence the processes occurring. Here we use a simple conceptual model (Figure 3.1) to illustrate how ecosystem function arises from this series of interactions. In agro-ecosystems, biological interactions and biochemical processes are the consequence of the interactions between soil, site (e.g. climate, hydrology) and management factors (e.g. crop choice, fertilisation, tillage intensity). Within the soil, interactions between soil organic matter, soil organisms and the soil habitat determine the processes occurring and their direction, rate, etc. Each of these aspects (habitat, organic matter, soil organisms) can be described, and sometimes

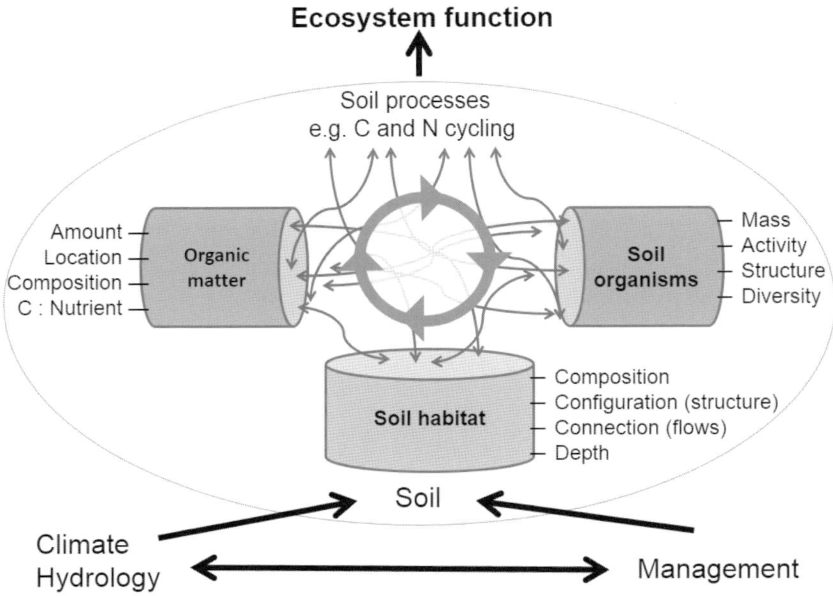

Figure 3.1 Conceptual model showing the ecosystem interactions between soil organisms, organic matter and soil habitat factors and their interactions with site factors (climate, hydrology) and management practices. The outcome of these interactions is a range of soil processes, which then underpin ecosystem function.

subdivided, according to which separable components are present (e.g. species, fractions, minerals) and in what proportions, together with information on their spatial and temporal locations and connectivity/activity.

In this chapter, we explore how the soil microbial biomass concept (and subsequent developments thereof) has supported improved understanding of interactions between the soil microbial community and soil organic matter and the role of interactions between these components in driving soil function in agro-ecosystems. Although the concepts presented in this chapter can be applied to the consideration of soil function in its broadest sense, the examples we present focus on the processes of the soil C and N cycle and their links to soil functions. From the earliest studies by Jenkinson and others (e.g. Jenkinson, 1966, 1971), these two interlocking cycles have remained the topics most studied in research on soil microbial biomass in agro-ecosystems, largely because soil organic C and N mineralisation are fundamental biogeochemical processes that underpin soil fertility and crop production. Nitrogen is usually the major yield-limiting nutrient and is consequently a major regulator of ecosystem function, and both C and N cycles have a central role in determining the greenhouse gas balance.

Size of the Microbial Biomass in Agro-Ecosystems — Understanding the Controlling Factors

The wide range of biological taxa occurring in soils and the relatively limited knowledge of their ecophysiology, in many cases, means that soil biota are often grouped by organism size (microorganisms, microfauna, mesofauna and macrofauna) or according to trophic or other functional groups (e.g. nematode functional guilds; Ferris *et al.*, 2001) to describe and/or model community and ecosystem processes (Wall, 2004). Since the development of the fumigation–incubation method (Jenkinson and Powlson, 1976 a, b), it has been common to treat the soil microbial community as a single, undifferentiated unit (black box), which is termed the soil microbial biomass and defined as all soil organisms $<5000\ \mu m^3$ volume. Significant differences in the amount, activity and diversity of the microbial biomass have been shown to result from the interactions of climate, soil and management factors. In his extensive review of the data then available on soil microbial biomass, Wardle (1992) identified how the size of the microbial biomass was influenced by plants, soil amendments (manure, residues and nutrients), soil pH, soil texture/structure and climatic factors (moisture and temperature). In particular, across a range of climates and ecosystems, Wardle (1992) showed a strong correlation between total organic C contents in soil and the size of the soil microbial biomass. Similarly, at a global scale, estimates of soil microbial biomass within biomes can be directly correlated with plant productivity and soil organic C content (Fierer *et al.*, 2009). Drawing on an updated geo-referenced dataset of soil microbial biomass estimates, Serna-Chavez *et al.* (2013) showed that at a global scale, soil microbial biomass patterns within and across biomes could largely be predicted by a climatic indicator of moisture availability (derived from both rainfall and evapotranspiration data). While these impacts may be direct, it is clear that water will also affect plant communities and their net primary productivity (and thus soil C status). However, the careful analysis of Serna-Chavez *et al.* (2013) shows that the direct effects are as important as those mediated via net primary productivity. Factors that affect the amount of energy supplied to the system (i.e. organic C inputs), but also the capacity of the microbial community to access and decompose them (e.g. access via water films in soil), are logically expected to provide key metabolic constraints to microbial activity and hence, over the medium- to long-term, population size.

At a more localised scale and especially within land-use classes, Thakur *et al.* (2015) showed that climatic factors have less impact on soil microbial biomass than local factors (such as plant community composition and soil properties). It is commonly observed that within biomes, and often co-correlated with variation in plant communities within them, unmanageable (fixed) site factors, such as slope, soil depth, mineralogy and soil texture, are also often shown to have a major effect on soil organisms in terms of both size and activity (Figure 3.2). These effects are both directly and/or indirectly mediated by impacts on soil water availability, plant growth, etc. Hence, there is potential for some sites to always have higher size, activity and diversity of microbial biomass than others as a result of the interaction of the almost limitless combinations of fixed site factors. For example, analysis of data on soil microbial biomass collected across the French soil quality monitoring network (2150 sites; Dequiedt *et al.*, 2011) showed that the main factors affecting the size of the soil microbial biomass at regional scale are soil texture, pH and soil organic matter content, as well as land use, with land under intensive management (arable cropping, horticulture, viticulture) having a smaller soil microbial biomass pool.

In an agricultural landscape in France (across 278 sites), Constancias *et al.* (2015) found that soil texture (small increases in clay content in loamy soils) had a larger impact on the composition of the microbial community (reducing richness, evenness) than on the size of the population. Other studies (at a range of scales) have found that increasing cell numbers and bacterial biomass in soil are found as the particle size decreases in soil (clay and silt) and hence with the proportion of pores 5–30 μm in size (Amato and Ladd, 1992; Hassink *et al.*, 1993; Kirchmann and Gerzabek, 1999). Sessitsch *et al.* (2001) showed that the bacterial community associated with clay and silt fractions was larger and more diverse than that associated with the sand fraction; it was suggested that bacteria found in larger pores operate in a relatively nutrient-poor environment and are more strongly affected by predation. Tiedje *et al.* (2001) suggested that spatial isolation limits competition between microbial species and Carson *et al.* (2010) showed that the observed links between soil texture and the diversity of microbial communities are often indirect and are related directly to soil water content and pore connectivity. There is also a very marked difference in active surface area in the soil with increasing clay content (Saggar *et al.*, 1996), which is likely to interact with the changes in pore size distribution and continuity. Thus, soil texture is often considered the key soil property driving the habitat parameters and their interactions with the soil microbial biomass (Figure 3.1).

A range of physicochemical properties also directly affect soil microbial biomass and its activity; pH is commonly identified as an important driver of soil processes (Cookson *et al.*, 2007, 2008). Constancias *et al.* (2015) found that

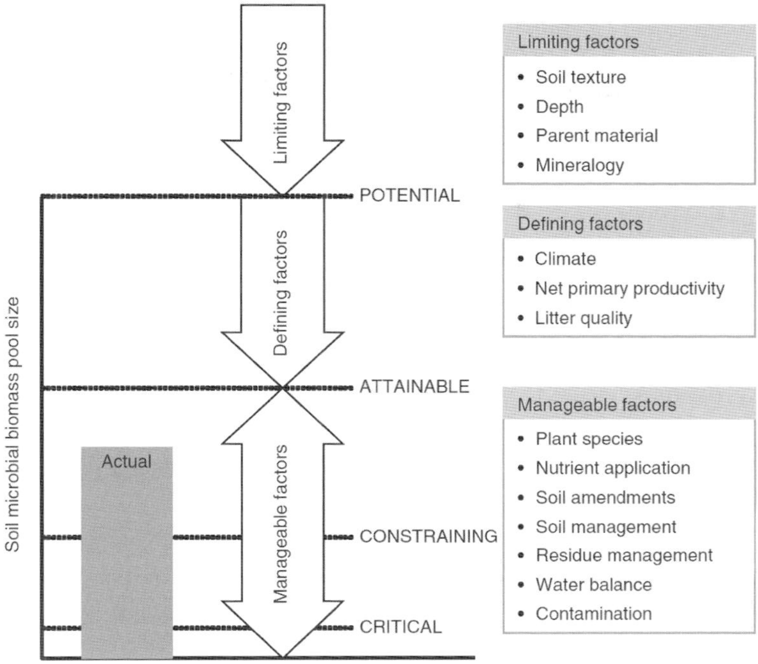

Figure 3.2 Conceptual diagram showing the size of the soil microbial biomass pool in relation to target values and their defining factors (Adapted from Gonzalez-Quiñones *et al.*, 2011). *Potential* soil microbial biomass is the maximum microbial population that could be sustained given otherwise non-limiting conditions and is constrained by the inherent site and soil characteristics which define the limits of the physicochemical habitable space. *Attainable* soil microbial biomass is then limited by climate and factors that regulate the input of organic C and nutrients into the ecosystem. Any improvement in climatic stressors or management practices that increase these inputs will tend to move the attainable value towards the potential. *Constraining* soil microbial biomass occurs when soil biological functions occur at the bottom of allowable limits, while *critical* soil microbial biomass defines the value at which a function is lost irreversibly. These lower limits are defined in relation to specific soil ecosystem functions. However, as little evidence is currently available to define the link between microbial biomass size and function (Sparling *et al.*, 2004), human judgement is most often used to define target values (Sparling *et al.*, 2003).

increasing soil pH stimulated bacterial community diversity (richness, evenness) and this has also been observed at other spatial scales (e.g. Fierer and Jackson, 2006; Green and Bohannan, 2006; Rousk *et al.*, 2010). More acidic soils ($pH < 4.5$) are often found to have a smaller soil microbial biomass and a change in community composition so that Gram +ve bacteria and fungi are much more dominant (Bååth *et al.*, 1995; Bardgett *et al.*, 2001; Aciego-Pietri and Brookes, 2009). However, these impacts are a result of both direct impacts of pH and also the interactions of a range of factors also affected by pH, including plant productivity and associated organic matter inputs and their quality, nutrient availability and toxicity effects arising from the increased availability of Al^{3+} and other heavy metals at low soil pH. Geisseler and Scow (2014) showed that the size of the soil microbial biomass increased with fertiliser application using data from long-term fertiliser trials in a major meta-analysis; however, this was concomitant with an increase in soil organic matter, which was considered to be the main driving factor, and was only seen where soil pH was maintained (pH > 5) whilst fertiliser was applied. In an elegant study, Aciego-Pietri and Brookes (2009) showed the importance of considering driving factors in combination; in arable systems, straw inputs markedly decreased the difference between microbial communities measured in both more acidic (pH 4) and more neutral (pH 6) soils.

Plant root systems are a dynamic and varied component of below-ground ecology. They form a key habitat component for a number of soil organisms including symbiotic bacteria, mycorrhizal fungi, root pathogens and herbivores (Brussaard, 1998). Lupwayi *et al.* (2004) showed that distinct populations of culturable endophytic bacteria were associated within the plant root for a range of crop plants (barley, wheat, oilseed rape), while Normander and Prosser (2000) also showed discrete endophytic bacterial community compositions using molecular methods. The composition of the community of root endophytic bacteria is determined by a combination of both plant- and soil-specific factors — endophytic bacterial communities show similarity to the soil community (Germida *et al.*, 1998; Seghers *et al.*, 2004) but root communities had much lower bacterial species richness than those associated with the root surface (rhizoplane) or soil associated with roots (rhizosphere). Hence crop choice and variety are important, not only in determining the likely inputs of organic C to the system through net primary production, but also through direct effects on the soil microbial biomass and its community composition.

Until the development of the fumigation–extraction method, it was not possible to quantify the sizes of the pools of nutrients (such as N, P, S) associated with the soil microbial biomass. It is now known that the amounts of nutrients, immobilised in the cells of the microbial biomass, are quite large though a small proportion (e.g. N and P between 0.02 and 0.04%) of the total soil pool (Wardle, 1992; Oehl *et al.*, 2001). Jenkinson and Rayner (1977) proposed a turnover time of 2.5 years for microbial biomass C measured in the Broadbalk Continuous Wheat Experiment under UK field conditions, and a turnover time for N of

1.5 years was proposed for the soil microbial biomass in soils of the same experiment, again measured under field conditions (Jenkinson and Parry, 1989). While turnover times are only estimates, as well as being site- and season-specific, these data clearly indicate that the microbial biomass is far from being a static component of the total soil organic matter pool. Wardle (1992) clearly showed that soil microbial biomass and hence microbial nutrient pools tend to be larger where soil organic matter is maintained at higher levels through increased annual inputs of C (e.g. where plant growth and associated residue/root returns are increased by fertilisers, or where external inputs of organic matter such as manures are applied, Table 3.2). While findings from Murphy *et al.* (2011) indicated that labile fractions of soil organic C are more strongly related to microbial community structure and function than is total soil C. These findings agree with those of Grayston *et al.* (2004), who suggest that although the size of the microbial biomass is altered by the quantity of resources available, microbial community structure and function are better related to the quality of available organic matter.

Table 3.2 Pools and rates of microbially mediated nutrient transformations in plots of the Broadbalk continuous wheat experiment, Rothamsted Research, Harpenden, UK.

	No N added	**Mineral fertiliser (144 kg N ha^{-1} year^{-1})**	**Farmyard manure (35 t ha^{-1} year^{-1})**
	Plot 3	**Plot 8**	**Plot 2.2**
Wheat yield (t ha^{-1} at 85% DM)	0.93	6.01	4.93
Return in stubble and chaff (t ha^{-1})	0.6	2.3	1.9
pH (H_2O)	8.2	7.3	7.8
Soil organic C (%)	0.76	1.06	3.00
Soil organic N (%)	0.09	0.12	0.31
Available P (Olsen) (mg kg^{-1})	7	78	102
Microbial biomass C (mg kg^{-1})	167	259	545
Microbial biomass P (mg kg^{-1})	6.0	5.3	28.9
Basal respiration (mg CO_2-C kg^{-1} day^{-1})	7.2	8.5	13.8
Gross N mineralisation (mg N kg^{-1} day^{-1})	0.08	0.23	0.89
Gross nitrification (mg N kg^{-1} day^{-1})	0.16	0.21	0.76

Source: The soil is a calcareous silty clay loam. Adapted from Murphy *et al.* (2003b).

Opening the Microbial "Black Box"

Soil is where most of the biodiversity on Earth exists (Hinsinger *et al.*, 2009) and is a challenging system to research as it comprises a variety of microhabitats with different physicochemical gradients and discontinuous environmental conditions. Microorganisms adapt to microhabitats and live together in consortia, interacting with each other and with other parts of the soil biota. Very little is known about the spatial organisation of soil microorganisms, beyond which organisms tend to be aggregated in relation to the spatial architecture of the soil system and that many biologically mediated processes exhibit high spatial variability across a wide range of scales. Developments of hybridisation techniques (see Sayler and Layton, 1990 for review) for nucleic acids (DNA and RNA) and the application of the polymerase chain reaction to amplify gene sequences in environmental studies have shown tremendous potential in opening the microbial "black box" to quantify the vast diversity and function of soil organisms (see Chapter 8). These methods have been applied in the detection of specific organisms in soil (e.g. Holben *et al.*, 1988) and to describe both the genetic and phenotypic diversity of the soil microbial biomass without the need for culturing (e.g. Torsvik *et al.*, 1990). These approaches are increasingly elucidating not just population structures, but also making clear links to ecosystem function (Banning *et al.*, 2015) and the geographical distribution of soil organisms (Martiny *et al.*, 2006).

High-throughput molecular techniques provide a means to quantify and monitor the response of key molecular genetic markers (Schwartz *et al.*, 2007) and can be linked to other data (e.g. geological, meteorological, land use and management history) to explore how ecosystems are structured and function. As such, many countries have embarked on extensive benchmarking exercises of the below-ground biota in order to define baselines against which future changes can be evaluated. In Australia, the national soil quality monitoring program (www.soilquality.org.au) uses a commercial soil DNA diagnostic service (Ophel-Keller *et al.*, 2008) to identify plant parasitic nematodes and soil-borne diseases. These findings are used to inform pre-seeding management decisions (e.g. crop choice) and contribute to identifying the below-ground soil constraints (biological, chemical and physical) that together prevent a crop from reaching the water-limited yield potential. Furthermore, the Biome of Australia Soil Environments project (www.bioplatforms.com/soil-biodiversity) is collecting DNA sequence information on microbial community composition to create a reference map of Australia's soil. The below-ground bacterial diversity map of England and Wales (on a 5-km grid) has already been completed (Griffiths *et al.*, 2011). The Earth Microbiome Project (www.earthmicrobiome.org) is more ambitious and seeks to map the microbial biome of the entire planet.

Most biochemical-based techniques for studying nutrient cycling and microorganism–plant nutrient transfers are applied at scales several orders of magnitude greater (i.e. cm and mm, grams of soil) than at the cellular scale at which the processes actually occur (Herrmann *et al.*, 2007). For example, the available nutrient concentration in the extraction of a few grams of soil may bear little relation to the concentrations of the nutrient that microorganisms may experience at the microscale, which could range from effectively zero in some microsites, to very high in the proximity of organic matter or fertiliser granules. Simultaneous analysis of identity (who is there?) and function (what are they doing?) of soil organisms with consideration for maintaining the soil three-dimensional structure (where are they located?) poses numerous technical challenges (Herrmann *et al.*, 2007). It requires study of the links between soil physiochemical heterogeneity and biochemical processes. Yet, this is the cellular scale at which the microbial biomass and plants interact, decompose organic matter, cycle nutrients and produce greenhouse gases. Understanding C and N cycling at the soil–plant interface still remains one of the greatest challenges limiting our ability to predict the extent by which climate and anthropogenic disturbance alter nutrient flows between soil, plant and atmosphere.

The use of isotopic tracers coupled with the ability to measure the nutrient pools in the soil microbial biomass has made it possible to trace with greater precision the cycling of nutrients and the fate of N applied in fertiliser through the complex and interacting microbial processes occurring in soil (e.g. Jenkinson and Parry, 1989; Murphy *et al.*, 2003a; Herrmann *et al.*, 2007). Increasingly nanoscale secondary ion mass spectrometry (NanoSIMS) which allows for precise, spatially explicit, elemental and isotopic analysis down to <100 nm resolution is being used to understand pathways involved in the transport of C and N (and other elements) into plant and microbial cells at scales and sensitivities not attainable by other methods. NanoSIMS enables simultaneous quantification of location (e.g. soil minerals, pores, organic matter, plant roots, microorganisms; Figure 3.3) and function (e.g. competition between microbial and plant cells for nutrients; Figure 3.4). The power of NanoSIMS for the study of soil microbial processes lies in the ability of the instrument to distinguish stable isotopes of elements with a high sensitivity (i.e. concentrations in parts per million). This can be achieved for all elements of importance to agro-ecosystem studies (e.g. carbon, ^{12}C and ^{13}C; nitrogen, ^{14}N and ^{15}N). In addition, the multi-collection capability allows for a simultaneous measurement of a number of ion species, thus enabling images to be obtained for different isotopes from the same region of interest (Figure 3.3; ^{28}Si for the soil silica sand matrix, ^{12}C for soil organic matter and the $^{12}C^{15}N/^{12}C^{14}N$ ratio for nitrogen fertiliser) from which quantitative isotope ratios

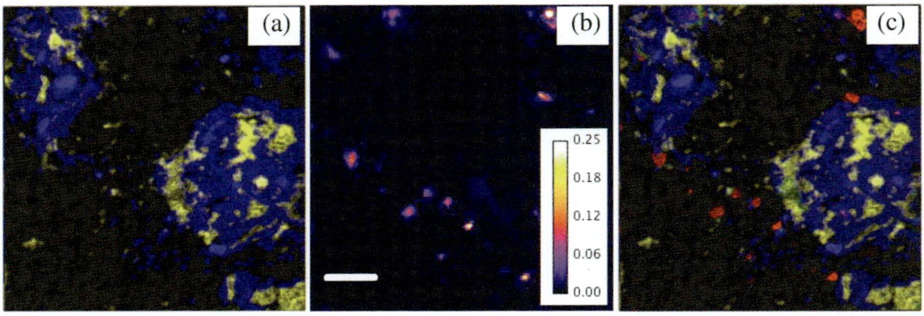

Figure 3.3 NanoSIMS ion images highlighting *in situ* spatial relationships between soil particles, organic matter and microorganisms within the soil matrix, coupled with levels of ^{15}N enrichment. (a) overlay of ^{28}Si (blue; defining quartz-dominated sand particles) and $^{12}C^{14}N$ (yellow; defining soil organic matter) ion images representative of the structure of the soil. (b) $^{15}N/^{14}N$ isotope ratio images from the same regions, showing areas of ^{15}N-ammonium fertiliser uptake into microbial cells where the colour ratio scale bar depicts the $^{15}N/^{14}N$ ratio abundance. White scale bar is 5 μm. (c) Overlay of ^{28}Si (blue), $^{12}C^{14}N$ (yellow) and $^{12}C^{15}N/^{12}C^{14}N$ (red; defining microorganisms) ion images illustrating the spatial location of soil microorganisms that were active (i.e. ^{15}N assimilation) within the soil matrix. (Adapted from Clode *et al.*, 2009.)

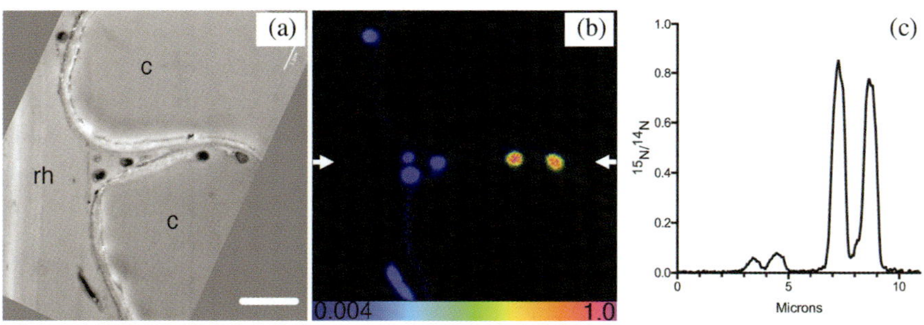

Figure 3.4 Correlative transmission electron microscopy (TEM) and NanoSIMS isotopic image analyses of a wheat rhizosphere (rh) exposed to ^{15}N-labelled ammonium fertiliser. (a) TEM image confirms the presence of rhizobacteria on and in the wheat root cells (c). White scale bar is 2 μm. (b) $^{15}N/^{14}N$ NanoSIMS isotope ratio image (10×10 μm) where the colour scale ranges from natural (0.004) $^{15}N/^{14}N$ ratio abundance (blue) to highly enriched (1.0) $^{15}N/^{14}N$ ratio abundance (pink). (c) Line scan across the NanoSIMS image (white arrows on B define the 10 μm line) illustrating the capacity of NanoSIMS to quantify ^{15}N isotopic data from individual labelled (i.e. uptake of the ^{15}N-ammonium) and non-labelled bacterial cells with the structural arrangement of the rhizosphere. (Adapted from Clode *et al.*, 2009.)

of individual components (e.g. bacterium; Figure 3.4) can be determined. Development of secondary ion mass spectrometry, *in situ* hybridisation (Orphan *et al.*, 2001; Behrens *et al.*, 2008; Li *et al.*, 2008) and single-cell extraction techniques (Eichorst *et al.*, 2015) provides opportunities to concurrently link molecular identification (microbial phylogeny) and function (metabolic activity). Furthermore, being able to link phylogenetic and functional microarrays with NanoSIMS (Chip-SIP; Mayali *et al.*, 2012) enables the design of a new generation of isotopic tracer experiments that study biochemical functions within complex microbial communities and associated with numerous functional genes. These and other such technological breakthroughs in the complementary fields of proteomics (analysis of protein structure and function) and metabolomics (analysis of intermediates and products of metabolism) open the microbial "black box" and offer endless opportunities to explore soil organisms within agro-ecosystems. Bioinformatics, the science of analysing and interpreting large-scale complex biological data such as genetic codes, goes hand-in-hand with these techniques. Here, the challenge is to identify the pathways of significance amongst these complex networks of ecological interactions (Almeida, 2002). Identifying the environmental stressors and management responses of critical soil processes that need to be modified to improve the sustainability and greenhouse gas footprint of agro-ecosystems should be a priority.

Role of the Soil Microbial Biomass in Organic Matter Decomposition and Nutrient Cycling

The fact that soil-derived organisms could decompose protein, cellulose and a wide range of organic substrates was well known by the end of the nineteenth century with distinct metabolic pathways shown under aerobic and anaerobic conditions. Carbon cycling and decomposition of organic matter in soils is driven mainly by biological mechanisms, with abiotic chemical oxidation likely to account for less than 5% of organic matter decomposition (Lavelle *et al.*, 1993). Since decomposition of organic matter results from large consortia of microorganisms processing a wide range of substrates, the development of holistic "black box" methods to measure the soil microbial biomass (Jenkinson and Powlson, 1976a, b) allowed for significant progress in the understanding of the processes of decomposition. A key finding of these early studies was that regulation of soil organic matter decomposition does not seem to result from the size of the soil microbial biomass or its ability to secrete enzymes (see Chapter 1). Initially research focussed on the importance of the chemical composition of any added organic materials and their decomposability (i.e. quality). The quality of plant material has a large influence on the "burst" of initial decomposition (Waksman and Tenney, 1927). Kaiser *et al.* (2014) showed that resource stoichiometry (C:N ratio) controls competition between microbial groups with different life-history traits (cellular C:N ratios and species-specific turnover rates). They also showed that this competition in turn governs changes in community dynamics and it is these dynamics which control the rate of litter decomposition. No wonder that Wardle *et al.* (1997) observed that increasing species richness of plant litter inputs (beyond two species) led to unpredictable responses in ecosystem processes and properties. Kaiser *et al.* (2014) suggest that there may well be decoupling between soluble bioavailable organic matter from litter (which is critical in regulating the microbial response to litter input) and the overall C:N ratio. Kaiser *et al.* (2015) further explore the social dynamics in microbial communities showing the interacting roles of microorganisms producing extracellular enzymes ("decomposers") and microorganisms exploiting the catalytic activities of others ("cheaters") in the regulation of organic matter turnover. As discussed earlier, the importance of physical protection mechanisms for the subsequent stabilisation of remaining organic matter in soil has been identified; Sollins *et al.* (1996) presented a conceptual model of decomposition and stabilisation of organic matter in soil showing interactions of a range of stabilisation mechanisms. Baldock and Skjemstad (2000) show the importance of interaction between organic matter and mineral particles in regulating stabilisation. Pore size and its architectures that result in significant portions of total pore space existing in pores <0.5 μm will limit microbial activity since microorganisms are not able to enter pores that small (Hassink *et al.*, 1993), resulting in protection of any organic materials associated with particles in these pores. The spatial location of soil organisms within the soil's three-dimensional matrix is also of importance as non-preferred soil spaces/pores (i.e. where biological function is low or absent) are more likely to stabilise soil organic matter (Ekschmitt *et al.*, 2008).

For soils in C-balance, the mineralisation flux to the atmosphere is equal to the deposition of plant-derived C to soil, but it is increasingly recognised that this balance may be disrupted by environmental and land-use changes, potentially resulting in significant (positive or negative) feedbacks to rising CO_2 concentrations in the atmosphere (Houghton, 2007; Langley *et al.*, 2009). However, uncertainties in measuring soil C-balance (Saby *et al.*, 2008) and a lack of knowledge of the drivers and controls of soil organic matter mineralisation (Bradford *et al.*, 2008) means that the role of soils in global C-cycling budgets is still poorly understood (Trumbore, 2006).

Understanding and management of the seasonality of the soil microbial biomass and the microbial processes controlling nutrient supply is as important as determining the annual flux of nutrients through the soil microbial biomass for management of the N cycle in practice. Climatic factors, in particular, the interaction of temperature and moisture, largely control the rate of biological processes and hence organic matter decomposition and N cycling (Nishio and Fujimoto, 1989; Recous *et al.*, 1999). Seasonal variation in plant growth also leads to temporal variability in organic matter inputs to soil (Murphy *et al.*, 2007). Spatial patterns and activity of soil organisms often show greater fluctuations than underlying patterns of soil resources (Ettema and Wardle, 2002), though mineral N variability is highly dependent on interactions between microbial activity and water movement through soil. Temporal variability in the biomass and activity of soil organisms can therefore be as significant as spatial variability. Hoyle and Murphy (2006) found that the size of the microbial biomass pool was strongly related to historical treatment differences in crop residue retention, whereas microbial activity (CO_2 production and gross N mineralisation) and microbial community structure showed seasonal dynamics strongly linked to variations in temperature and soil moisture. In the UK, Murphy *et al.* (2007) showed that while the size of the soil microbial biomass was significantly larger in a ley-arable than continuous arable rotation, there was no clear seasonal pattern. In contrast, soil CO_2 production and gross N mineralisation showed strong seasonality, which was linked to soil temperature and moisture. Normalisation of soil CO_2 production and gross N mineralisation with respect to these environmental regulators enabled the underlying influence of the incorporation of fresh plant material into soil on these processes to be studied. There was no simple relationship between soil CO_2 production and gross N mineralisation. This was attributed to changes in the C:N ratio of the mineralising pool(s) through the season, for example, increased proportion of root exudates during stem elongation (Murphy *et al.*, 2007).

Nitrogen is the primary nutrient-limiting plant production globally and is consequently a major regulator of ecosystem function. In the case of N, and organic forms, in particular, there is strong competition between soil microorganisms and plant roots and the flow of N is intrinsically linked to the flow of C from root exudates, which is typically 20–40% of total net primary productivity (Jones *et al.*, 2004). Cleveland and Liptzin (2007) showed that despite the high spatial heterogeneity in soils, there were well-constrained C:N:P ratios (60:7:1) in the soil microbial biomass; this microbial element stoichiometry has been shown to provide a useful tool to understand and model the way in which nutrient limitations may affect microbial biomass size and activity (Manzoni and Porporato, 2009).

Nutrient cycling in soil is often measured by quantification of available pools of soil nutrients (and crop uptake) through time; however, it is well known that such data gives an indication only of the net outcome of a complex set of interacting biochemical processes (Jarvis *et al.*, 1996). The ^{15}N isotope pool dilution technique has been widely applied to the study of N cycling (Murphy *et al.*, 2003a; Booth *et al.*, 2005) and has more completely revealed the complex interacting processes at the heart of the soil N cycle (e.g. Schimel *et al.*, 1989; Davidson *et al.*, 1992; Hart *et al.*, 1994; Cookson *et al.*, 2006). Kirkham and Bartholomew (1954, 1955) first formulated the differential equations to estimate gross processes in soils that form the basic concepts of the isotope pool dilution techniques that have been developed to measure the gross (total) rates of dissimilatory processes. For example, processes such as gross N mineralisation (i.e. ammonification) and gross nitrification are estimated by enriching the product pool with ^{15}N (^{15}N-ammonium for mineralisation and ^{15}N-nitrate for nitrification) and by measuring the changes in the product pool size and dilution of ^{15}N in this pool over time (Table 3.2). The ^{15}N isotopic pool dilution approach also indicates the importance of assimilatory processes such as gross N immobilisation in the soil N cycle. However, gross N immobilisation rates in soils are challenging to estimate at the meso-scale and studies are fraught with difficulty. Gross N immobilisation rates are usually estimated indirectly by measuring ^{15}N tracers that move into the microbial biomass using the fumigation–extraction method (Ledgard *et al.*, 1998; Hatch *et al.*, 2000) or by determination of residual ^{15}N in soils after KCl extractions in combination with numerical modelling of N processes (Mary *et al.*, 1998; Recous *et al.*, 1999; Andersen and Jensen, 2001).

The biogeochemical cycling of N in soils has a pivotal role both in ensuring continued sustainable plant production and in determining the environmental impact of land use, since both nitrate (NO_3^-) leaching and N_2O emissions can result from N cycling in soil. Synchronising N supply (soil) and N demand (crop) processes in both time and space will optimise crop N use efficiency and hence also minimise losses to water and air. Stockdale *et al.* (2002) showed that different fertiliser histories did not affect the gross mineralisation–immobilisation turnover in grassland; however, measured rates of gross nitrification increased with increasing previous ammonium (NH_4^+) fertiliser addition and there was a higher NO_3^--to-NH_4^+ ratio. There was no correlation in any of these measured processes, considered individually, with losses of NO_3^- by leaching. However, where the relative importance of microbial pathways competing for NH_4^+ was expressed as a nitrification to immobilisation ratio (*N/I*), this index was shown to change with management and was a key indicator of leaching risk (Stockdale *et al.*, 2002). While the *N/I* index is an integrating value that cannot be directly connected to any specific biological or mechanistic processes, the index highlights the importance of understanding the interactions and competition between different biochemical pathways for N when considering both N supply and loss in agro-ecosystems.

Recently, mathematical modelling has begun to integrate our understanding of the soil–plant N cycle and the soil, plant, environmental and management factors that govern it. Models are usually first developed in response to a particular question and at a particular scale and draw together current knowledge and hypotheses about biological systems, their subsystems and interactions and are able to integrate scattered data into a coherent whole (Jenkinson *et al.*, 1990). Models represent simplifications of the system; we describe processes in models with various levels of exactitude, all of them imperfect (Addiscott *et al.*, 1991). The complexity of the cycle and the large number of interacting factors which control it mean that even for temperate agricultural soils our models do not closely approach reality (de Willigen, 1991). Process-based models of N supply usually consider soil organic matter as one or more pools of material, where each pool is made up of material, which is considered to behave similarly and can be described by a set of unique characteristics (Jarvis *et al.*, 1996). The number and type of pools described varies enormously between models and reflects both the system being modelled and the hypotheses of the modellers. Many models are modular in structure, allowing modules to be simply replaced by more sophisticated (or even simpler) modules as the model evolves in response to increased scientific understanding of the underlying processes in the crop and soil. Consequently, many of the complex models appear to be similar; the C and N cycles are closely linked and microbial biomass, readily available and recalcitrant pools of organic matter, are often included (e.g. Parton *et al.*, 1987; Bradbury *et al.*, 1993). The rate of release of N from organic matter is affected by external factors such as moisture and temperature and most of the multi-compartmental models use first-order kinetics to approximate the decomposition of pools. Monod kinetics have also been introduced to describe microbial growth (Molz *et al.*, 1986). Models of the soil N cycle (e.g. ANIMO: Rijtema and Kroes, 1991; DAISY: Hansen *et al.*, 1991; SUNDIAL: Smith *et al.*, 1996) explicitly incorporate descriptions of the processes of soil N turnover and enable the effects of seasonal variations and timing of management practices to be simulated.

For many models, the long-term aim is not only to provide an adequate scientific description of the underlying processes, but also to provide simulations that support policy-making and/or decisions on-farm. For example, one of the main aims underpinning the development of the SUNDIAL model has been to provide fertiliser recommendations for farmers, that is, to specify how much fertiliser N is required to grow a particular crop for a particular yield in a particular field and when this N should be applied. Such recommendations are required in early spring, though they can be updated for subsequent fertiliser applications, and the SUNDIAL model is driven, in part, by the farmer's estimated target marketable yield (grain, tubers, etc.). The model requires inputs describing the soil, cropping history management and the preceding weather, and the information required was chosen to reflect the information, which is likely to be readily available to farmers (Smith *et al.*, 1996). In SUNDIAL, soil N dynamics are driven by the C cycle. Nitrogen and C are returned to the soil by growing crops, not only at harvest as stubble and straw, but also throughout the growing season as root exudates, dead leaves and fragments of roots. The model first simulates the decomposition of organic C according to the soil type as it moves through the various compartments (released as CO_2, incorporated into microbial biomass or into humus) and then calculates the N content of these compartments from their C:N ratios (for humus and biomass, these are assumed to remain constant). If the C:N ratio rises, due to a higher C:N composition of the crop debris, N is immobilised first from NH_4^+ and then from NO_3^- in the soil. If the C:N ratio falls, N is mineralised to NH_4^+. Ammonium may then be nitrified to NO_3^-, and NO_3^- may be lost by denitrification or leaching. To provide a fertiliser recommendation for a particular crop, the model is run with the field information and local weather data. The model simulates the potential soil N supply and any losses of N over the growing period. The amount and timing of fertiliser applications is optimised within the model by increasing the application from zero to a level where the response of crop N uptake to added fertiliser is negligible. In a series of trials carried out specifically to validate the use of the model as a fertiliser recommendation system, the SUNDIAL recommendation was not significantly different from the measured optimum (Table 3.3). However, it is not realistic to expect that fertiliser recommendations for the current crop will ever exactly match the measured optima due to the influence of the (largely) unpredictable weather on crop growth and microbial processes. When SUNDIAL was used retrospectively, that is, with the actual weather and the target yield set at the actual optimum achieved, the SUNDIAL recommendation almost exactly predicted the measured optimum (Table 3.3). It is important that SUNDIAL achieves good retrospective predictions as this demonstrates that the model can be used reliably to carry over information from 1 year to the next without introducing significant errors. SUNDIAL can also be used to assist in designing rotations to optimise the synchrony of N supply from the soil and crop demand; the leaks can then be plugged (Smith and Glendining, 1996).

Table 3.3 Performance of two N fertiliser recommendation systems compared to the optimum measured in a field trial.

			Sundial	
Method	**Optimum**[a]	**RB209**	**Default weather and yield**	**Actual weather and yield**
Recommendation (kg N ha^{-1})	139	182	162	141
Calculated yield[a] (t ha^{-1})	9.1	9.5	9.3	9.1
% difference from optimum yield		4	2	0

Note: RB209 is the standard reference book providing fertiliser recommendations for crops in the UK (in this case, the recommendations were made using the 8th edition Defra, 2000). SUNDIAL is a process-based dynamic model system developed using long-term data at Rothamsted Research.
[a]Calculated from linear plus exponential curve fitted to trial data.

Managing the Soil Microbial Biomass — Developing Sustainable Agro-Ecosystems

Taken overall, the role of soil organisms in supporting food production has been increasingly marginalised with the intensification of farming systems during the twentieth century (Giller *et al.*, 1997). Technological advances in agriculture, particularly in the second half of the twentieth century, in tillage, agrochemicals and plant breeding, have enabled food crops to be grown with decreasing regard for naturally occurring soil processes. Intensification of farming practices has also led to less variation among farming systems (e.g. monocultures) and their associated landscapes (Benton *et al.*, 2003). In addition, most agricultural practices also affect a range of soil properties (both directly and indirectly) and hence also impact soil biota. Consequently, farmed land is usually associated with simplified ecological systems and soil food webs compared to semi-natural systems (e.g. Culman *et al.*, 2010). Significant differences in the amount, activity and diversity of soil organisms have also been shown in comparisons of agricultural systems. For example, significant differences in both abundance and community composition of soil biota have been shown in dominantly pastoral compared with arable agricultural systems (van der Putten *et al.*, 2004; Jangid *et al.*, 2008). Postma-Blaauw *et al.* (2010) found that conversion of grassland into arable land negatively affected both the abundance and functional diversity of soil biota with larger impacts on the abundance and diversity of taxonomic groups with larger body size (earthworms, enchytraeids, microarthropods and nematodes) than smaller-sized taxonomic groups (protozoans, bacteria and fungi). Dequiedt *et al.* (2011) showed that land under intensive management (arable cropping, horticulture, viticulture) has a smaller soil microbial biomass pool than other agricultural land uses. However, despite the increasing intensity of the disturbance to soils caused by agricultural practice (tillage, fertilisers, pesticides, manipulation of organic residues and crop species), the soil microbial biomass continues to function effectively as the basic consumer trophic level in the decomposer subsystem controlling the breakdown and cycling of organic matter in soils.

Change in the size of the soil microbial biomass has long been recognised as more sensitive to management practices (thus an early warning) than total soil organic matter (Powlson *et al.*, 1987). It might be argued that, without the development of methods for measuring the size of the microbial biomass (and its nutrient content), we would not have been fully equipped to judge the impacts of various agricultural management practices on soil organic matter or nutrient cycling (e.g. Powlson and Jenkinson, 1981; Saffigna *et al.*, 1989). Increasingly, measurements of the biodiversity of the microbial biomass are also being used to assess the impact of particular agricultural practices or systems (e.g. Clegg *et al.*, 2003; De Goede *et al.*, 2003; Dunfield and Germida, 2004). However, currently there is no reliable method to predict the impact of management on the soil microbial biomass and its whole range of activity due to the complexity of the interacting indirect (through changes in soil physical and chemical conditions) and direct effects of management on soil microbial biomass, and the ability of the soil microbial biomass itself to alter the soil habitat in response to management changes. Stockdale *et al.* (2013) showed that the resistance and resilience of soil function might be increased through the selection of crops and organic matter inputs. Differences in the response of soil microbial communities to stress may be related to the relative proportions of active and dormant organisms (Stockdale *et al.*, 2013). However, redundancy in the microbial population with regard to many soil processes means that measured changes in functional diversity do not necessarily have a direct impact on ecosystem function.

Measurement of soil microbial biomass has been widely recommended for use as a soil quality indicator (e.g. Insam, 2001; Schloter *et al.*, 2003; Bending *et al.*, 2004) and there are several soil quality monitoring programmes where it has been included (Carter *et al.*, 1999; Nielsen and Winding, 2002; Winding *et al.*, 2005; www.soilquality.org.au). For example, Carter *et al.* (1999) proposed that although soil microbial biomass could not be used as a sole indicator of either crop productivity or aggregate stability, it could be used as an indicator of energy/C cycling in soils. However, the approach required to measure and interpret soil microbial biomass values for soil quality monitoring is not universal and should be considered in the specific context of climate, land use, management systems (as illustrated in Figure 3.2) and objective of the monitoring program. Increasing on-farm uptake of agricultural practices and systems to enhance soil microbial biomass and biologically supported soil processes require not only measurements of soil biological quality, but also interpretation of the data collected to provide a guide to practice. While it is useful to measure trends through time for the same site/field or compare values between fields with different management, to support farmer decision-making with regard to changes in practice it is important to know at least the direction of the change required (Is more or less of the measured characteristic better?). Wienhold *et al.* (2009) describe on-going work in the USA on a soil management assessment framework, which is aimed at producing scoring curves as a way of interpreting soil indicators. Within broader soil quality frameworks for on-farm use, which include physical and chemical indicators, a simple scoring approach has been developed in the UK but the biological measures currently only include observations of earthworms and crop residues (simply sustainable soils accessed via www.leafuk.org).

Typically, an increase in soil microbial biomass is considered beneficial and an improvement to the system, while a decline is considered detrimental, but this level of interpretation is not sufficient if measurements of soil microbial biomass are to be included in a soil monitoring framework. Merrington *et al.* (2006) established a tiered risk-based procedure for users to select indicators for soil sampling and monitoring in the UK and used an expert group approach to identify trigger values or workable ranges for a number of soil chemical measures and soil bulk density; they indicated that biological indicators were still only available as research tools. Wander *et al.* (2002) report a participatory approach to developing such a soil quality benchmarking system as a result of farmer and researcher interaction in cropping systems in Illinois, which highlighted that farmers often have the desire to take soil quality into account in decision-making but lack tools which actually enable this to happen in the field. However, little evidence is available of the links between soil microbial biomass levels and ecological function to allow for these limits to be identified (Sparling *et al.*, 2004; Gonzalez-Quiñones *et al.*, 2011).

While it is often argued that maintenance and enhancement of the diversity, activity and biomass of soil biota is of benefit within all agricultural systems (e.g. Doran and Smith, 1987; Beauchamp and Hume, 1997; Clapperton *et al.*,

2003), conclusions of scientific reviews often indicate only in the most general way how this can be practically enacted at a farming system level. For example, Clapperton *et al.* (2003) in a review of the role of soil microbial biomass in controlling nutrient release and plant uptake conclude: "Ideally agro-ecosystems should be managed to maintain the structural integrity of the [soil] habitat, increase soil organic matter (OM) and optimise the C:N ratios in soil OM using cover crops and/or crop sequence." There are indications that changes in agricultural management practices that result in "reducing the intensity of use of mechanical and manufactured inputs and (re)-discovering cost-effective ways to integrate biological inputs, will benefit below-ground biodiversity, particularly in lowland grassland and cropping systems" (Stockdale *et al.*, 2006). For example, restoration of a less intensive management system (re-establishment of grassland on arable fields) was shown to increase species abundance (Postma-Blaauw *et al.*, 2010). However, the original community composition was not restored within 4 years. While there is limited evidence available, it seems likely that increasing benefit for the soil biota (biomass, activity and diversity) accrues where a range of systems-oriented approaches are adopted in combination, such as required by organic farming standards or the practices of no-till (conservation agriculture) systems. These principles, particularly increased organic matter inputs and reduced tillage intensity, are common in the literature discussing options for increasing the sustainability of agro-ecological systems more generally. Farmers have adopted these system-oriented practices, which benefit soil biota, for a range of other reasons including, but not only, fuel reduction, C sequestration and conservation of above-ground biodiversity. However, such recommendations are general and system-oriented; individual management practices are not recommended. This seems to leave farmers without answers to a range of pertinent and practically important questions such as "how many cover crops and which ones, where the right balance (economic as well as ecological) is between minimising tillage and optimising weed control …" (Stockdale *et al.*, 2006).

Future Challenges

One in nine people in the world (795 million of the 7.3 billion) are estimated to be suffering from chronic undernourishment, where the vast majority live in developing countries (Food and Agriculture Organisation, 2015). Of the global challenges that society currently faces, the basic requirement to feed the increasing world population from finite resources is of critical importance; failure to maintain soil and water security will critically undermine food security. As the twenty-first century advances, alongside the changes to the climate itself, it seems almost certain that fossil fuel-based resources, including N fertiliser and pesticides, will become increasingly expensive and scarce. Increasing scarcity of other mineral resources, for example, phosphate rock, also seems likely. Consequently, a search for appropriate agronomic techniques may mean a renewed focus on biologically based approaches where an increased capacity to manage the complexity within the soil microbial biomass is required (Abbott and Murphy, 2003). Within the context of this chapter, increasing fertiliser-use efficiency (Cassman *et al.*, 2002) and managing greenhouse gas emissions (Paustian *et al.*, 2016) require in part a better understanding of the key groups of soil organisms involved and how their functioning responds to external factors (climate and management).

Managing agro-ecosystems in an environmentally and economically sustainable manner necessitates that benefits from applied organic and inorganic fertilisers are maximised while off-site impacts, including nutrient flow to waterways and greenhouse gas emissions, are minimised. As an example, China has more than 20% of the world population but only 8% of the total world arable land (Food and Agriculture Organization, 2010). Increased availability and use of inorganic fertilisers have played important roles in feeding China's population. China is now the largest user of inorganic fertilisers, accounting for 90% of the global increase in use (Liu and Diamond, 2005) and consuming one-third of global production (Hvistendahl, 2010). Between 1980 and 2010, there was a six-fold increase in inorganic fertiliser use in China (National Bureau of Statistics of China, 2012) during which time grain yields doubled (e.g. 1980: wheat 1.9 t ha^{-1}; 2010: wheat 4.7 t ha^{-1}; National Bureau of Statistics of China, 2012). However, over the same time period, soil organic C stocks only changed slightly, with a general decrease in arid/semi-arid regions and increase in humid/semi-humid regions (Jiang *et al.*, 2014). Application of organic amendments to agricultural soils has declined in many parts of the world as a direct consequence of increased availability and use of inorganic fertilisers. Excessive addition of soluble inorganic nutrients to crops can uncouple C and nutrient cycles in space and time. The soils of the Loess Plateau in northwest China are used mainly for agricultural production and have formed over thousands of years through regular inputs of organic manures. Here the decomposition rate of soil organic matter was measured to be faster when inorganic fertiliser was applied alone compared to with manure (Liang *et al.*, 2012). Similarly, modelling scenarios for northern China predicted that continued application of inorganic fertilisers without organic amendments will decline soil organic C stores and increase CO_2 emissions under current farming systems practices (Jiang *et al.*, 2014). If China continues to maintain self-sufficiency in food production, then arable lands will need to increase productivity without causing loss of soil fertility. This challenge also resonates with other countries that have shifted from low-input to high-input synthetic-based farming systems.

Expectations of shortage in global reserves of rock phosphate over the next 50–100 years (Cordell *et al.*, 2009) is predicted to result in erratic supply of P fertilisers and increased cost but decreased quality. The fixation of P at soil mineral surfaces is well recognised as a constraint to P fertiliser efficiency. In Australia alone, the "bank" of fixed P is estimated to be worth AUS$10 billion or 100 kg P per hectare of arable land (Cornish, 2009). A number of mechanisms have been proposed whereby increased soil organic matter and soil microbial activity decreases sorption of P, potentially leading to increase plant uptake (Guppy *et al.*, 2005). In particular, microorganisms release enzymes that mineralise organic P and they enhance dissolution of minerals containing P through a combination of soil acidification and the release of metal-complexing agents (predominantly organic acids and siderophores). In highly P-limited environments, such as those in Australia, this strategy is important for mobilising insoluble mineral-bound P. Also P incorporated into microbial cells may become available for plant uptake upon microbial death. Finding a way to manage the microbial biomass P pool could increase crop uptake from the P bank in soil and produce the same yield with smaller inputs of P fertiliser. However, several barriers remain to be overcome. First, a mechanistic understanding of microbial release of fixed P and of interactions between plants, microorganisms and minerals in the rhizosphere is required. Second, the effect of specific agronomic practices that can be used to proactively manage microbial release and plant uptake of fixed P need to be quantified.

While new analytical methods and science findings will continue to increase our understanding of the soil microbial biomass and the factors that affect its size, activity and diversity, the challenge has, and may always be, to translate this knowledge into information that can be applied by farmers to increase the sustainability of farming system practices. Both top-down (e.g. government legislation) and bottom-up (e.g. land-holder participatory action research) approaches have been used to implement change, both have strengths and weaknesses but both are predicated on effective knowledge exchange with the research community. Given the increasing interest and active engagement of the farming industry with the concepts of soil health/quality, we believe that it is appropriate and timely to take the existing underpinning research knowledge and translate it into practical tools to support decision-making on-farm. However, we also recognise that, to fully deliver on the ambition to effectively manage sustainable agro-ecosystems, there is an associated need to continue to identify research questions and undertake the underpinning basic scientific research that can provide step changes in the capacity of productive farm systems to manage the soil microbial biomass which underpins sustainable agro-ecosystems.

Acknowledgements

The authors are grateful for the inspiration of David Jenkinson during their post-doctoral years at Rothamsted. We thank Professor Peta Clode, Centre for Microscopy Characterisation and Analysis at the University of Western Australia for the preparation of NanoSIMS figures. Daniel Murphy contributed to this chapter while supported by the Australian Research Council Future Fellowship Scheme (FT110100246).

References

Abbott, L. K. and Murphy, D. V. (2003). In Abbott, L. K. and Murphy, D. V. (eds.), *Soil Biological Fertility. A Key to Sustainable Land Use in Agriculture*. Kluwer Academic Publishers, London, pp. 1–254.

Aciego-Pietri, J. C. and Brookes, P. C. (2009). Substrate inputs and pH as factors controlling microbial biomass, activity and community structure in an arable soil. *Soil Biol. Biochem.* 41, 1396–1405.

Addiscott, T. M., Whitmore, A. P. and Powlson, D. S. (1991). *Farming, Fertilisers and the Nitrate Problem.* CAB International,Wallingford.

Almeida, J. S. (2002). Predictive non-linear modeling of complex data by artificial neural networks. *Curr. Opin. Biotechnol.* 13, 72–76.

Amato, M. and Ladd, J. N. (1992). Decomposition of ^{14}C labelled glucose and legume material in soils: properties influencing the accumulation of organic residue C and microbial biomass C. *Soil Biol. Biochem.* 24, 455–464.

Andersen, M. K. and Jensen, L. S. (2001). Low soil temperature effects on short-term gross N mineralisation–immobilisation turnover after incorporation of a green manure. *Soil Biol. Biochem.* 33, 511–521.

Bååth, E., Frostegård, Å., Pennanen, T. and Fritze, H. (1995). Microbial community structure and pH response in relation to soil organic matter quality in wood-ash fertilized, clear-cut or burned coniferous forest soils. *Soil Biol. Biochem.* 27, 229–240.

Baldock, J. A. and Skjemstad, J. O. (2000). Role of the soil matrix and minerals in protecting natural organic materials against biological attack. *Org. Geochem.* 31, 697–710.

Banning, N. C., Maccarone, L. D., Fisk, L. M. and Murphy, D. V. (2015). Ammonia-oxidising bacteria not archaea dominate nitrification activity in semi-arid agricultural soil. *Sci. Rep.* 5, 11146.

Bardgett, R. D., Jones, A. C., Jones, D. L., Kemmitt, S. J., Cook, R. and Hobbs, P. J. (2001). Soil microbial community patterns related to the history and intensity of grazing in sub-montane ecosystems. *Soil Biol. Biochem.* 33, 1653–1664.

Bardgett, R. D. and Wardle, D. A. (2010). *Above-ground–Below-ground Linkages. Biotic Interactions, Ecosystem Processes, and Global Change.* Oxford University Press, Oxford, pp. 1–301.

Beauchamp, E. G. and Hume, D. J. (1997). Agricultural soil manipulation: the use of bacteria, manuring and plowing. In van Elsas, J. D., Trevors, J. T. and Wellington, E. M. H. (eds.), *Modern Soil Microbiology.* Marcel Dekker, New York, pp. 643–664.

Behrens, S., Lösekann, T., Pett-Ridge, J., Weber, P. K., Ng, W. O., Stevenson, B. S., *et al.* (2008). Linking microbial phylogeny to metabolic activity at the single-cell level by using enhanced element labeling-catalyzed reporter deposition fluorescence *in situ* hybridization (EL-FISH) and NanoSIMS. *Appl. Environ. Microbiol.* 74, 3143–3150.

Bending, G. D., Turner, M. K., Rayns, F., Marx, M. C. and Wood, M. (2004). Microbial and biochemical soil quality indicators and their potential for differentiating areas under contrasting agricultural management regimes. *Soil Biol. Biochem.* 36, 1785–1792.

Benton, T. G., Vickery, J. A. and Wilson, J. D. (2003). Farmland biodiversity: is habitat heterogeneity the key? *Trends Ecol. Evol.* 18, 182–188.

Booth, M. S., Stark, J. M. and Rastetter, E. (2005). Controls on nitrogen cycling in terrestrial ecosystems: a synthetic analysis of literature data. *Ecol. Monogr.* 75, 139–157.

Bradford, M. A., Davies, C. A., Frey, S. D., Maddox, T. R., Melillo, J. M., Mohan, J. E., *et al.* (2008). Thermal adaptation of soil microbial respiration to elevated temperature. *Ecol. Lett.* 11, 1316–1327.

Bradbury, N. J., Whitmore, A. P., Hart, P. B. S. and Jenkinson, D. S. (1993). Modelling the fate of nitrogen in crop and soil in the years following application of ^{15}N-labelled fertilizer to winter wheat. *J. Agric. Sci.* 121, 363–379.

Brussaard, L. (1998). Soil fauna, guilds, functional groups and ecosystem processes. *Appl. Soil Ecol.* 9, 123–135.

Carson, J. K., Gonzalez-Quiñones, V., Murphy, D. V., Hinz, C., Shaw, J. A. and Gleeson, D. B. (2010). Low pore connectivity increases bacterial diversity in soil. *Appl. Environ. Microbiol.* 76, 3936–3942.

Carter, M. R., Gregorich, E. G., Angers, D. A., Beare, M. H., Sparling, G. P., Wardle, D. A., *et al.* (1999). Interpretation of microbial biomass measurements for soil quality assessment in humid temperate regions. *Can. J. Soil Sci.* 79, 507–520.

Cassman, K. G., Dobermann, A. and Walters, D. T. (2002). Agroecosystems, nitrogen-use efficiency, and nitrogen management. *AMBIO J. Hum. Environ.* 31, 132–140.

Clapperton, M. J., Chan, K. Y. and Larney, F. J. (2003). Managing the soil habitat for enhanced biological fertility. In: Abbott, L. K. and Murphy, D. V. (eds.), *Soil*

Biological Fertility — A Key to Sustainable Land Use in Agriculture. Kluwer Academic Publishers, Dordrecht, The Netherlands, pp. 203–224.

Clegg, C. D., Lovell, R. D. and Hobbs, P. J. (2003). The impact of grassland management regime on the community structure of selected bacterial groups in soils. *FEMS Microbiol. Ecol.* 43, 263–270.

Cleveland, C. C. and Liptzin, D. (2007). C:N:P stoichiometry in soil: is there a "Redfield ratio" for the microbial biomass? *Biogeochemistry* 85, 235–252.

Clode, P. L., Kilburn, M. R., Jones, D. L., Stockdale, E. A., Cliff, J. B., Herrmann, A. M., *et al.* (2009). *In situ* mapping of nutrient uptake in the rhizosphere using nanoscale secondary ion mass spectrometry. *Plant Physiol.* 151, 1751–1757.

Coleman, D. C., Ingham, R. E., McClellan, J. F. and Trofymow, J. A. (1984). Soil nutrient transformations in the rhizosphere via animal–microbial interactions. In: Anderson, J. M., Rayner, A. D. M. and Walton, D. W. H. (eds.), *Invertebrate–Microbial Interactions.* Cambridge University Press, Cambridge, UK, pp. 35–58.

Constancias, F., Terrat, S., Saby, N., Horrigue, W., Villerd, J., Guillemin, J. P., *et al.* (2015). Mapping and determinism of soil microbial community distribution across an agricultural landscape. *Microbiol. Open* 4, 505–517.

Cookson, W. R., Müller, C., O'Brien, P. A., Murphy, D. V. and Grierson, P. F. (2006). Nitrogen dynamics in an Australian semiarid grassland soil. *Ecology* 87, 2047–2057.

Cookson, W. R., Murphy, D. V. and Roper, M. M. (2008). Characterizing the relationships between soil organic matter components and microbial function and composition along a tillage disturbance gradient. *Soil Biol. Biochem.* 40, 763–777.

Cookson, W. R., Osman, M., Marschner, P., Abaye, D. A., Clark, I., Murphy, D. V., *et al.* (2007). Controls on soil nitrogen cycling and microbial community composition across land use and incubation temperature. *Soil Biol. Biochem.* 39, 744–756.

Cordell, D., Drangert, J .O. and White, S. (2009). The story of phosphorus: global food security and food for thought. *Glob. Environ. Chang.* 19, 292–305.

Cornish, P. S. (2009). Research directions: improving plant uptake of soil phosphorus, and reducing dependency on input of phosphorus fertiliser. *Crop Pasture Sci.* 60, 190–196.

Culman, S. W., Young-Mathews, A., Hollander, A. D., Ferris, H., Sánchez-Moreno, S., O'Geen, A. T., *et al.* (2010). Biodiversity is associated with indicators of soil ecosystem functions over a landscape gradient of agricultural intensification. *Landsc. Ecol.* 25, 1333–1348.

Davidson, E. A., Hart, S. C. and Firestone, M. K. (1992). Internal cycling of nitrate in soils of a mature coniferous forest. *Ecology* 73, 1148–1156.

Defra. (2000). *Fertiliser Manual RB209.* Department for Environment, Food and Rural Affairs, London, UK.

De Goede, R. G. M., Brussaard, L. and Akkermans, A. D. L. (2003). On-farm impact of cattle slurry manure management on biological soil quality. *NJAS-Wagen. J. Life Sci.* 51, 103–133.

Dequiedt, S., Saby, N. P. A., Lelievre, M., Jolivet, C., Thioulouse, J., Toutain, B., *et al.* (2011). Biogeographical patterns of soil molecular microbial biomass as influenced by soil characteristics and management. *Global Ecol. Biogeogr.* 20, 641–652.

de Willigen, P. (1991). Nitrogen turnover in the soil-crop system: comparison of fourteen simulation models. *Fertil. Res.* 27, 141–149.

Diamond, J. (2005). *Collapse: How Societies Choose to Fail or Succeed.* Penguin Books, London, pp. 1–571.

Doran, J. W. and Parkin, T. B. (1994). Defining and assessing soil quality. In *Defining Soil Quality for a Sustainable Environment.* Soil Science Society of America Special Publication 35. Soil Science Society of America and American Society of Agronomy, Madison, WI, pp. 1–21.

Doran, J. W. and Smith, M. S. (1987). Organic matter management and utilization of soil and fertilizer nutrients. In: Follett, R. F., Stewart, J. W. B. and Cole, C. V. (eds.), *Soil Fertility and Organic Matter as Critical Components of Production Systems.* Soil Science Society of America Special Publication 19. Soil Science Society of America and American Society of Agronomy, Madison, WI, pp. 53–72.

Dunfield, K. E. and Germida, J. J. (2004). Impact of genetically modified crops on soil-and plant-associated microbial communities. *J. Environ. Qual.* 33, 806–815.

Eichorst, S. A., Strasser, F., Woyke, T., Schintlmeister, A., Wagner, M. and Woebken, D. (2015). Advancements in the application of NanoSIMS and Raman microspectroscopy to investigate the activity of microbial cells in soils. *FEMS Microbiol. Ecol.* 91, fiv106.

Ekschmitt, K., Kandeler, E., Poll, C., Brune, A., Buscot, F., Friedrich, M., *et al.* (2008). Soil-carbon preservation through habitat constraints and biological limitations on decomposer activity. *J. Plant Nut. Soil Sci.* 171, 27–35.

Ettema, C. H. and Wardle, D. A. (2002). Spatial soil ecology. *Trends Ecol. Evol.* 17, 177–183.

Ferris, H., Bongers, T. and De Goede, R. G. M. (2001). A framework for soil food web diagnostics: extension of the nematode faunal analysis concept. *Appl. Soil Ecol.* 18, 13–29.

Fierer, N. and Jackson, R. B. (2006). The diversity and biogeography of soil bacterial communities. *Proc. Natl. Acad. Sci. USA* 103, 626–631.

Fierer, N., Strickland, M. S., Liptzin, D., Bradford, M. A. and Cleveland, C. C. (2009). Global patterns in below-ground communities. *Ecol. Lett.* 12, 1238–1249.

Food and Agriculture Organization. (2010). *FAO Statistical Yearbook 2010.* Available at http://www.fao.org/economic/ess/syb/en/.

Food and Agriculture Organization. (2015). *The State of Food Insecurity in the World. Meeting the 2015 International Hunger Targets: Taking Stock of Uneven Progress.* Available at http://www.fao.org/3/a-i4646e.pdf.

Geisseler, D. and Scow, K. M. (2014). Long-term effects of mineral fertilizers on soil microorganisms — a review. *Soil Biol. Biochem.* 75, 54–63.

Germida, J. J., Siciliano, S. D., de Freitas, J. R. and Seib, A. M. (1998). Diversity of root-associated bacteria associated with field-grown canola (*Brassica napus* L.) and wheat (*Triticum aestivum* L.). *FEMS Microbiol. Ecol.* 26, 43–50.

Giller, K. E., Beare, M. H., Lavelle, P., Izac, A. M. and Swift, M. J. (1997). Agricultural intensification, soil biodiversity and agroecosystem function. *Appl. Soil Ecol.* 6, 3–16.

Gonzalez-Quiñones, V., Stockdale, E. A., Banning, N. C., Hoyle, F. C., Sawada, Y., Wherrett, A. D., *et al.* (2011). Soil microbial biomass — interpretation and consideration for soil monitoring. *Aust. J. Soil Res.* 49, 287–304.

Grayston, S. J., Campbell, C. D., Bardgett, R. D., Mawdsley, J. L., Clegg, C. D., Ritz, K., *et al.* (2004). Assessing shifts in microbial community structure across a range of grasslands of differing management intensity using CLPP, PLFA and community DNA techniques. *Appl. Soil Ecol.* 25, 63–84.

Green, J. and Bohannan, B. J. (2006). Spatial scaling of microbial biodiversity. *Trends Ecol. Evol.* 21, 501–507.

Griffiths, R. I., Thomson, B. C., James, P., Bell, T., Bailey, M. and Whiteley, A. S. (2011). The bacterial biogeography of British soils. *Environ. Microbiol.* 13, 1642–1654.

Guppy, C. N., Menzies, N. W., Moody, P. W. and Blamey, F. P. C. (2005). Competitive sorption reactions between phosphorus and organic matter in soil: a review. *Soil Res.* 43, 189–202.

Hansen, S., Jensen, H. E., Nielsen, N. E. and Svendsen, H. (1991). Simulation of nitrogen dynamics and biomass production in winter wheat using the Danish simulation model DAISY. *Fertil. Res.* 27, 245–259.

Hart, S. C., Stark, J. M., Davidson, E. A. and Firestone, M. K. (1994). Nitrogen mineralization, immobilisation, and nitrification. *Methods of Soil Analysis: Part 2 — Microbiological and Biochemical Properties*, Soil Science Society of America, Madison, Wisconsin, USA, pp. 985–1018.

Hassink, J., Bouwman, L. A., Zwart, K. B. and Brussaard, L. (1993). Relationships between habitable pore space, soil biota and mineralization rates in grassland soils. *Soil Biol. Biochem.* 25, 47–55.

Hatch, D. J., Lovell, R. D., Antil, R. S., Jarvis, S. C. and Owen, P. M. (2000). Nitrogen mineralization and microbial activity in permanent pastures amended with nitrogen fertilizer or dung. *Biol. Fertil. Soils* 30, 288–293.

Herrmann, A. M., Ritz, K., Nunan, N., Clode, P. L., Pett-Ridge, J., Kilburn, M. R., *et al.* (2007). Nano-scale secondary ion mass spectrometry — a new analytical tool in biogeochemistry and soil ecology: a review article. *Soil Biol. Biochem.* 39, 1835–1850.

Hinsinger, P., Bengough, A. G., Vetterlein, D. and Young, I. M. (2009). Rhizosphere: biophysics, biogeochemistry and ecological relevance. *Plant Soil* 321, 117–152.

Holben, W. E., Jansson, J. K., Chelm, B. K. and Tiedje, J. M. (1988). DNA probe method for the detection of specific microorganisms in the soil bacterial community. *Appl. Environ. Microbiol.* 54, 703–711.

Houghton, R. A. (2007). Balancing the global carbon budget. *Annu. Rev. Earth Planet. Sci.* 35, 313–347.

Hoyle, F. C. and Murphy, D. V. (2006). Seasonal changes in microbial function and diversity associated with stubble retention versus burning. *Soil Res.* 44, 407–423.

Hvistendahl, M. (2010). China's push to add by subtracting fertilizer. *Science* 327, 801.

Insam, H. (2001). Developments in soil microbiology since the mid 1960s. *Geoderma* 100, 389–402.

Jangid, K., Williams, M. A., Franzluebbers, A. J., Sanderlin, J. S., Reeves, J. H., Jenkins, M. B., *et al.* (2008). Relative impacts of land-use, management intensity and fertilization upon soil microbial community structure in agricultural systems. *Soil Biol. Biochem.* 40, 2843–2853.

Jarvis, S. C., Stockdale, E. A., Shepherd, M. A. and Powlson, D. S. (1996). Nitrogen mineralization in temperate agricultural soils: processes and measurement. *Adv. Agron.* 57, 187–235.

Jenkinson, D. S. (1966). Studies on the decomposition of plant material in soil. *J. Soil Sci.* 17, 280–302.

Jenkinson, D. S. (1971). Studies on the decomposition of C14 labelled organic matter in soil. *Soil Sci.* 111, 64–70.

Jenkinson D. S. (1977). The soil microbial biomass. *NZ Soil News* 25, 213–218.

Jenkinson, D. S., Andrew, S. P. S., Lynch, J. M., Goss, M. J. and Tinker, P. B. (1990). The turnover of organic carbon and nitrogen in soil [and discussion]. *Philos. Trans. R. Soc. Lond. B* 329, 361–368.

Jenkinson, D. S. and Parry, L. C. (1989). The nitrogen cycle in the Broadbalk wheat experiment: a model for the turnover of nitrogen through the soil microbial biomass. *Soil Biol. Biochem.* 21, 535–541.

Jenkinson, D. S. and Powlson, D. S. (1976a). The effects of biocidal treatments on metabolism in soil — I. Fumigation with chloroform. *Soil Biol. Biochem.* 8, 167–177.

Jenkinson, D. S. and Powlson, D. S. (1976b). The effects of biocidal treatments on metabolism in soil — V. A method for measuring soil biomass. *Soil Biol. Biochem.* 8, 209–213.

Jenkinson, D. S. and Rayner, J. H. (1977). The turnover of soil organic matter in some of the Rothamsted classical experiments. *Soil Sci.* 123, 298–305.

Jiang, G., Xu, M., He, X., Zhang, W., Huang, S., Yang, X., *et al.* (2014). Soil organic carbon sequestration in upland soils of northern China under variable fertilizer management and climate change scenarios. *Glob. Biogeochem. Cycles* 28, 319–333.

Jones, D. L., Hodge, A. and Kuzyakov, Y. (2004). Plant and mycorrhizal regulation of rhizodeposition. *New Phytol.* 163, 459–480.

Kaiser, C., Franklin, O., Dieckmann, U. and Richter, A. (2014). Microbial community dynamics alleviate stoichiometric constraints during litter decay. *Ecol. Lett.* 17, 680–690.

Kaiser, C., Franklin, O., Richter, A. and Dieckmann, U. (2015). Social dynamics within decomposer communities lead to nitrogen retention and organic matter build-up in soils. *Nat. Commun.* 6, Article number: 8960.

Kirchmann, H. and Gerzabek, M. H. (1999). Relationship between soil organic matter and micropores in a long-term experiment at Ultuna, Sweden. *J. Plant Nutr. Soil Sci.* 162, 493–498.

Kirkham, D. O. N. and Bartholomew, W. V. (1954). Equations for following nutrient transformations in soil, utilizing tracer data. *Soil Sci. Soc. Am. J.* 18, 33–34.

Kirkham, D. O. N. and Bartholomew, W. V. (1955). Equations for following nutrient transformations in soil, utilizing tracer data. II. *Soil Sci. Soc. Am. J.* 19, 189–192.

Langley, J. A., McKinley, D. C., Wolf, A. A., Hungate, B. A., Drake, B. G. and Megonigal, J. P. (2009). Priming depletes soil carbon and releases nitrogen in a scrub-oak ecosystem exposed to elevated CO_2. *Soil Biol. Biochem.* 41, 54–60.

Lavelle, P., Blanchart, E., Martin, A., Martin, S. and Spain, A. (1993). A hierarchical model for decomposition in terrestrial ecosystems: application to soils of the humid tropics. *Biotropica* 25, 130–150.

Ledgard, S. F., Jarvis, S. C. and Hatch, D. J. (1998). Short-term nitrogen fluxes in grassland soils under different long-term nitrogen management regimes. *Soil Biol. Biochem.* 30, 1233–1241.

Li, T., Wu, T. D., Mazéas, L., Toffin, L., Guerquin-Kern, J. L., Leblon, G., *et al.* (2008). Simultaneous analysis of microbial identity and function using NanoSIMS. *Environ. Microbiol.* 10, 580–588.

Liang, B., Yang, X., He, X., Murphy, D. V. and Zhou, J. (2012). Long-term combined application of manure and NPK fertilizers influenced nitrogen retention and stabilization of organic C in Loess soil. *Plant Soil* 353, 249–260.

Liu, J. G. and Diamond, J. (2005). China's environment in a globalizing world. *Nature* 435, 1179–1186.

Lupwayi, N. Z., Clayton, G. W., O'Donovan, J. T., Harker, K. N., Turkington, T. K. and Rice, W. A. (2004). Soil microbiological properties during decomposition of crop residues under conventional and zero tillage. *Can. J. Soil Sci.* 84, 411–419.

Manzoni, S. and Porporato, A. (2009). Soil carbon and nitrogen mineralization: theory and models across scales. *Soil Biol. Biochem.* 41, 1355–1379.

Martiny, J. B. H., Bohannan, B. J., Brown, J. H., Colwell, R. K., Fuhrman, J. A., Green, J. L., *et al.* (2006). Microbial biogeography: putting microorganisms on the map. *Nat. Rev. Microbiol.* 4, 102–112.

Mary, B., Recous, S. and Robin, D. (1998). A model for calculating nitrogen fluxes in soil using ^{15}N tracing. *Soil Biol. Biochem.* 30, 1963–1979.

Mayali, X., Weber, P. K., Brodie, E. L., Mabery, S., Hoeprich, P. D. and Pett-Ridge, J. (2012). High-throughput isotopic analysis of RNA microarrays to quantify microbial resource use. *ISME J.* 6, 1210–1221.

Merrington, G., Fishwick, S., Barraclough, D., Morris, J., Preedy, N., Boucard, T., *et al.* (2006). The development and use of soil quality indicators for assessing the role of soil in environmental interactions. Report to UK SIC, SC030265, Environment Agency.

Molz, F. J., Widdowson, M. A. and Benefield, L. D. (1986). Simulation of microbial growth dynamics coupled to nutrient and oxygen transport in porous media. *Water Resour. Res.* 22, 1207–1216.

Murphy, D. V., Cookson, W. R., Braimbridge, M., Marschner, P., Jones, D. L., Stockdale, E. A., *et al.* (2011). Relationships between soil organic matter and the soil microbial biomass (size, functional diversity, and community structure) in crop and pasture systems in a semi-arid environment. *Soil Res.* 49, 582–594.

Murphy, D. V., Recous, S., Stockdale, E. A., Fillery, I. R. P., Jensen, L. S., Hatch D. J., *et al.* (2003a). Gross nitrogen fluxes in soil: theory, measurement and application of ^{15}N pool dilution techniques. *Adv. Agron.* 79, 69–118.

Murphy, D. V., Stockdale, E. A., Brookes, P. C. and Goulding, K. W. T. (2003b). Impact of microorganisms on chemical transformations in soil. In Abbott, L. K. and Murphy, D. V. (eds.), *Soil Biological Fertility. A Key to Sustainable Land Use in Agriculture.* Kluwer Academic Publishers, London, pp. 37–59.

Murphy, D. V., Stockdale, E. A., Poulton, P. R., Willison, T. W. and Goulding, K. W. T. (2007). Seasonal dynamics of carbon and nitrogen pools and fluxes under continuous arable and ley-arable rotations in a temperate environment. *Eur. J. Soil Sci.* 58, 1410–1424.

National Bureau of Statistics of China. (2012). *China Statistical Year Book* [in Chinese], China Statistics Press, Beijing, China.

Nielsen, M. N. and Winding, A. (2002). Microorganisms as indicators of soil health. National Environmental Research Institute, Denmark. NERI Technical Report, No. 388.

Nishio, T. and Fujimoto, T. (1989). Mineralization of soil organic nitrogen in upland fields as determined by a $^{15}NH_4^+$ dilution technique, and absorption of nitrogen by maize. *Soil Biol. Biochem.* 21, 661–665.

Normander, B. and Prosser, J. I. (2000). Bacterial origin and community composition in the barley phytosphere as a function of habitat and presowing conditions. *Appl. Environ. Microbiol.* 66, 4372–4377.

Oehl, F., Oberson, A., Probst, M., Fliessbach, A., Roth, H. R. and Frossard, E. (2001). Kinetics of microbial phosphorus uptake in cultivated soils. *Biol. Fertil. Soils* 34, 31–41.

Ophel-Keller, K., McKay, A., Hartley, D., Curran, H. and Curran, J. (2008). Development of a routine DNA-based testing service for soilborne diseases in Australia. *Austral. Plant Pathol.* 37, 243–253.

Orphan, V. J., House, C. H., Hinrichs, K. U., McKeegan, K. D. and DeLong, E. F. (2001). Methane-consuming archaea revealed by directly coupled isotopic and phylogenetic analysis. *Science* 293, 484–487.

Parton, W. J., Schimel, D. S., Cole, C. V. and Ojima, D. S. (1987). Analysis of factors controlling soil organic matter levels in Great Plains grasslands. *Soil Sci. Soc. Am. J.* 51, 1173–1179.

Paustian, K., Lehmann, J., Ogle, S., Reay, D., Robertson, G. P. and Smith, P. (2016). Climate-smart soils. *Nature* 532, 49–57.

Postma-Blaauw, M. B., de Goede, R. G. M., Bloem, J., Faber, J. H. and Brussaard, L. (2010). Soil biota community structure and abundance under agricultural intensification and extensification. *Ecology* 91, 460–473.

Powlson, D. S., Brookes, P. C. and Christensen, B. T. (1987). Measurement of soil microbial biomass provides an early indication of changes in total soil organic matter due to straw incorporation. *Soil Biol. Biochem.* 19, 159–164.

Powlson, D. S. and Jenkinson, D. S. (1981). A comparison of the organic matter, biomass, adenosine triphosphate and mineralizable nitrogen contents of ploughed and direct-drilled soils. *J. Agric. Sci.* 97, 713–721.

Raes, J. and Bork, P. (2008). Molecular eco-systems biology: towards an understanding of community function. *Nat. Rev. Microbiol.* 6, 693–699.

Recous, S., Aita, C. and Mary, B. (1999). *In situ* changes in gross N transformations in bare soil after addition of straw. *Soil Biol. Biochem.* 31, 119–133.

Rijtema, P. E. and Kroes, J. G. (1991). Some results of nitrogen simulations with the model ANIMO. *Fertil. Res.* 27, 189–198.

Rousk, J., Bååth, E., Brookes, P. C., Lauber, C. L., Lozupone, C., Caporaso, J. G., *et al.* (2010). Soil bacterial and fungal communities across a pH gradient in an arable soil. *ISME* 4, 1340–1351.

Saby, N., Bellamy, P. H., Morvan, X., Arrouays, D., Jones, R. J., Verheijen, F. G., *et al.* (2008). Will European soil-monitoring networks be able to detect changes in topsoil organic carbon content? *Global Change Biol.* 14, 2432–2442.

Saffigna, P. G., Powlson, D. S., Brookes, P. C. and Thomas, G. A. (1989). Influence of sorghum residues and tillage on soil organic matter and soil microbial biomass in an Australian Vertisol. *Soil Biol. Biochem.* 21, 759–765.

Saggar, S., Parshotam, A., Sparling, G. P., Feltham, C. W. and Hart, P. B. S. (1996). ^{14}C-labelled ryegrass turnover and residence times in soils varying in clay content and mineralogy. *Soil Biol. Biochem.* 28, 1677–1686.

Sayler, G. S. and Layton, A. C. (1990). Environmental application of nucleic acid hybridization. *Annu. Rev. Microbiol.* 44, 625–648.

Schimel, J. P., Jackson, L. E. and Firestone, M. K. (1989). Spatial and temporal effects on plant-microbial competition for inorganic nitrogen in a California annual grassland. *Soil Biol. Biochem.* 21, 1059–1066.

Schloter, M., Dilly, O. and Munch, J. C. (2003). Indicators for evaluating soil quality. *Agric. Ecosyst. Environ.* 98, 255–262.

Schwartz, M. K., Luikart, G. and Waples, R. S. (2007). Genetic monitoring as a promising tool for conservation and management. *Trends Ecol. Evol.* 22, 25–33.

Seghers, D., Wittebolle, L., Top, E. M., Verstraete, W. and Siciliano, S. D. (2004). Impact of agricultural practices on the *Zea mays* L. endophytic community. *Appl. Environ. Microbiol.* 70, 1475–1482.

Serna-Chavez, H. M., Fierer, N. and Bodegom, P. M. (2013). Global drivers and patterns of microbial abundance in soil. *Global Ecol. Biogeogr.* 22, 1162–1172.

Sessitsch, A., Weilharter, A., Gerzabek, M. H., Kirchmann, H. and Kandeler, E. (2001). Microbial population structures in soil particle size fractions of a long-term fertilizer field experiment. *Appl. Environ. Microbiol.* 67, 4215–4224.

Smith, J. U., Bradbury, N. J. and Addiscott, T. M. (1996). SUNDIAL: a PC-based system for simulating nitrogen dynamics in arable land. *Agron. J.* 88, 38–43.

Smith, J. U. and Glendining, M. J. (1996). A decision support system for optimising the use of nitrogen in crop rotations. *Asp. Appl. Biol.* 47, 103–110.

Sollins, P., Homann, P. and Caldwell, B. A. (1996). Stabilization and destabilization of soil organic matter: mechanisms and controls. *Geoderma* 74, 65–105.

Sparling, G. P., Parfitt, R. L., Hewitt, A. E. and Schipper, L. A. (2003). Three approaches to define desired soil organic matter contents. *J. Environ. Qual.* 32, 760–766.

Sparling, G. P., Schipper, L. A., Bettjeman, W. and Hill, R. (2004). Soil quality monitoring in New Zealand: practical lessons from a 6-year trial. *Agric. Ecosyst. Environ.* 104, 523–534.

Stockdale, E. A., Banning, N. C. and Murphy, D. V. (2013). Rhizosphere effects on functional stability of microbial communities in conventional and organic soils following elevated temperature treatment. *Soil Biol. Biochem.* 57, 56–59.

Stockdale, E. A. and Brookes, P. C. (2006). Detection and quantification of the soil microbial biomass — impacts on the management of agricultural soils. *J. Agric. Sci.* 144, 285–302.

Stockdale, E., Hatch, D., Murphy, D., Ledgard, S. and Watson, C. (2002). Verifying the nitrification to immobilisation ratio (N/I) as a key determinant of potential nitrate loss in grassland and arable soils. *Agronomie* 22, 831–838.

Stockdale, E. A., Watson, C. A., Black, H. I. J. and Philipps, L. (2006). Do farm management practices alter below-ground biodiversity and ecosystem function? Implications for sustainable land management. JNCC Report No. 354. ISSN 0963 8901. Peterborough: Joint Nature Conservation Committee.

Thakur, M. P., Milcu, A., Manning, P., Niklaus, P. A, Roscher, C., Power, S., *et al.* (2015). Plant diversity drives soil microbial biomass carbon in grasslands irrespective of global environmental change factors. *Global Change Biol.* 21, 4076–4085.

Tiedje, J. M., Cho, J. C., Murray, A., Treves, D., Xia, B., Zhou, J., *et al.* (2001). Soil teeming with life: new frontiers for soil science. In Rees, R. M., Ball, B. C., Campbell, C. D. and Watson, C. A. (eds.), *Sustainable Management of Soil Organic Matter*. CABI Publishing, Wallingford, Oxon, UK, pp. 393–425.

Torsvik, V., Goksøyr, J. and Daae, F. L. (1990). High diversity in DNA of soil bacteria. *Appl. Environ. Microbiol.* 56, 782–787.

Trumbore, S. (2006). Carbon respired by terrestrial ecosystems — recent progress and challenges. *Global Change Biol.* 12, 141–153.

Van der Putten, W. H., Anderson, J. M., Bardgett, R. D., Behan-Pelletier, V., Bignell, D. E., Brown, G. G., *et al.* (2004). The sustainable delivery of goods and services provided by soil biota. In Wall, D. H. (ed.), *Sustaining Biodiversity and Ecosystem Services in Soils and Sediments*. SCOPE 64. Island Press, Washington, DC, pp. 15–43.

Waksman, S. A. and Tenney, F. G. (1927). The composition of natural organic materials and their decomposition in the soil: II. Influence of age of plant upon the rapidity and nature of its decomposition-Rye plants. *Soil Sci.* 24, 317–334.

Wander, M. M., Walter, G. L., Nissen, T. M., Bollero, G. A., Andrews, S. S. and Cavanaugh-Grant, D. A. (2002). Soil quality. *Agron. J.* 94, 23–32.

Wardle, D. A. (1992). A comparative assessment of factors which influence microbial biomass carbon and nitrogen levels in soil. *Biol. Rev.* 67, 321–358.

Wardle, D. A., Bonner, K. I. and Nicholson, K. S. (1997). Biodiversity and plant litter: experimental evidence which does not support the view that enhanced species richness improves ecosystem function. *Oikos* 79, 247–258.

Wardle, D. A. and Giller, K. E. (1996). The quest for a contemporary ecological dimension to soil biology. *Soil Biol. Biochem.* 28, 1549–1554.

Wienhold, B. J., Karlen, D. L., Andrews, S. S. and Stott, D. E. (2009). Protocol for indicator scoring in the soil management assessment framework (SMAF). *Renew. Agric. Food Syst.* 24, 260–266.

Winding, A., Hund-Rinke, K. and Rutgers, M. (2005). The use of microorganisms in ecological soil classification and assessment concepts. *Ecotoxicol. Environ. Saf.* 62, 230–248.

Young, I. M. and Ritz, K. (1998). Can there be a contemporary ecological dimension to soil biology without a habitat? *Soil Biol. Biochem.* 30, 1229–1232.

4

Microbial Biomass and Functions in Paddy Soil

Kazuyuki Inubushi* and Hirohiko Nagano†

Chiba University, Matsudo, Chiba 271-8510, Japan
**inubushi@faculty.chiba-u.jp*
†Hnagano@alaska.edu

Introduction

Paddy fields are important agro-ecosystems for rice production, as rice is the staple food for more than two billion people. Paddy rice production is mainly practised in East, South and Southeast Asia under the monsoon climate. Paddy soils are unique in that they are mostly managed under waterlogged conditions during the rice-growing season. Although these soils may have similar properties to natural wetland soils, such as partially anaerobic conditions, they also differ from natural wetlands because they are managed for agricultural production and are subject to fertilisation, monoculture cultivation and weeding. In this chapter, we first describe a modification of the original method established by David Jenkinson and his group in the UK for determining microbial biomass C and N, as well as the adenosine 5′ triphosphate (ATP) method for use under waterlogged soil conditions. Second, we examine the differences in microbial biomass between aerobic and waterlogged soils using this modified method, with special reference to soil nitrogen dynamics, exchange between soil and rice plants, and carbon dynamics, including methane production. In Japan, about half of all arable lands are paddy fields, while the other half are upland fields. A quarter of these total arable lands are covered by Andosols. As these soils are derived from volcanic parent materials, we also compare the microbial biomass in Andosols (or Andisols) and non-Andosols in this chapter.

Microbial Biomass C and N Determination in Paddy Soil

The original method for determining microbial biomass by chloroform fumigation was based on the autolysis of microbial cells exposed to chloroform vapour. However, as chloroform is an organic solvent that is not water-soluble, we expected that wetland soils from paddy fields might not be effectively fumigated following the original method. Also, if water is removed from wet soil by drying, the microbial biomass and soil organic matter would be changed during the drying process. So, we first attempted to introduce chloroform vapour into wet soil spread thinly on a Petri dish by evacuating the soil using two desiccators (Figure 4.1).

Effective fumigation of wet paddy soil was confirmed when more than 95% of the original microbial population was killed (Inubushi *et al.*, 1984). Some easily decomposable organic matter, rather than microbial cell components, was solubilised during fumigation, but when the fumigation time was limited to less than 24 h (Inubushi *et al.*, 1985; Inubushi and Wada, 1988), only a small proportion was solubilised. We then used this modified chloroform fumigation–incubation method with an ^{15}N tracer to investigate the nitrogen dynamics between soil and rice plants via the microbial biomass in paddy soils. This research began with laboratory incubation, but was later conducted in a tropical paddy field at the International Rice Research Institute in the Philippines. In the incubation experiment, ^{15}N from both labelled ammonium sulphate and $^{15}N_2$ gas was incorporated into the microbial biomass over a short period (<1 week) and

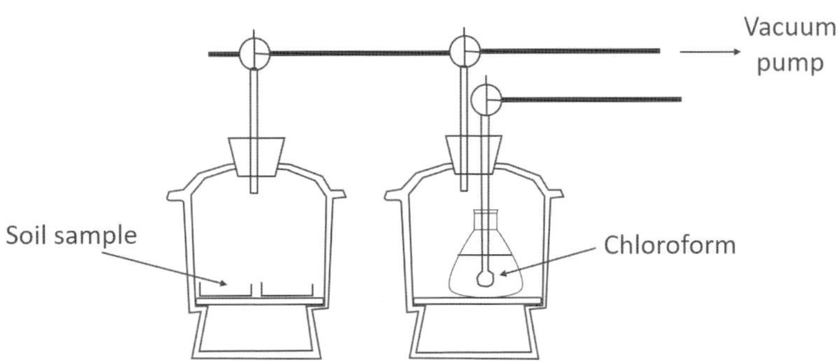

Figure 4.1 Schematic diagram showing the fumigation of wet paddy field soils with chloroform.

then re-mineralised by the end of the incubation (4 weeks) (Inubushi and Watanabe, 1987). In the field experiment, we found that ammonium sulphate derived from fertiliser was the initial source of N for rice plants, but in the middle to latter stages, the microbial biomass was a more important source of N for rice. This was based on periodic measurements of the ^{15}N abundance during N uptake by rice and microbial biomass (Inubushi and Watanabe, 1986; Watanabe and Inubushi, 1986).

Microbial Biomass C, N, ATP and AEC Determinations in Paddy Soil for Comparison with Upland Soil

Another series of laboratory experiments was conducted at Rothamsted Experimental Station, UK, to modify the fumigation–extraction method for use with paddy soils (Inubushi *et al.*, 1991). Measurements of the soil ATP content and adenylate energy change (AEC) were modified for use in waterlogged soils, as these indices also provide valuable insights into the effects that agricultural management has on soil microbial processes and organic matter turnover (Inubushi *et al.*, 1989a, b). The effects of long-term fertiliser application on microbial biomass were then investigated under aerobic and anaerobic conditions at both the Rothamsted long-term wheat and grassland experiment in the UK, and at the Konosu long-term paddy experiment in eastern Japan. Dr. Jenkinson suggested the use of liquid chloroform dripped onto moist soil to ensure microbial autolysis and efficient fumigation, as chloroform is only slightly soluble in water. Apart from this change, no other modifications to the fumigation and extraction times were made. Our results show that chloroform fumigation by this modified method was effective. ATP was effectively extracted from waterlogged soils, and the long-term application of organic manure increased the microbial biomass C, N and ATP contents (Inubushi *et al.*, 1989a). However, ATP decreased rapidly after waterlogged incubation, especially in Rothamsted soils, while aeration increased ATP. Soil ADP and AMP levels were also measured to calculate the AEC, which decreased during waterlogged incubation, but they recovered rapidly after short-term aeration. This indicates that soil ATP was converted into ADP and AMP during anaerobic incubation, but was quickly re-converted into ATP after aeration. AEC recovery was faster in the Konosu paddy soil than in the Rothamsted soils (Inubushi *et al.*, 1989a). The microbial community in paddy soil may be better adapted to soil conditions under flooded/aerated cycling for rice cultivation management.

Microbial Biomass in Local Paddy Field Experiments

In Japan, heavy metal contamination (e.g. Cd, Hg and As) of paddy soils became a serious problem starting in the 1880s, mostly due to mining activities. In Mie prefecture, central Japan, however, indigenous soil parent materials already contain some heavy metals. When heavy metal-contaminated paddy soil was reclaimed by covering it with uncontaminated mountain soil, the microbial biomass proved to be a good indicator for evaluating how soil fertility recovers in reclaimed soil and is stabilised by organic matter amendment (Yoshikawa and Inubushi, 1995).

In Shiga prefecture, western Japan, Shibahara and Inubushi (1995, 1997) also investigated the microbial biomass and N dynamics in soil microbial biomass–rice plant interactions in long-term paddy field experiments. Under field conditions, the N turnover time in the soil biomass was estimated to be as rapid as under tropical conditions in the Philippines (Inubushi *et al.*, 1997; Shibahara *et al.*, 1998), indicating that soil microbial N turnover is well adapted to hot summer conditions during the rice-growing season. Additional studies are needed to compare such features under broader conditions, not only for N but also for other nutrients.

Microbial Biomass in Paddy Soil under Elevated Atmospheric CO_2 Conditions

Several studies have monitored changes in microbial biomass under elevated atmospheric CO_2 conditions, first in incubation experiments, then in growth chambers and finally under field conditions (Inubushi *et al.*, 1999, 2001a, 2003; Cheng *et al.*, 2000, 2001; Hoque *et al.*, 2001, 2002a, b) in relation to rice growth, methane production (Cheng *et al.*, 2008) and oxidation (Inubushi *et al.*, 2011). Under free-air CO_2 enrichment (FACE) conditions, the microbial biomass in waterlogged paddy field soil first increased due to algal growth in flooded water and then from increased rice root growth. Elevated atmospheric CO_2 conditions influenced the algal and rice root biomass directly, and the soil microbial biomass indirectly. The response of microbial biomass C (MBC) was more obvious than that of microbial biomass N under a fixed N application rate, indicating that N is a limiting factor under elevated CO_2 conditions for both soil microbes and rice plants. A comparison of the results from the Japanese and Chinese FACE experiments would be interesting, as the latter also involved heavy N application (Zheng *et al.*, 2006). The results of studies investigating methane oxidation in relation to rice roots and aquatic weeds suggest a close relationship between plants and soil methanotrophs (Gilbert and Frenzel, 1998; Inubushi *et al.*, 2001b).

Microbial Biomass as Affected by Fertiliser and Soil Type

In Japanese paddy soils, controlled-release N fertiliser, especially coated urea (CRCU), is commonly used to reduce labour costs because only a single basal application is needed at transplantation time. This practice also has environmental benefits due to reduced N loss by leaching to groundwater and denitrification to the atmosphere. Microbial biomass N can be regarded as another controlled-release N source after immobilisation/re-mineralisation. Therefore, the dynamics of microbial biomass N in four types of paddy soil were studied during the rice-growing period using both an ^{15}N tracer and microbial community analysis based on phospholipid fatty acids (PLFAs; Acquaye *et al.*, 2000; Inubushi *et al.*, 2002; Acquaye and Inubushi, 2004a, b; Inubushi and Acquaye, 2004) (Figures 4.2 and 4.3). A significant linear relationship was observed between MBC, measured by the chloroform fumigation–extraction method, and total PLFAs in the surface and subsurface soils of the paddy soils. The differences in the effect of fertiliser type on microbial biomass N were small, even with a 20% reduced application rate of coated compared to uncoated urea. These differences were also small among soil types, but obviously different between soil layers [either surface (0–1 cm) or subsurface (1–10 cm)] (Figure 4.2).

A similar result was obtained for microbial diversity, namely the differences in the effects of fertiliser and soil types were small, but those of soil layers and timing were significant, indicating that soil diversity was impacted more heavily by soil layer (surface or subsurface) and timing (flooding or pre-flooding) than by the type of fertiliser or soil. This is probably because the subsurface soil became more anaerobic than the surface soil during rice cultivation, while the surface soil remained oxidative and became a more favourable environment for phototrophic microorganisms (Figure 4.3).

Several studies have also investigated microbial diversity in paddy soils using DNA/RNA techniques, including denitrifiers (Yoshida *et al.*, 2009), methanogens (Watanabe *et al.*, 2006) and the microbial biomass K in paddy soil (Yamashita *et al.*, 2014). Their results suggested that better soil management was required to reduce N loss via denitrification and C loss via methane emission, as well as better fertiliser management of K in paddy field soils for environmental sustainability. The diversity of methanogens in paddy soil has also been reviewed (Dubey, 2005) and specifically examined in Indian paddy soils (Dubey *et al.*, 2014).

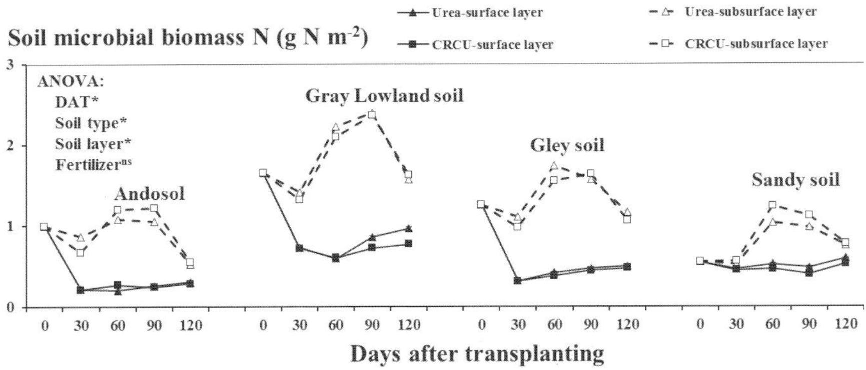

Figure 4.2 Changes in microbial N after rice transplantation in four Japanese paddy soils from depths of 0–1 cm (surface layer) and 1–10 cm (subsurface layer). Andosol was sampled from Kashiwa, Grey Lowland soil from Matsumoto, Gley soil from Abiko, and Sandy soil from Kujukuri, Japan. The total amounts of soil carbon were 55.8, 37.8, 28.7 and 7.8 g kg^{-1}, respectively. Standard errors of the means; * and ns represent significant and non-significant differences at $p < 0.05$, respectively.

Long-Term Rice–Wheat Rotation in Japan and India

In southern and some central parts of Japan, rice–wheat rotation is practised because of warmer and drier conditions during the winter season compared with other regions. The microbial biomass and soil quality were investigated in long-term rice–wheat experiments in Kyushu, southern Japan, for comparison with Indian rice–wheat experiments (Tirol-Padre *et al.*, 2005, 2007). Based on these relationships, organic manure was determined to be essential for maintaining sustainable soil quality. The amount of organic manure needed to supplement inorganic fertilisers must be optimised to increase C and N accumulation in the soil without negative effects on crop yield.

Biochar and Biogas Digested Slurry Affect the Microbial Biomass in Paddy Soil

Biogas production is used worldwide, including Japan, as a source of renewable bioenergy. After biogas production, the remaining biogas digester slurry is an unused bioresource, or another waste material, if it is not utilised. A project was started in Japan to develop effective uses for these materials when applied to soil. The process first involves liquid–solid separation, after which the dehydrated

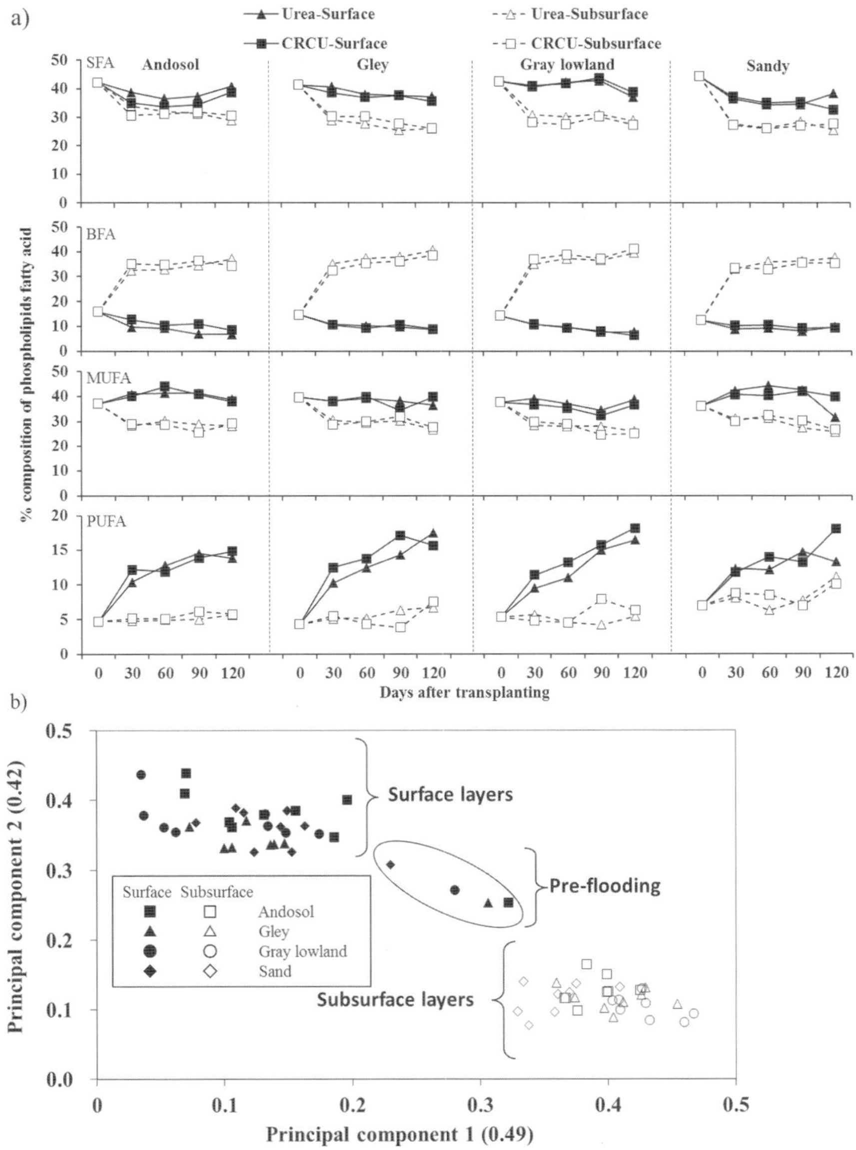

Figure 4.3 Changes in the (a) PLFA composition of the soil microbial biomass in four Japanese paddy field soils after urea or coated-urea (CRCU) fertilisation. The fatty acids were saturated (SFA, all living microorganisms), branched (BFA, Gram-positive bacteria), monounsaturated (MUFA, Gram-negative bacteria) and polyunsaturated (PUFA, eukaryotes). (b) Principal component analysis of the temporal changes in PLFAs (using data from (a)). The cumulative contributions of PC1 and PC2 were >0.60.

solid cake is used to make biochar and the liquid component used to make liquid fertiliser (Nakamura *et al.*, 2014). The impact of these materials on microbial biomass in paddy soil was then investigated in microcosm and pot experiments to examine the effect on soil fertility and crop production. Methane emissions in paddy soil were also measured to assess the impact on the global environment (Singla *et al.*, 2014, 2015). The application of biochar significantly increased the soluble organic C, MBC and soluble total N contents of the soils. Methanogenic activity was also increased, without affecting the microbial biomass N content, rice yield or methanogenic diversity, suggesting that biogas digester slurry can be used to sustainably manage paddy soil fertility and minimise the impact on the global environment, provided that appropriate application rates and methods are used.

Land-Use Change and Amendments Affect the Microbial Biomass in Tropical Wetlands

The sustainability of land-use/cover change is an important issue not only in Japan, but also in developing countries. For example, a land-use change from wetland forests to fertilised vegetable fields in Central Kalimantan, Indonesia, was shown to affect the size of the microbial biomass and emission of greenhouse gases, including CO_2 and N_2O (Arai *et al.*, 2014). Several amendments were tested in an irrigated paddy field in Bangladesh, and significant positive effects were found, including an increase in microbial biomass and rice productivity, as well as improved soil quality (Ali *et al.*, 2014). The drying of wetlands to saline soils near the Aral Sea decreased the MBC, but increased the extractable C, in the soil, indicating that salt affected the microbial activity balance (Mamilov *et al.*, 2004). Forest soils near a volcano can be extremely acidic, and in such situations the soil microbial biomass can be a good indicator of the effects on plant growth (Watanabe *et al.*, 2010). The microbial biomass and greenhouse gas production/consumption potentials of soils under different land-use types were compared in central Japan and eastern Hungary, and it was found that not only climatic conditions, but also soil type and land-use were the principal factors influencing soil microbial properties and greenhouse gas dynamics (Kong *et al.*, 2013a, b). Another field study in relation to land-use history was conducted in former pasture and croplands in semi-dry Eurasian steppe soil. The lands had been left fallow and restored with native grasses for 18 years. The effects of land-use change history on microbial biomass and activity were examined by taking soil and gas samples from nearby sites with a similar soil type but different land-use history. The post-restoration MBC content was significantly larger in the pasture than in the cropland, although there was no significant difference in CO_2 efflux between the two sites, indicating that microbial biomass was affected not only by current land-use, but also by past (until 18 years prior to the study) land use (Nagano *et al.*, 2012b). Soil microbial biomass can therefore be a useful environmental index, especially under such critical/extreme conditions.

Microbial Biomass in Andosols and Other Soils

In Japan, about half of the upland is covered by volcanic soils (Andosols). The MBC/total C ratio in these soils was found to be much lower than in non-volcanic soils (Inubushi *et al.*, 2005), indicating slower soil organic matter turnover and/or stability in Andosols. The effect of organic fertiliser on microbial biomass was also examined in Andosols under vegetable fields in relation to crop growth and greenhouse gas emissions (Goyal *et al.*, 1999, 2000, 2006a, b; Inubushi *et al.*, 2000; Nagano *et al.*, 2012a). Several organic treatments over a short period (<2 years) increased both the microbial biomass and crop growth to some extent. The environmental impact of recycling organic residues has also been investigated in grasslands (Ushiwata *et al.*, 2007). Such effects can be seen not only in the surface soil, but also in subsurface soils (Zaman *et al.*, 2002a, b, 2004; Matsushima *et al.*, 2009). Soil MBC and soil organic C (SOC) were compared in Andosol and non-Andosol (Sand-dune Regosol) soils from Japan and at Rothamsted Experimental Station, UK (Inubushi and Kong, 2014). A significant linear correlation was observed for Japanese Andosols, but the slope of the correlation (MBC/SOC) was significantly lower for Andosols than for non-Andosols. This indicates that the MBC was lower at the same SOC level in the Andosol compared with the non-Andosol. A subsurface Andosol in buried paleosol had a higher MBC than the upper subsoil layer, indicating that accumulated soil C may be a substrate for microorganisms, even after thousands of years of eruptions and surface horizon formation. These findings highlight the unique features of microbial biomass and C turnover in Andosols. Even when Andosols contain large amounts of soil organic matter due to acidic soil conditions, microbial biomass and activity are sensitive indicators that show how soil organic matter changes.

The Spatial Distribution of Microbial Biomass in Relation to Other Soil Parameters

The spatial distribution of soil microbial biomass and N_2O emissions were investigated extensively in an onion field on Fluvaquent in Hokkaido, northern Japan, and were found to be correlated significantly with other soil chemical parameters, including the soil organic matter content and pH (Yanai *et al.*, 2003). Soil heterogeneity is critical to properly manage agricultural fields, as it improves soil fertility with minimum environmental risks. Soil microbial biomass is one of the most sensitive parameters needed to monitor such heterogeneity.

The Production/Consumption of Methane and Other Greenhouse Gases in Wetland and Forest Soils

Several studies have investigated the influence of vegetation types and soil properties on MBC and metabolic quotients in temperate volcanic and tropical forest soils (Xu *et al.*, 2006, 2007; Nagano *et al.*, 2013). Multiple timescale variations and controls of soil/root respiration and soil microbial biomass were also investigated in a tropical, dry Dipterocarp forest in western Thailand, and it was found that soil microbial biomass plays a crucial role in the magnitude of and temporal variations in soil respiration. In this forest ecosystem, large seasonal variations in soil respiration were observed and were mainly attributed to the responses of soil microbial biomass to soil moisture (Hanpattanakit *et al.*, 2015).

Future Research Challenges

In paddy soil ecosystems, modern technologies such as heavy machinery use and low-cost soil management are used in Japan and many other rice-growing countries, but their effect on soil microbial biomass and diversity is not clearly understood. Consequently, the effectiveness of these technologies for sustainable agriculture and land management is not yet proved. Other global issues such as soil C sequestration, greenhouse gas emission/consumption, land-use/cover changes, desertification and deforestation have become ever more critical issues in many regions. Further research is urgently needed using soil microbial parameters and their controlling variables to assess sustainable practices and to provide long-term perspectives.

References

Acquaye, S., Hadi, A., Cheng, W. and Inubushi, K. (2000). Effect of fertilizers on soil microbial biomass in two types of paddy fields. *Tech. Bull. Fac. Hort., Chiba Univ.* 54, 1–11.

Acquaye, S. and Inubushi, K. (2004a). Comparative effects of application of coated and non-coated urea in clayey and sandy paddy soil microcosms examined by the ^{15}N

tracer technique. I. Effects on growth, N uptake and yield of rice crop. *Soil Sci. Plant Nutr.* 50, 205–213.

Acquaye, S. and Inubushi, K. (2004b). Comparative effects of application of coated and non-coated urea in clayey and sandy paddy soil microcosms examined by the ^{15}N tracer technique. II. Effects on soil microbial biomass N and microbial ^{15}N immobilization. *Soil Sci. Plant Nutr.* 50, 215–223.

Ali, M., Sattar, A., Islam, M. N. and Inubushi, K. (2014). Integrated effects of organic, inorganic and biological amendments on methane emission, soil quality and rice productivity in irrigated paddy ecosystem of Bangladesh: field study of two consecutive rice growing seasons. *Plant Soil* 378, 239–252.

Arai, H., Hadi, A., Darung, U., Limin, S. H., Takahashi, H., Hatano, R. and Inubushi, K. (2014). Land use change affects microbial biomass and fluxes of carbon dioxide and nitrous oxide in tropical peatlands. *Soil Sci. Plant Nutr.* 60, 423–434.

Cheng, W., Chander, K. and Inubushi, K. (2000). Effect of elevated CO_2 and temperature on microbial biomass nitrogen and nitrogen mineralization in submerged soil microcosms. *Soil Microorganisms* 54, 51–59.

Cheng, W., Inubushi, K., Hoque, M. M., Sasaki, H., Kobayashi, K., Yagi, K., Okada, M. and Hasegawa, T. (2008). Effect of elevated [CO_2] on soil bubble and CH_4 emission from a rice paddy: a test by ^{13}C pulse-labeling under free-aid CO_2 enrichment. *Geomicrobiol. J.* 25(7–8), 396l.

Cheng, W., Inubushi, K., Yagi, K., Sakai, H. and Kobayashi, K. (2001). Effect of elevated CO_2 on biological nitrogen fixation, nitrogen mineralization and carbon decomposition in submerged rice soil. *Biol. Fert. Soils* 34, 7–13.

Dubey, S. K. (2005). Microbial ecology of methane emission in rice agroecosystem. *Appl. Ecol. Environ. Res.* 3(2), 1–27.

Dubey, S. K., Singh, A., Watanabe, T., Asakawa, S., Singla, A., Arai, H. and Inubushi, K. (2014). Methane production potential and methanogenic archaeal community structure in tropical dryland irrigated Indian paddy soils. *Biol. Fert. Soils* 50, 369–379.

Gilbert, B. and Frenzel, P. (1998). Rice roots and CH_4 oxidation: the activity of bacteria, their distribution and the microenvironment. *Soil Biol. Biochem.* 30(4), 1903–1916.

Goyal, S., Inubushi, K., Kato, S., Xu, H. L. and Umemura, H. (1999). Effect of anaerobically fermented manure on the soil organic matter, microbial properties and growth of spinach under greenhouse conditions. *Indian J. Microb.* 39, 211–216.

Goyal, S., Sakamoto, K. and Inubushi, K. (2000). Microbial biomass and activities along an andosol profile in relation to soil organic matter level. *Microb. Environ.* 15, 143–150.

Goyal, S., Sakamoto, K. and Inubushi, K. (2006b). Decomposition of sewage sludge compost and its effect on soil microbial biomass and growth of spinach. *Res. Crops* 7, 517–521.

Goyal, S., Sakamoto, K., Inubushi, K. and Kamewada, K. (2006a). Long-term effects of inorganic fertilization and organic amendments on soil organic matter and soil microbial properties in Andisols. *Arch. Agron. Soil Sci.* 52, 617–625.

Hanpattanakit, P., Monique, Y., Leclerc, M. Y., McMillan, A. M. S., Limtong, P., Maeght, J. L., Panuthai, S., Inubushi, K. and Chidthaisong, A. (2015). Multiple timescale variations and controls of soil respiration in a tropical dry dipterocarp forest, western Thailand. *Plant Soil* 390(1–2), 167–181. doi: 10.1007/s11104-015-2386-8.

Hoque, M. M., Inubushi, K., Miura, S., Kobayashi, K., Kim, H.-Y., Okada, M. and Yabashi, S. (2001). Biological dinitrogen fixation and soil microbial biomass carbon as influenced by free-air carbon dioxide enrichment (FACE) at three levels of nitrogen fertilization in a paddy field. *Biol. Fert. Soils* 34, 453–459.

Hoque, M. M., Inubushi, K., Miura, S., Kobayashi, K., Kim, H. Y., Okada, M. and Yabashi, S. (2002b). Nitrogen dynamics in paddy field as influenced by free-air CO_2 enrichment (FACE) at three levels of nitrogen fertilization. *Nutr. Cycl. Agroecosyst.* 63, 301–308.

Hoque, M. M., Inubushi, K. and Yabashi, S. (2002a). Nitrogen dynamics in paddy field under free-air CO_2 enrichment (FACE) examined by the ^{15}N tracer technique. *Tech. Bull. Fac. Hortic., Chiba Univ.* 56, 65–75.

Inubushi, K and Acquaye, S. (2004). Role of microbial biomass in biogeochemical processes in paddy soil environments. *Soil Sci. Plant Nutr.* 50(6), 793–805.

Inubushi, K., Acquaye, S., Tsukagoshi, S., Shibahara, F. and Komatsu, S. (2002). Effects of controlled-release coated urea (CRCU) on soil microbial biomass N in paddy fields examined by the ^{15}N tracer technique. *Nutr. Cycl. Agroecosyst.* 63(2–3), 291–300.

Inubushi, K., Brookes, P. C. and Jenkinson, D. S. (1989a). Adenosine 5′ triphosphate and adenylate energy charge in waterlogged soil. *Soil Biol. Biochem.* 21, 733–739.

Inubushi, K., Brookes, P. C. and Jenkinson, D. S. (1989b). Influence of paraquat on the extraction of adenosine triphosphate from soil by trichloroacetic acid. *Soil Biol. Biochem.* 21, 741–742.

Inubushi, K., Brookes, P. C. and Jenkinson, D. S. (1991). Soil microbial biomass C, N and ninhydrin-N in aerobic and anaerobic soils measured by the fumigation–extraction method. *Soil Biol. Biochem.* 23, 737–741.

Inubushi, K., Cheng, W., Aonuma, S., Hoque, M. M., Kobayashi, K., Miura, S., Kim, H. Y. and Okada, M. (2003). Effects of free-air CO_2 enrichment (FACE) on CH_4 emission from a rice paddy field. *Global Change Biol.* 9(10), 1458–1464.

Inubushi, K., Cheng, W. and Chander, K. (1999). Carbon dynamics in submerged soil microcosms as influenced by elevated CO_2 and temperature. *Soil Sci. Plant Nutr.* 45, 863–872.

Inubushi, K., Cheng, W., Mizuno, T., Lou, Y., Hasegawa, T., Sakai, H. and Kobayashi, K. (2011). Microbial biomass carbon and methane oxidation influenced by rice cultivars and elevated CO_2 in a Japanese paddy soil. *Eur. J. Soil Sci.* 62, 69–73.

Inubushi, K., Goyal, S., Sakamoto, K., Wada, Y., Yamakawa, K. and Arai, T. (2000). Influence of application of sewage sludge compost on N_2O production in soils. *Chemosphere-Global Change Sci.* 2, 329–334.

Inubushi, K., Hoque, M. M., Miura, S., Kobayashi, K., Kim, H. Y., Okada, M. and Yabashi, S. (2001a). Effect of free-air CO_2 enrichment (FACE) on microbial biomass in paddy field soil. *Soil Sci. Plant Nutr.* 47, 737–745.

Inubushi, K. and Kong, Y. (2014). Soil microbial biomass and C storage of an Andosol. In Hartemink, A. E., McSweeney, K. (eds.), *Soil Carbon, Progress in Soil Science*. Springer, Switzerland, pp. 173–178.

Inubushi, K., Sakamoto, K. and Sawamoto T. (2005). Properties of microbial biomass in acid soils and their turnover. *Soil Sci. Plant Nutr.* 51(5), 605–608.

Inubushi, K., Shibahara, F., Hasegawa, K. and Yamamuro, S. (1997). Effects of added organic matter on microbial biomass nitrogen dynamics and plant uptake in paddy soil. *Dev. Plant Soil Sci.* 78, 777–778.

Inubushi, K., Sugii, H., Nishino, S. and Nishino, E. (2001b). Effect of aquatic weeds on methane emission from submerged paddy soil. *Am. J. Bot.* 88(6), 975–979.

Inubushi, K. and Wada, H. (1988). Mineralization of carbon and nitrogen in chloroform-fumigated paddy soil under submerged conditions. *Soil Sci. Plant Nutr.* 34, 287–291.

Inubushi, K., Wada, H. and Takai, Y. (1984). Determination of microbial biomass-nitrogen in submerged soil. *Soil Sci. Plant Nutr.* 30, 455–459.

Inubushi, K., Wada, H. and Takai, Y. (1985). Factors influencing the determination of microbial biomass-nitrogen in submerged soil. *Soil Sci. Plant Nutr.* 31, 481–485.

Inubushi, K. and Watanabe, I. (1986). Dynamics of available nitrogen in paddy soils. II. Mineralized N of chloroform-fumigated soil as a nutrient source for rice. *Soil Sci. Plant Nutr.* 32, 561–577.

Inubushi, K. and Watanabe, I. (1987). Microbial biomass nitrogen in anaerobic soil as affected by N-immobilization and N_2-fixation. *Soil Sci. Plant Nutr.* 33(2), 213–224.

Kong, Y., Nagano, H., Kátai, J., Vágó, I., Oláh, Á. Z., Yashima, M. and Inubushi, K. (2013a). CO_2, N_2O and CH_4 production/consumption potentials of soils under different land-use types in central Japan and eastern Hungary. *Soil Sci. Plant Nutr.* 59, 455–462.

Kong, Y., Watanabe, M., Nagano, H., Watanabe, K., Yashima, M. and Inubushi, K. (2013b). Effects of land-use type and nitrogen addition on nitrous oxide and carbon dioxide production potentials in Japanese Andosols. *Soil Sci. Plant Nutr.* 59, 790–799.

Mamilov, A., Dilly, O. M., Mamilov, S. and Inubushi, K. (2004). Microbial eco-physiology of degrading Aral Sea wetlands: consequences for C-cycling. *Soil Sci. Plant Nutr.* 50(6), 839–842.

Matsushima, M., Choi, W. J. and Inubushi, K. (2009). Nitrification inhibitor reduces N_2O production from different soil profiles of an Andosol. *Com. Soil Sci. Plant Anal.* 40, 3194–3206.

Nagano, H., Kato, S., Ohkubo, S. and Inubushi, K. (2012a). Emissions of carbon dioxide, methane, and nitrous oxide from short- and long-term organic farming Andosols in central Japan. *Soil Sci. Plant Nutr.* 58, 793–801.

Nagano, H., Kasahara, T., Takahashi, M., Yoshioka, R., Kong, Y., Yashima, M., Okazaki, M., *et al.* (2013). Temporal and spatial variation in physiochemical and microbial properties in Chiba University's Environmental Horticulture and Forestry Farm: Comparative research of results in 2002, 2003 and 2011. *Hort. Res.* 67, 21–27 (in Japanese with English summary).

Nagano, H., Sugihara, S., Matsushima, M., Okitsu, S., Prikhodka, V. E., Ikhodko, R., Manakhova, E., *et al.* (2012b). Carbon and nitrogen contents and greenhouse gas fluxes of the Eurasian steppe soils with different land-use histories located in the Arkaim museum reserve of South Ural, Russia. *Soil Sci. Plant Nutr.* 58, 238–244.

Nakamura, M., Yuyama, Y., Yamaoka, M. and Shimizu, N. (2014). Global warming impacts of the process to utilize digested slurry from methane fermentation as a fertilizer: case study of the Yamada Biomass Plant. *Paddy Water Environ.* 12(Suppl. 2), S295.

Shibahara, F. and Inubushi, K. (1995). Measurements of microbial biomass C and N in paddy soils by the fumigation–extraction method. *Soil Sci. Plant Nutr.* 41, 681–689.

Shibahara, F. and Inubushi, K. (1997). Effects of organic matter application on microbial biomass and available nutrients in various types of paddy soils. *Soil Sci. Plant Nutr.* 43, 191–203.

Shibahara, F., Yamamuro, S. and Inubushi, K. (1998). Dynamics of microbial biomass nitrogen as influenced by organic matter application in paddy fields. I. Fate of fertilizer and soil organic N determined by ^{15}N tracer technique. *Soil Sci. Plant Nutr.* 44, 167–178.

Singla, A., Dubey, S. K., Ali, M. A. and Inubushi, K. (2015). Methane flux from paddy vegetated soil: a comparison between biogas digested liquid and chemical fertilizer. *Wetlands Ecol. Manage.* 23, 139–148.

Singla, A., Dubey, S. K., Singh, A. and Inubushi, K. (2014). Effect of biogas digested slurry-based biochar on methane flux and methanogenic archaeal diversity in paddy soil. *Agric. Ecosyst. Environ.* 197, 278–287.

Tirol-Padre, A., Ladha, J. K., Regmi, A. P., Bhandari, A. L. and Inubushi, K. (2007). Organic amendments affect soil parameters in two long-term rice–wheat experiments. *Soil Sci. Soc. Am. J.* 71, 442–452.

Tirol-Padre, A., Tsuchiya, K., Inubushi, K. and Ladha, J. K. (2005). Enhancing soil quality through residue management in a rice–wheat system in Fukuoka, Japan. *Soil Sci. Plant Nutr.* 51, 849–860.

Ushiwata, S., Sasa, H. and Inubushi, K. (2007). Influence of steam-treated grass clipping on grass growth, drainage water quality and soil microbial properties in a simulation of green course. *Soil Sci. Plant Nutr.* 53(4), 489–498.

Watanabe, I. and Inubushi, K. (1986). Dynamics of available nitrogen in paddy soils. I. Changes in available N during rice cultivation and origin of N. *Soil Sci. Plant Nutr.* 32, 37–50.

Watanabe, T., Kimura, M. and Asakawa, S. (2006). Community structure of methanogenic archaea in paddy field soil under double cropping (rice–wheat). *Soil Biol. Biochem.* 38, 1264–1274.

Watanabe, M., Yamamura, S., Takamatsu, T., Koshikawa, M. K., Hayashi, S., Murata, T., Saito, S., Inubushi, K. and Sakamoto, K. (2010). Microbial biomass and nitrogen transformations in volcanic acid surface soils strongly acidified by volcanic hydrogen sulfide deposition in Osorezan, Japan. *Soil Sci. Plant Nutr.* 56, 123–132.

Xu, X., Han, L., Wang, Y. and Inubushi, K. (2007). Influence of vegetation types and soil properties on microbial biomass carbon and metabolic quotients in temperate volcanic and tropical forest soils. *Soil Sci. Plant Nutr.* 53, 430–440.

Xu, X., Inubushi, K. and Sakamoto, K. (2006). Effect of vegetations and temperature on microbial biomass carbon and metabolic quotients of temperate volcanic forest soils. *Geoderma* 136, 310–319.

Yamashita, K., Honjo, H., Nishida, M., Kimura, M. and Asakawa, S. (2014). Estimation of microbial biomass potassium in paddy field soil. *Soil Sci. Plant Nutr.* 60, 512–519.

Yanai, J., Sawamoto, T., Oe, T., Kusa, K., Yamakawa, K., Sakamoto, K., Naganawa, T., Inubushi, K., *et al.* (2003). Spatial variability of N_2O emissions and their soil-related determining factors in an agricultural field. *J. Environ. Qual.* 32, 1965–1977.

Yoshida, M., Ishii, S., Otsuka, S. and Senoo, K. (2009). Temporal shifts in diversity and quantity of *nirS* and *nirK* in a rice paddy field soil. *Soil Biol. Biochem.* 41, 2044.

Yoshikawa, S. and Inubushi, K. (1995). Characteristics of microbial biomass and soil organic matter in newly reclaimed wetland rice soils. *Biol. Fert. Soils* 19, 292–296.

Zaman, M., Cameron, K. C., Di, H. J. and Inubushi, K. (2002b). Changes in mineral N, microbial biomass and enzyme activities in different soil depths after surface applications of dairy shed effluent and chemical fertilizer. *Nutr. Cycl. Agroecosyst.* 63, 275–290.

Zaman, M., Di, H. J., Sakamoto, K., Goto, S., Hayashi, H. and Inubushi, K. (2002a). Effect of sewage sludge compost and chemical fertilizer applications on microbial biomass and N mineralization rates. *Soil Sci. Plant Nutr.* 48, 195–201.

Zaman, M., Matsushima, M., Chang, S. X., Inubushi, K., Nguyen, L., Goto, S., Kaneko, F. and Yoneyama, T. (2004). Nitrogen mineralization, N_2O production and soil microbiological properties as affected by long-term applications of sewage sludge composts. *Biol. Fert. Soils* 40, 101–109.

Zheng, X., Zhou, Z., Wang, Y., Zhu, J., Wang, Y., Yue, J., Shi, Y., Kobayashi, K., Inubushi, K., Huang, Y., Han, S., Xu, Z., Xie, B., Butterbach-Bahl, K. and Yang, L. (2006). Nitrogen-regulated effects of free-air CO_2 enrichment on methane emissions from paddy rice fields. *Global Change Biol.* 12, 1717–1732.

5

Soil Biodiversity and Ecosystem Functioning

Richard D. Bardgett*,§, Franciska T. de Vries* and Wim H. van der Putten†,‡

**School of Earth and Environmental Sciences, Michael Smith Building, The University of Manchester, Manchester M13 9PT, UK*
†Department of Terrestrial Ecology, Netherlands Institute of Ecology (NIOO-KNAW), P.O. Box 50, 6700 AB Wageningen, The Netherlands
‡Laboratory of Nematology, Wageningen University, PO Box 8123, 6700 ES, Wageningen, The Netherlands
§r.bardgett@manchester.ac.uk

Introduction

A striking feature of soil is that it hosts a vast diversity of life; soils of most ecosystems contain literally hundreds of thousands of species of bacteria and fungi, and alongside them live thousands of animal species, such as nematodes, microarthropods, enchytraieids, earthworms, ants and moles. Not only do these soil-dwelling species cover virtually every taxonomic group found in terrestrial ecosystems, but also it has been estimated that they could represent more than 25% of the described living species on Earth (Decaëns, 2010). Scientific interest in this rich diversity of soil life has boomed in recent years, with an ever-increasing number of studies being published on the distribution of soil biodiversity across different spatial and temporal scales and its functional importance for vegetation dynamics and biogeochemical cycles. Policy-makers, land managers and farmers are also becoming more interested in soil biodiversity, as awareness of its importance for sustainable soil management, human health and ecosystem services grows (Wall *et al.*, 2012, 2015).

In a similar way that the advent of new methods for measuring soil microbial biomass opened the way for a spate of studies on the role of soil microorganisms in nutrient and carbon cycling (Jenkinson and Powlson, 1976), a new generation of tools for interrogating the diversity and functional role of complex belowground communities has helped to spawn a new era of soil biodiversity research. This new research has not only extended the concept of soil microbial biomass to consider the rich diversity of microbial form and function in soil, but also it has revealed the existence of highly complex networks of biotic interactions in soil that play a major role in shaping plant communities and the functioning of terrestrial ecosystems (Bardgett and van der Putten, 2014). This is what this chapter is about: summarising how understanding of the functional role of soil biodiversity has advanced over recent decades, in terms of ecosystem processes, vegetation dynamics and ecosystem resilience to global change and identifying future challenges for soil biodiversity research.

A Short History of Soil Biodiversity Research

The importance of soil organisms for soil fertility has long been known. Aristotle referred to earthworms as 'the intestines of the earth', and Cleopatra, the last pharaoh of Egypt, declared them to be sacred because of their contribution to Egyptian agriculture. In his last book, published in 1881, Darwin detailed the importance of earthworms for soil fertility and biogeochemical cycles, and in the late 1800s and early 1900s, pioneering work of the Russian microbiologist Winogradsky established some of the fundamental roles that soil microbes play in the nitrogen and sulphur cycles. Despite all this historic wisdom, it was not until the 1980s when ecologists began to seriously explore the functional role of ecological interactions in soil. Initial work in this area focused on the role of trophic interactions in soil food webs for processes of decomposition and nutrient cycling (Anderson *et al.*, 1983; Coleman *et al.*, 1983; Clarholm, 1985), and plant nutrient acquisition and growth (Ingham *et al.*, 1985; Setälä and Huhta, 1991), and of root-associated organisms, especially mycorrhizal fungi and root-feeding fauna, as drivers of nutrient cycling (Read, 1991) and vegetation dynamics (Grime *et al.*, 1987; Brown and Gange, 1989). Around the same time, information on soil food webs was starting to be incorporated into ecosystem models of carbon and nutrient cycling (Hunt *et al.*, 1987), and models were also developed to examine functional consequences of shifts in food web energy channels in soil (Moore and

Hunt, 1988). These, and other studies, laid the foundation for subsequent work that showed below-ground multi-trophic interactions to serve as important drivers of ecosystem processes of decomposition and nutrient cycling and that the rate and magnitude of ecosystem processes depends on the architecture of soil food webs (Bardgett, 2002; Mikola *et al.*, 2002; Scheu and Setälä, 2002; Wardle, 2002).

Soil biodiversity research took off during the 1990s and early 2000s. It is hard to pin point what triggered this; several factors simultaneously played a role. The development of new DNA-based techniques during the early 1990s opened the door to the exploration of microbial diversity in soil and provided new insights into its true extent and distribution. As already mentioned, the pioneering research of soil ecologists in the 1980s laid the foundation for a new generation of studies in the 1990s and early 2000s exploring how the diversity and architecture of soil communities influenced ecosystem processes (e.g. De Ruiter *et al.*, 1995; van der Heijden *et al.*, 1998; Laakso and Setälä, 1999; Cragg and Bardgett, 2001; Hunt and Wall, 2002); international conferences and workshops were held to explore how to 'go beyond the biomass' (Ritz *et al.*, 1994) and consider the significance and regulation of soil biodiversity (e.g. Freckman, 1994; Collins *et al.*, 1995; Bardgett *et al.*, 2005); and the study of biodiversity–function relationships was fast becoming a dominant theme in ecology due to global concerns about the rapid pace of species loss and its consequences for ecosystem function (Schulze and Mooney, 1994; Loreau *et al.*, 2002). Together, these factors prompted a growing number of soil ecologists to start exploring how diverse soil communities are distributed, how they respond to global change, and the causes and consequences of soil biodiversity for ecosystem processes and vegetation dynamics. Since then, soil biodiversity research has exploded: the ISI Web of Science reveals a marked increase in publications on soil biodiversity since the early 2000s, a pattern which remains on a steep upward trajectory; global initiatives, such as the Global Soil Biodiversity Initiative (GSBI), have been established to promote awareness of soil biodiversity and its integration into policy and land management decisions; an increasing number of major, multinational funding initiatives are being focused on soil biodiversity research; and a new generation of soil ecologists, plant scientists and microbiologists are entering the field, ensuring that it is well poised to tackle future challenges in soil biodiversity research.

Soil Biodiversity and Ecosystem Functioning

The diversity of soil life is highly vulnerable to anthropogenic disturbances (Table 5.1). The conversion of natural forests and grasslands to agriculture causes dramatic declines in soil biodiversity and the interactions within the soil

Table 5.1 Soil biodiversity loss and change can result from a broad array of anthropogenic impacts on ecosystems worldwide.

Disturbance	Remarks
Land conversion	Conversion of tropical forest to agriculture strongly reduces the diversity of soil fauna, including nematodes and termites, and mycorrhizal fungi. The conversion of natural grasslands to agriculture reduces the diversity of many groups of soil fauna, enhances the abundance of plant parasitic nematodes and causes a shift from fungal to bacterial-dominated food webs
Tillage	Soil disturbance, compaction and erosion caused by tillage negatively impact the abundance and diversity of most soil faunal groups, especially larger bodied organisms such as earthworms and mycorrhizal fungi. It also causes a shift from fungal to bacterial-dominated food webs
Invasive species	Invasive soil species impact the diversity of native soil communities via predation and competition and through changing the soil environment. Invasive plant species change the soil physicochemical environment and may have resistance to native pathogens and mycorrhizal fungi, thereby affecting their diversity
Climate change	Changes in the means, extremes and variability of climate can induce shifts in the abundance of diversity of soil biota. Effects of climate change on soil biota are both direct, via effects of changing temperature and moisture availability, and indirect through changing plant growth and plant community structure in response to altered climate and elevated carbon dioxide concentration
Pollution	Pesticides and herbicides, which affect both target and non-target organisms, and chronic pollutant stresses, such as those caused by heavy metal contamination reduce the diversity and abundance of microbes and animals in soils, generally favouring bacteria over fungi
Nutrient enrichment	Chronic nitrogen enrichment generally suppresses the abundance and diversity of decomposer and mycorrhizal fungi by altering quantity, quality and diversity of organic matter entering soil and by acidifying the soil. Effects of nitrogen enrichment on soil animals are variable, whereas it generally causes a shift from fungal to bacteria-dominated food webs

(*Continued*)

Table 5.1 (*Continued*)

Disturbance	Remarks
Desertification	Desertification causes loss of soil biodiversity and of below-ground biomass in general. Disturbances such as trampling, which are commonly associated with desertification, destroys vegetation and biocrusts in drylands, causing significant declines in soil biodiversity
Fire	Severe and frequent wildfires negatively affect soil animal populations, especially litter dwelling and less mobile species that cannot escape, and soil fungal and mycorrhizal communities. Fire impacts operate by changing soil pH and hydrological characteristics, and through encouraging vegetation change and plant invasions
Sealing	Construction of roads, housing and other urban developments effectively suffocates the soil, severing links that soils biota have with plants and their roots and preventing infiltration of water and gaseous exchange. This all causes declines in soil diversity at all taxonomic levels
Mining	Surface mining completely removes soil biodiversity, such as in the case of sand mining, and can cause acute soil pollution, for instance, with heavy metals and oil

The major drivers of soil biodiversity change are listed, and while not exhaustive, they have all been associated with species loss in at least some regions of the world.

food web (Bloemers *et al.*, 1997; Eggleton *et al.*, 2002; Dlamini and Haynes, 2004), as does the intensification of agriculture (Mulder *et al.*, 2005; Postma-Blaauw *et al.*, 2010; Tsiafouli *et al.*, 2015). Climate change and, especially, extreme climate events pose a significant threat to soil biodiversity (Bardgett *et al.*, 2008), as does nitrogen enrichment, which negatively impact soil fungi and dramatically changes soil bacterial communities and their functioning (Treseder, 2008; Ramirez *et al.*, 2012). Also, soil sealing, which effectively ceases natural soil functioning (Gardi *et al.*, 2013), and invasive plant species, which impact native soil organisms via a range of mechanisms, including altered resource supply, competitive interactions and predation, and physical and chemical modification of soil (Wardle *et al.*, 2011). Given the sensitivity of soil biodiversity to these anthropogenic disturbances, and the rapid rate of soil change resulting from human activity, an obvious question is whether the resulting loss of biodiversity and simplification of soil food webs matters for soil and ecosystem functioning?

It is well known that soil species differ in their effects on soil functioning, but what is less clear is how changes in the diversity, composition and complexity of below-ground communities influence the functioning of soil. However, a recent synthesis of experimental studies on soil diversity–function relationships concluded that diversity effects on processes of nutrient and carbon cycling are highly variable, but that effects of species loss are greatest at the low end of the diversity spectrum (Nielsen *et al.*, 2011). These authors also concluded that changes in the composition of below-ground communities, and the variety of organism functional traits and biotic interactions, are of more importance for soil functioning than are changes in species richness *per se* (Nielsen *et al.*, 2011). As an example, microcosm studies have shown that functional dissimilarity among detritivores (i.e. the degree to which component species differ in their effects on ecosystem processes) is a superior predictor of litter decomposition to species richness (Heemsbergen *et al.*, 2004) and that the selective removal of certain groups of organisms from complex soil communities impairs multiple soil functions (Wagg *et al.*, 2014). It has also been shown that rare soil microbes can have an important functional role, for instance in plant defence (Hol *et al.*, 2010), and that changes in the functional complexity of soil communities elicit widely divergent responses of individual ecosystem processes, which arise through below-ground effects on above-ground communities that also impact ecosystem functioning (Bradford *et al.*, 2014). This latter finding suggests that effects of changes in soil community composition vary for different ecosystem processes and that the net effect on ecosystem functioning is an aggregate of both positive and negative relationships for different processes with differing underlying causes (Bradford *et al.*, 2014).

All of the aforementioned studies were done under laboratory or glasshouse conditions, often in experimental systems with structurally simplistic soils and narrow diversity gradients that bear little resemblance to those found in the field (but see Bradford *et al.*, 2014). However, a new generation of studies is attempting to overcome these limitations by testing for soil biodiversity–function relationships *in situ* in real soil systems. A recent cross-biome experiment, spanning sites from the subarctic to the tropics, for example, showed that the loss of key components of the decomposer communities consistently slowed rates of litter decomposition and carbon and nitrogen cycling, pointing to negative effects of soil biodiversity loss on these functions on broad spatial scales (Handa *et al.*, 2014). Also, in a study that encompassed four locations across Europe with different climatic and soil conditions, soil food web properties were found to strongly and consistently predict processes of carbon and nitrogen cycling and they were a better predictor of these processes than land use (De Vries *et al.*,

2013). In particular, nitrogen cycling processes were explained by soil food web properties such as biomass of arbuscular mycorrhizal fungi and the bacterial-energy channel of the soil food web, whereas carbon cycling processes were related to earthworm biomass and the fungal to bacterial energy channel ratio. This study not only provided the first quantification of relationships between soil biodiversity and ecosystem processes on a broad spatial scale, but also the results suggested that soil food web parameters should be included in carbon and nitrogen cycling models to enhance their predictive power (De Vries *et al.*, 2013).

A spate of recent field-based studies has also combined the use of high-throughput molecular approaches with measures of biogeochemical processes to demonstrate how complex soil microbial communities regulate ecosystem function *in situ*. As an example, Zhou *et al.* (2012) used metagonomic approaches linked to field-based measures of soil carbon cycling to show that soil microorganisms regulate impacts of experimental warming on tall-grass prairie on soil carbon and nitrogen dynamics via differential stimulation of microbial populations and genes involved in decomposition and nitrogen cycling. Likewise, molecular approaches were used to demonstrate that rapid and dynamic responses of microbial communities to permafrost thaw relate to methane emissions (Mackelprang *et al.*, 2011) and that root-associated fungi play an important role in regulating soil carbon dynamics in boreal forest (Clemmensen *et al.*, 2013). And in a study of soil taken from many locations across Scotland and a network of global drylands, Delgado-Baquerizo *et al.* (2016) discovered that soil microbial diversity positively relates to multi-functionality, even when accounting for effects of climate, soil abiotic factors and spatial predictors on ecosystem processes. Although these examples did not explicitly test for soil diversity effects *per se*, they do point to an important functional role of complex soil communities in biogeochemical cycling in the real world.

As a final point, changes in soil biodiversity are likely to impact soil physical properties, with consequences for ecosystem services related to soil formation and water regulation. Diversity effects on soil physical properties have not been explicitly studied, but they are likely to be important given the potential for different soil organisms and species to differentially impact soil structure through different routes. For example, fungi promote soil aggregate stability through the physical enmeshment of soil particles by their extensive networks of mycelia, whereas bacteria produce metabolic products, mainly polysaccharides, which bind soil particles together (Hallett *et al.*, 2009). Mycorrhizal infection can also influence soil aggregate stability through physical enmeshment of soil particles by their extensive networks of mycelium, but also through the binding of soil particles via the production of extracellular polysaccharides and proteins, including the protein glomalin, which alters the wetting behaviour of soil (Rillig and Mummey, 2006). And soil animals, especially ecosystem engineers such as earthworms and termites, impact soil structure by creating macropores and channels, thereby improving water movement through soil. As already mentioned, our understanding of diversity effects on soil physical properties is very limited, which we argue needs to be redressed given the crucial importance of soil structure for soil functioning and ecosystem services.

Soil Biodiversity and Resilience

Against a background of ongoing climate change and increasing frequency of extreme weather events, there is a growing interest in how soil biodiversity responds to disturbances and the extent to which complex soil communities and their functioning can resist and recover from disturbances. The relationship between soil biodiversity and stability, and particularly whether greater soil biodiversity increases the resistance and resilience of soil functioning to disturbances, is poorly understood. While there is limited evidence that soil biodiversity *per se* increases stability of functioning, a number of modelling and empirical studies on soil food webs have found consistent patterns in the attributes of soil communities and food webs that convey stability. For example, the first modelling studies on soil food webs showed that the patterning of interaction strengths in soil food webs is important for stability (De Ruiter *et al.*, 1995; Neutel *et al.*, 2002). In their description of the first complete soil food web, Hunt *et al.* (1987) already identified the presence of two distinct soil food web compartments, namely the fungal and the bacterial energy channel: the bacterial energy channel was made up by organisms with a higher turnover rate that mineralised more C and N than those of the fungal energy channel. Almost 20 years later, Rooney *et al.* (2006) combined this compartmentalisation with the findings by De Ruiter *et al.* (1995) and Neutel *et al.* (2002) to show that the fungal energy channel, with its low turnover rate and many weak interactions, is more stable than the bacterial energy channel with its high turnover rate and strong interactions. In particular, they argued that the coupling of these two energy channels by top predators conveyed stability to the soil food web.

Empirical studies on the relationship between soil biodiversity and stability after disturbance initially focused on diversity in its strictest sense, namely within-trophic-level species number. There is some evidence that species richness of soil communities increases their stability following disturbance. One of the first studies to examine how soil biodiversity and functioning are linked under disturbance was done by Griffiths *et al.* (2000), who experimentally reduced soil microbial diversity through chloroform fumigation before imposing both a transient and a persistent perturbation. They found that greater microbial diversity increased the recovery of respiration after the transient perturbation, as well as the resistance of respiration to the persistent stress. Similarly, Wittebolle *et al.* (2009) found that evenness of denitrifying communities increased functional resistance to salinity stress, although Liiri *et al.* (2002) found no evidence that more species-rich microarthropod communities were less affected by drought than were more species-poor communities.

More recent work suggest that functional traits are important predictors of the response of soil organisms to disturbance, for example, higher dispersal ability would increase community recovery after disturbance (Figure 5.1; Lindberg and Bengtsson, 2006; De Vries and Shade, 2013). It has also been suggested that fast growing organisms, such as those that make up the bacterial energy channel and copiotrophic bacteria, invest more in growth, but less in defense strategies and would therefore be less resistant, but more resilient, to disturbances in general (Figure 5.1). This idea is in line with the theoretical work described above and is supported by empirical work that shows that soil food webs with a prominent fungal energy channel, consisting of slower growing organisms, are more resistant to drought, in terms of both their composition and functioning, than bacterial-based soil food webs that are dominated by faster-growing organisms (De Vries *et al.*, 2012a). In addition, the ability of soil organisms to form spores or to enter dormancy would confer resistance to a wide range of disturbances (Figure 5.1; De Vries and Shade, 2013). But specific functional traits are probably more important for determining the resistance to a specific type of disturbance (De Vries and Shade, 2013). For example, it has been shown repeatedly that fungi are more resistant to drought than bacteria (Bapiri *et al.*, 2010; De Vries *et al.*, 2012a, b) and this observation has recently been attributed to the ability of fungi to hydraulically redistribute water in their hyphal networks (Guhr *et al.*, 2015). More specifically, the ability to use specific substrates can affect the response of soil organisms to disturbances because many disturbances alter the form and availability of below-ground resources (De Vries *et al.*, 2013). The selection for specific traits can also alter the ability of a soil community to resist and recover from multiple disturbances. For example, when subsequent disturbances select for or favour similar traits, the soil community might be better able to resist the second disturbance, as has been shown for multiple droughts, or a heat wave followed by a drought (Berard *et al.*, 2012; De Vries *et al.*, 2012a). In contrast, when two subsequent disturbances select for contrasting traits, the second disturbance might hit a soil community much harder than without the first disturbance, for example, when a flood follows a drought.

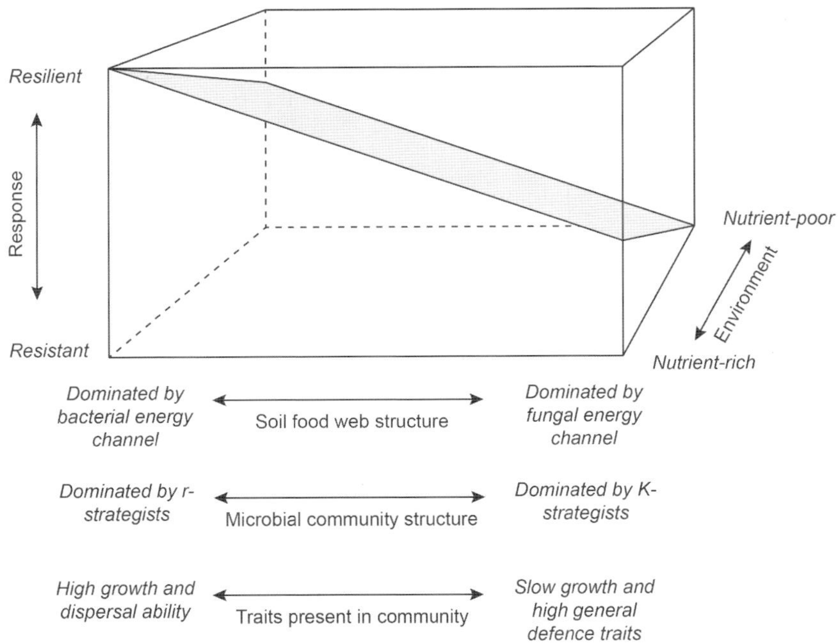

Figure 5.1 Conceptual framework predicting soil community response to disturbance (Adapted from De Vries and Shade, 2013). The response of the soil community is predicted by the relative abundance of the bacterial and fungal energy channels, the relative abundance of r- vs. K-strategists and the relative abundance of general growth, dispersal and defence traits. The presence and abundance of these community properties is correlated with environmental properties, and environmental properties themselves will influence the ability of soil organisms to re-grow and/or re-disperse. A soil community dominated by the fungal energy channel, oligotrophic K-strategists and slow growth and high abundance of general resistance traits will be resistant, but not resilient, to disturbance. In contrast, a community dominated by the bacterial energy channel, copiotrophic r-strategists and high growth and dispersal rate will have low resistance, but high resilience under disturbance. In addition, the presence of specific traits will affect the response of the soil community to a specific disturbance (not indicated here).

Another important driver of soil biodiversity response to disturbance is the availability of resources. It has been suggested that soils rich in carbon are more resistant and/or resilient to disturbances (Orwin *et al.*, 2006), and in line with the ideas above, higher resource availability would favour soil community resilience by increasing growth rates (Figure 5.1). Thus, altered plant physiological processes, primary productivity and community composition, which all influence the quality and quality of organic inputs into the soil, can have cascading effects on the response of below-ground communities. Indeed, plant presence, species identity and functional richness have been shown to affect the response of soil organisms to disturbances (Orwin and Wardle, 2005; De Vries *et al.*, 2012b). Similarly, the response of soil organisms to disturbances can feedback to plant growth and community composition, for example, by altering the availability of nutrients during and after a disturbance (De Vries *et al.*, 2012a), or via direct antagonistic and facilitative effects on individual plant species (Pellkofer *et al.*, 2016), such as the positive effect of mycorrhizal fungi on the drought resistance of their host plant (Raviv, 2010).

Soil Biodiversity, Plant Growth and Community Dynamics

Interest in the role of soil biodiversity as a driver of plant community dynamics has increased dramatically in recent years. Soil biota can influence vegetation dynamics via direct interactions between plant roots and root-associated pathogens, herbivores and symbiotic organisms, but also via the activities of free-living soil organisms that regulate nutrient availability and supply to plants. Some of the first examples of soil organisms acting as determinants of plant community dynamics emerged in the late 1980s and early 1990s, with studies demonstrating the role of mycorrhizal fungi, pathogens and root-feeding invertebrates as drivers of vegetation succession (Grime *et al.*, 1987; Brown and Gange, 1989; van der Putten *et al.*, 1993; Bever *et al.*, 1997). These studies helped to set in motion a major research effort that has now placed plant–soil feedback at the heart of the study of plant community dynamics (van der Putten *et al.*, 2013) and shown that particular groups of soil organisms, such as ecosystem engineers (e.g. ants and termites), impact vegetation dynamics via their influence on physical properties and nutrient availability in soil (van der Heijden *et al.*, 2008; Bardgett and Wardle, 2010).

Many studies have examined how changes in plant diversity influence soil communities and the processes that they govern, but few have examined the reverse, that is, how soil biodiversity influences plant diversity. Moreover, those studies that have explored how soil biodiversity influences plant diversity have focussed on the role of single taxonomic groups, such as mycorrhizal fungi (van der Heijden *et al.*, 1998), or on selective manipulation of taxonomic groups, largely on the basis of body size (Bradford *et al.*, 2002; Wagg *et al.*, 2014). These studies show highly variable results, with increasing complexity and diversity of soil communities having both positive (van der Heijden *et al.*, 1998; Wagg *et al.*, 2014) and negative effects (Bradford *et al.*, 2002) on plant diversity. This points to the wide range of mechanisms involved and the likelihood that the net effect of changes in soil biodiversity on plant diversity depends on the balance between positive and negative effects of different taxa. However, such variability in the response of plant communities to soil biodiversity is not surprising given that they operate via many mechanisms, including direct effects of herbivory, symbiosis or pathogenesis, and indirect effects of changing resource availability and multi-trophic interactions between above- and below-ground communities (Wardle *et al.*, 2004; Bezemer and van Dam, 2005).

Recent research points to other mechanisms by which soil biodiversity can affect plant communities, such as the role of common mycorrhizal networks, which interconnect roots of individual plants. These common mycorrhizal networks connect the roots of two or more plants (Selosse *et al.*, 2006), and hence have potential to be supplied by carbon from several individual plants and to distribute mineral nutrient resources from soil to host plants unequally, thereby influencing plant competitive interactions. One potential consequence of such asymmetrical resource transfer is the establishment of 'cheaters', whereby a plant may invest relatively little carbon into a mycorrhizal network, but obtain a disproportionately large amount of mineral nutrients (Walder *et al.*, 2012). Also, it has been suggested that plants might 'reward' AM fungi with carbon and these fungi then enforce the cooperation by transferring more phosphorus to the plant (Kiers *et al.*, 2011), thereby potentially affecting plant competitive interactions. However, evidence for these roles of common mycorrhizal networks comes from laboratory studies, so their impact on plant community dynamics in the field is yet to be explored.

An area that has attracted considerable recent attention is the role of complex rhizosphere microbial communities in plant disease suppression. Not only has it been shown that there is an enormous diversity of microorganisms associated with plant roots, but also studies reveal that this plant-associated microbial community is crucial for plant health (Mendes *et al.*, 2011; Berendsen *et al.*, 2012). As noted by Mendes *et al.* (2013), the rhizosphere provides a 'frontline defence' for plant roots against attack by soil-borne pathogens via a variety of mechanisms, including the production of metabolites that inhibit competing microbes, such as antimicrobial compounds and volatile organic compounds, and through modulating the plant immune system. There is also accumulating evidence that upon pathogen or insect attack, plants can recruit protective microorganisms and enhance microbial activity to suppress pathogens in the rhizosphere (Berendsen *et al.*, 2012). The mechanisms involved are complex, but include the excretion of compounds in root exudates which act as attractants to consortia of beneficial microorganisms and activate antimicrobial activity of rhizosphere microflora

(Berendsen *et al.*, 2012). This is a fast-developing area of research of high relevance to crop protection; hence, a major challenge for the future is to identify the key players in the rhizosphere microbiome involved in plant defence and the mechanisms that enable plants to recruit and stimulate beneficial microorganisms in order to protect plant health.

Soil Biodiversity and Invasion

Biological invasions are a major threat for biodiversity and ecosystem functioning and there is now evidence that invasion of exotic species into new territories may both affect and be affected by soil biodiversity. Many studies have considered how invasive exotic plants influence soil processes and soil biodiversity. In general, three groups of soil biota can be distinguished for their specific role in plant invasiveness: pathogens and root herbivores, symbiotic mutualists and decomposers (Van der Putten *et al.*, 2007). Many invasive plants are early successional species that are often controlled by soil-borne enemies in their native range, whereas they make little use of symbiotic mutualists (Kardol *et al.*, 2006). Moreover, invasive plant species often produce high-quality litter that is easily decomposable, which accelerates rates of nutrient cycling (Kourtev *et al.*, 2002). Therefore, in the new range, introduced exotic plant species may have multiple benefits that promote their growth; not only do they escape their specialized soil-borne enemies and make less use of symbiotic mutualists, but also they increase nutrient availability (Blumenthal *et al.*, 2009). Such responses may also explain why invasive plants can reduce soil biodiversity, without being negatively affected themselves.

Besides invasive plants, there are also invasive animals, such as introduced earthworms (Hendrix and Bohlen, 2002), or rats that can change the entire functioning of ecosystems by influencing nutrient cycles and making nutrients more or less available (Fukami *et al.*, 2006). Introduced soil-borne pathogens can have devastating effects in invaded ecosystems. For example, the oomycete pathogen *Pythium cinnamomi* that has been introduced in Australia can attack entire Eucalypt forest communities (Peters and Weste, 1997; Weste, 1981) and the introduced soil pathogen *Armillaria luteobubalina* killed off 38% of all species in coastal Australian ecosystems (Shearer *et al.*, 1998). Also, *Phytophthora ramorum*, which causes sudden oak death, has infected over 70 plant species in Californian natural forests (Venette and Cohen, 2006). There are fewer examples of introduced symbionts. For example, in Australia and New Zealand, introduced ectomycorrhizal fungi, such as *Amanita muscaria* (Bougher, 1996), might influence invaded ecosystems; however, there is little understanding of the consequences for soil biodiversity, although they are likely to be profound (Van der Putten *et al.*, 2007).

A major effect of invasive plant species on local soil biodiversity may operate via the introduction of different litter quality. Several studies have demonstrated such effects. For example, the invasive annual grass species *Bromus tectorum* has been shown to reduce the abundance and diversity of soil fungi and invertebrates, but promote the abundance of bacteria in soil food webs in perennial grassland, which was attributed to changes in the amount and quality of resources entering soil (Belnap and Philips, 2001). In another study on invasive forest species in the eastern United States, it was shown that two introduced understory species changed soil microbial community structure and functioning underneath invaded patches, which was also attributed to changes in both the quantity and quality of resource input to soil (Kourtev *et al.*, 2002), and the invasive bamboo in native broadleaf forests reduced soil microbial diversity and altered microbial functioning (Xu *et al.*, 2015). In this latter case, however, the effect of the invasive bamboo on soil biodiversity was largely indirect, via its effect on native plant biodiversity (Xu *et al.*, 2015).

Relatively few studies have considered effects of soil biodiversity on biological invasions. In one study, however, the survival of the invading bacterial pathogen *Escherichia coli* in soil was found to be negatively related to microbial diversity, suggesting that soil microbial diversity is a key factor controlling the extent to which bacterial invaders can establish (van Elsas *et al.*, 2012). Moreover, it was recently proposed that the network architecture of complex root-associated bacterial communities determines pathogen invasion: diverse rhizosphere communities with low nestedness and high connectance, and a clear niche overlap with the pathogen, were found to reduce pathogen invasion success and lower disease levels in plants (Wei *et al.*, 2015). Soil biodiversity might also control invasiveness of introduced exotic plant species, although solid evidence is lacking. One such mechanism could be the control of introduced plant species by local soil pathogens or root herbivores. There is little experimental evidence for this, but it is reasonable to assume that a wider diversity of local soil-borne pathogens could enhance the chance that introduced plant species will become controlled by local pathogen strains that are present in the invaded range. For instance, one study on the black cherry tree *Prunus serotina* showed that the difference between the native and novel range might be the absence of virulent strains of an otherwise abundant soil-borne enemy (Reinhart *et al.*, 2010).

Conclusions and Outlook

Research on soil biodiversity has come a long way in recent years and new discoveries are constantly emerging. What is abundantly clear from this recent research effort is that below-ground communities are more diverse than previously thought and they also play a major role in shaping the structure and functioning of terrestrial ecosystems and their resilience to invasion and disturbance events associated with global change. Despite this new understanding, challenges remain. These include the need for improved understanding of the mechanisms that shape complex soil biological communities at different spatial and temporal scales, and the need for new integrative frameworks that encompass the many ecological interactions that operate within soil, and between below- and above-ground communities. Moreover, these integrative frameworks also need to take account of the structural and chemical complexity of soil. As noted elsewhere (Bardgett and van der Putten, 2014), theory in soil ecology has lagged behind experimental work; hence the need for new theoretical models to explain the causes and consequences of soil biodiversity for the structure and functioning of terrestrial ecosystem, and their responses to global change, is pressing.

Another major challenge is to extend soil biodiversity knowledge to the field. To date, most studies on relationships between soil biodiversity and ecosystem functioning have been done in simplified model systems, which do not reflect the true complexity of below-ground communities or the soil systems in which they live. Also, despite recent advances in our understanding of the role of soil biodiversity in regulating vegetation dynamics, relatively little is known about how such processes contribute to vegetation dynamics in the field. Of relevance here are recent molecular tools, such as metagenomics, which offer potential to detect changes in the structure and functional attributes of below-ground communities in response to environmental perturbations and link these changes to measurements of ecosystem process in the real world (e.g. Mackelprang *et al.*, 2011; Zhou *et al.*, 2012; Clemmensen *et al.*, 2013). Also, trait-based approaches are increasingly used to understand how community change above-ground influences the structure and functioning of soil microbial communities and consequences for ecosystem processes (e.g. De Deyn *et al.*, 2008; De Vries *et al.*, 2012c; Grigulis *et al.*, 2013; Bardgett *et al.*, 2014; Barberán *et al.*, 2015). These approaches also provide a means of simplifying complex soil food webs by identifying the key functional traits of below-ground communities that most strongly control ecosystem functions. Combined with new statistical and modelling tools, such approaches could help to explain how current and future changes in soil biodiversity resulting from human activities impact ecosystem functioning at scales relevant to the real world.

References

Anderson, J. M., Ineson, P. and Huish, S. A. (1983). Nitrogen and cation mobilization by soil fauna feeding on leaf litter and soil organic-matter from deciduous woodlands. *Soil Biol. Biochem.* 15, 463–467.

Bapiri, A., Bååth, E. and Rousk, J. (2010). Drying–rewetting cycles affect fungal and bacterial growth differently in an arable soil. *Microbial Ecol.* 60, 419–428.

Barberán, A., McGuire, K. L., Wolf, J. A., Jones, F. A., Wright, S. J., Turner, B. L., *et al.* (2015). Relating below-ground microbial composition to the taxonomic, phylogenetic, and functional trait distributions of trees in a tropical forest. *Ecol. Lett.* 18, 1397–1405.

Bardgett, R. D. (2002). Causes and consequences of biological diversity in soil. *Zoology* 105, 367–374.

Bardgett, R. D., Freeman, C. and Ostle, N. J. (2008). Microbial contributions to climate change through carbon-cycle feedbacks. *ISME J.* 2, 805–814.

Bardgett, R. D., Mommer, L. and De Vries, F. T. (2014). Going underground: root traits as drivers of ecosystem processes. *Trends Ecol. Evol.* 29, 692–699.

Bardgett, R. D., Usher, M. B. and Hopkins, D. W. (Eds.) (2005). *Biological Diversity and Function in Soils.* Cambridge University Press, UK.

Bardgett. R. D. and van der Putten, W. H. (2014). Below-ground biodiversity and ecosystem functioning. *Nature* 515, 505–511.

Bardgett, R. D. and Wardle, D. A. (2010). *Above-ground–Below-ground Linkages: Biotic Interactions, Ecosystem Processes, and Global Change.* Oxford Series in Ecology and Evolution. Oxford University Press, UK.

Belnap, J. and Phillips, S. L. (2001). Soil biota in an ungrazed grassland: response to annual grass (*Bromus tectorum*) invasion. *Ecol. Appl.* 11, 1261–1275.

Berard, A., Sassi, M.B., Renault, P. and Gros, R. (2012). Severe drought-induced community tolerance to heat wave. An experimental study on soil microbial processes. *J. Soils Sediments* 12, 513–518.

Berendsen, R. L., Pieterse, C. M. and Bakker, P. A. (2012). The rhizosphere microbiome and plant health. *Trends Plant Sci.* 17, 478–486.

Bever, J. D., Westover, K. M. and Antonovics, J. (1997). Incorporating the soil community into plant population dynamics: the utility of the feedback approach. *J. Ecol.* 85, 561–573.

Bezemer, T. M. and van Dam, N. M. (2005). Linking above-ground and below-ground interactions via induced plant defenses. *Trends Ecol. Evol.* 20, 617–624.

Bloemers, G. F., Hodda, M., Lambshead, P. J. D., Lawton, J. H. and Wanless, F. R. (1997). The effects of forest disturbance on diversity of tropical soil nematodes. *Oecologia* 111, 575–582.

Blumenthal, D., Mitchell, C. E., Pyšek, P. and Jarošík, V. (2009). Synergy between pathogen release and resource availability in plant invasion. *Proc. Natl. Acad. Sci. USA* 106, 7899–7904.

Bougher, N. L. (1996). Diversity of ectomycorrhizal fungi associated with eucalypts in Australia. In Brundrett, M., Dell, B., Malajczuk, N., Minquin, G. (eds.), *Mycorrhizas for Plantation Forestry in Asia.* Australian Centre for International Research, Canberra, Australia, pp 8–15.

Bradford, M. A., Jones, T. H., Bardgett, R. D., Black, H. I., Boag, B., Bonkowski, M., *et al.* (2002). Impacts of soil faunal community composition on model grassland ecosystems. *Science* 298, 615–618.

Bradford, M. A., Wood, S. A., Bardgett, R. D., Black, H. I. J., Bonkowski, M., Eggers, T., *et al.* (2014). Discontinuity in the responses of ecosystem processes and multifunctionality to altered soil community composition. *Proc. Natl. Acad. Sci. USA* 111, 14478–14483.

Brown, V. K. and Gange, A. C. (1989). Differential effects of above- and below-ground insect herbivory during early plant succession. *Oikos* 54, 67–76.

Clarholm, M. (1985). Interactions of bacteria, protozoa and plants leading to mineralization of soil nitrogen. *Soil Biol. Biochem.* 17, 181–187.

Clemmensen, K. E., Bahr, A., Ovaskainen, O., Dahlberg, A., Ekblad, A., Wallander, H., *et al.* (2013). Roots and associated fungi drive long-term carbon sequestration in boreal forest. *Science* 339, 1615–1618.

Coleman, D. C., Reid, C. P. P. and Cole, C. V. (1983). Biological strategies of nutrient cycling in soil systems. *Adv. Ecol. Res.* 13, 1–55.

Collins, H. P., Robertson, G. P. and Klug, M .J. (Eds.) (1995). *The Significance and Regulation of Soil Biodiversity.* Kluwer Academic Publishers, The Netherlands.

Cragg, R. G. and Bardgett, R. D. (2001). How changes in soil faunal diversity and composition within a trophic group influence decomposition processes. *Soil Biol. Biochem.* 33, 2073–2081.

De Deyn, G. B., Cornelissen, J. H. C. and Bardgett, R. D. (2008). Plant functional traits and soil carbon sequestration in contrasting biomes. *Ecol. Lett.* 11, 516–531.

De Ruiter, P. C., Neutel, A. M. and Moore, J. C. (1995). Energetics, patterns of interactions strengths, and stability in real ecosystems. *Science* 269, 1257–1260.

Decaëns, T. (2010). Macroecological patterns in soil communities. *Glob. Ecol. Biogeogr.* 19, 287–302.

Delgado-Baquerizo, M., Maestre, F. T., Reich, P. B., Jeffries, T. C., Gaitan, J. J., Encinar, D., *et al.* (2016). Microbial diversity drives multifunctionality in terrestrial ecosystems. *Nat. Commun.* 7, 10541.

De Vries, F. T., Liiri, M., Bjørnlund, L., Bowker, M., Christensen, S., Setälä, H. and Bardgett, R. D. (2012a). Land use alters the resistance and resilience of soil food webs to drought. *Nat. Clim. Chang.* 2, 276–280.

De Vries, F. T., Liiri, M., Bjørnlund, L., Setälä, H., Christensen, S. and Bardgett, R. D. (2012b). Legacy effects of drought on plant growth and the soil food web. *Oecologia* 170, 821–833.

De Vries, F., Manning, P., Tallowin, J., Mortimer, S., Pilgrim, E., Harrison, K., *et al.* (2012c). Abiotic drivers and plant traits explain landscape-scale patterns in soil microbial communities. *Ecol. Lett.* 15, 1230–1239.

De Vries, F. T. and Shade, A. (2013). Controls on soil microbial community stability under climate change. *Front. Microbiol.* 4, 265.

De Vries, F. T., Thébault, E., Liiri, M., Birkhofer, K., Tsiafouli, M. A., Bjørnlund, L., *et al.* (2013). Soil food web properties explain ecosystem services across European land use systems. *Proc. Natl. Acad. Sci. USA* 110, 14296–14301.

Dlamini, T. C. and Haynes, R. J. (2004). Influence of agricultural land use on the size and composition of earthworm communities in northern KwaZulu-Natal, South Africa. *Appl. Soil Ecol.* 27, 77–88.

Eggleton, P., Bignell, D. E., Hauser, S., Dibog, L., Norgrove, L. and Madong, B. (2002). Termite diversity across an anthropogenic disturbance gradient in the humid forest zone of West Africa. *Agric. Ecosyst. Environ.* 90, 189–202.

Freckman, D. W. (1994). *Life in the Soil. Soil Biodiversity: Its Importance to Ecosystem Processes.* Report of a workshop held at the Natural Resource History Museum, London, UK.

Fukami, T., Wardle, D. A., Bellingham, P. J., Mulder, C. P. H., Towns, D. R., Yeates, G. W., *et al.* (2006). Above- and below-ground impacts of introduced predators in seabird-dominated island ecosystems. *Ecol. Lett.* 9, 1299–1307.

Gardi, C., Jeffery, S. and Saltelli, A. (2013). An estimate of potential threats levels to soil biodiversity in EU. *Glob. Chang. Biol.* 195, 1538–1548.

Griffiths, B. S., Ritz, K., Bardgett, R. D., Cook, R., Christensen, S., Ekelund, F., *et al.* (2000). Ecosystem response of pasture soil communities to fumigation-induced microbial diversity reductions: an examination of the biodiversity–ecosystem function relationship. *Oikos* 90, 279–294.

Grigulis, K., Lavorel, S., Krainer, U., Legay, N., Baxendale, C., Dumont, M., *et al.* (2013). Relative contributions of plant traits and soil microbial properties to mountain grassland ecosystem services. *J. Ecol.* 101, 47–57.

Grime, J. P., Mackey, J. M. L., Hillier, S. H. and Read, D. J. (1987). Floristic diversity in a model system using experimental microcosms. *Nature* 328, 420–422.

Guhr, A., Borken, W., Spohn, M. and Matzner, E. (2015). Redistribution of soil water by a saprotrophic fungus enhances carbon mineralization. *Proc. Natl. Acad. Sci. USA* 112, 14647–14651.

Hallett, P. D., Feeney, D. S., Bengough, A. G., Rillig, M. C., Scrimgeour, C. M. and Young, I. M. (2009). Disentangling the impact of AM fungi versus roots on soil structure and water transport. *Plant Soil* 314, 183–196.

Handa, I. T., Aerts, R., Berendse, F., Berg, M. P., Bruder, A., Butenschoen, O., *et al.* (2014). Consequences of biodiversity loss for litter decomposition across biomes. *Nature* 509, 218–221.

Heemsbergen, D. A., Berg, M. P., Loreau, M., Van Hal, J. R., Faber, J. H. and Verhoef, H. A. (2004). Biodiversity effects on soil processes explained by interspecific functional dissimilarity. *Science* 306, 1019–1020.

Hendrix, P. F. and Bohlen, P. J. (2002). Exotic earthworm invasions in North America: ecological and policy implications. *Bioscience* 52, 801–811.

Hol, W. H., De Boer, W., Termorshuizen, A. J., Meyer, K. M., Schneider, J. H., Van Dam, N. M., *et al.* (2010). Reduction of rare soil microbes modifies plant–herbivore interactions. *Ecol. Lett.* 13, 292–301.

Hunt, H. W., Coleman, D. C., Ingham, E. R., Ingham, R. E., Elliott, E. T., Moore, J. C., *et al.* (1987). The detrital food web in a shortgrass prairie. *Biol. Fert. Soils* 3, 57–68.

Hunt, H. W. and Wall, D. H. (2002). Modelling the effects of loss of soil biodiversity on ecosystem function. *Glob. Chang. Biol.* 8, 33–50.

Ingham, R. E., Trofymow, J. A., Ingham, E. R. and Coleman, D. C. (1985). Interactions of bacteria, fungi, and their nematode grazers: effects on nutrient cycling and plant growth. *Ecol. Monogr.* 55, 119–140.

Jenkinson, D. S. and Powlson, D. S. (1976). The effects of biocidal treatments on metabolism in soil. V: a method for measuring soil biomass. *Soil Biol. Biochem.* 8, 209–213.

Kardol, P., Bezemer, T. M. and van der Putten, W. H. (2006). Temporal variation in plant–soil feedback controls succession. *Ecol. Lett.* 9, 1080–1088.

Kiers, E. T., Duhamel, M., Beesetty, Y., Mensah, J. A., Franken, O., Verbruggen, E., *et al.* (2011). Reciprocal rewards stabilize cooperation in the mycorrhizal symbiosis. *Science* 333, 880–882.

Kourtev, P. S., Ehrenfeld, J. G. and Haggblom, M. (2002). Exotic plant species alter the microbial community structure and function in the soil. *Ecology* 83, 3152–3166.

Laakso, J. and Setälä, H. (1999). Sensitivity of primary production to changes in the architecture of below-ground food webs. *Oikos* 87, 57–64.

Liiri, M., Setälä, H., Haimi, J., Pennanen, T. and Fritze, H. (2002). Relationship between soil microarthropod species diversity and plant growth does not change when the system is disturbed. *Oikos* 96, 137–149.

Lindberg, N. and Bengtsson, J. (2006). Recovery of forest soil fauna diversity and composition after repeated summer droughts. *Oikos* 114, 494–506.

Loreau, M., Naeem, S. and Inchausti, P. (Eds.) (2002). *Biodiversity and Ecosystem Functioning: Synthesis and Perspectives.* Oxford University Press, UK.

Mackelprang, R., Waldrop, M. P., De Angelis, K. M., David, M. M., Chavarria, K. L., Blazewicz, S. J., *et al.* (2011). Metagenomic analysis of a permafrost microbial community reveals a rapid response to thaw. *Nature* 480, 368–371.

Mendes, R., Garbeva, P. and Raaijmakers, J. M. (2013). The rhizosphere microbiome: significance of plant beneficial, plant pathogenic, and human pathogenic microorganisms. *FEMS Microbiol. Rev.* 37, 634–663.

Mendes, R., Kruijt, M., de Bruijn, I., Dekkers, E., van der Voort, M., Schneider, J. H., *et al.* (2011). Deciphering the rhizosphere microbiome for disease-suppressive bacteria. *Science* 332, 1097–1100.

Mikola, J., Bardgett, R. D. and Hedlund, K. (2002). Biodiversity, ecosystem functioning and soil decomposer food webs. In Loreau, M., Naeem, S. and Inchausti, P. (eds.), *Biodiversity and Ecosystem Functioning: Synthesis and Perspectives.* Oxford University Press, UK, pp. 169–180.

Moore, J. C. and Hunt, W. H. (1988). Resource compartmentation and the stability of real ecosystems. *Nature* 333, 261–263.

Mulder, C., Cohen, J. E., Setälä, H., Bloem, J. and Breure, A. M. (2005). Bacterial traits, organism mass, and numerical abundance in the detrital soil food web of Dutch agricultural grasslands. *Ecol. Lett.* 8, 80–90.

Neutel, A. M., Heesterbeek, J. A. P. and De Ruiter, P. C. (2002). Stability in real food webs: weak links in long loops. *Science* 296, 1120–1123.

Nielsen, U., Ayres, E., Wall, D. and Bardgett, R. D. (2011). Soil biodiversity and carbon cycling: a review and synthesis of studies examining diversity–function relationships. *Eur. J. Soil Sci.* 62, 105–116.

Orwin, K. H. and Wardle, D. A. (2005). Plant species composition effects on below-ground properties and the resistance and resilience of the soil microflora to a drying disturbance. *Plant Soil* 278, 205–221.

Orwin, K. H., Wardle, D. A. and Greenfield, L. G. (2006). Context-dependent changes in the resistance and resilience of soil microbes to an experimental disturbance for three primary plant chronosequences. *Oikos* 112, 196–208.

Pellkofer, S., van der Heijden, M. G. A., Schmid, B. and Wagg, C. (2016). Soil communities promote temporal stability and species asynchrony in experimental grassland communities. *PLoS ONE* 11, e0148015.

Peters, D. and Weste, G. (1997). The impact of *Phytophthora cinnamomi* on six rare native tree and shrub species in the Brisbane Ranges, Victoria. *Aust. J. Bot.* 45, 975–995.

Postma-Blaauw, M. B., de Goede, R. G. M., Bloem, J., Faber, J. H. and Brussaard, L. (2010). Soil biota community structure and abundance under agricultural intensification and extensification. *Ecology* 91, 460–473.

Ramirez, K. S., Craine, J. and Fierer N. (2012). Consistent effects of nitrogen amendments on soil microbial communities and processes across biomes. *Glob. Chang. Biol.* 18, 1918–1927.

Raviv, M. (2010). The use of mycorrhiza in organically-grown crops under semi arid conditions: a review of benefits, constraints and future challenges. *Symbiosis* 52, 65–74.

Read, D. J. (1991). Mycorrhizas in ecosystems. *Experientia* 47, 376–391.

Reinhart, K. O., Tytgat, T., Van der Putten, W. H. and Clay, K. (2010). Virulence of soil-borne pathogens and invasion by *Prunus serotina*. *New Phytol.* 186, 484–495.

Rillig, M. C. and Mummey, D. L. (2006). Mycorrhizas and soil structure. *New Phytol.* 171, 41–53.

Ritz, K., Dighton, J. and Giller, K. E. (Eds.) (1994). *Beyond the Biomass: Compositional and Functional Analysis of Soil Microbial Communities*. Wiley, UK.

Rooney, N., McCann, K., Gellner, G. and Moore, J. C. (2006). Structural asymmetry and the stability of diverse food webs. *Nature* 442, 265–269.

Scheu, S. and Setälä, H. (2002). *Multitrophic Interactions in Decomposer Food-Webs. Multitrophic Level Interactions.* Cambridge University Press, Cambridge, UK.

Schulze, E.-D. and Mooney, H. A. (Eds.). (1994). *Biodiversity and Ecosystem Function.* Springer, Berlin, Heidelberg.

Selosse, M. A., Richard, F., He, X. and Simard, S. W. (2006). Mycorrhizal networks: des liaisons dangereuses? *Trends Ecol. Evol.* 21, 621–628.

Setälä, H. and Huhta, V. (1991). Soil fauna increase *Betula pendula* growth: laboratory experiments with coniferous forest floor. *Ecology* 72, 665–671.

Shearer, B. L., Crane, C. E., Fairman, R. G. and Grant, M. J. (1998). Susceptibility of plant species in coastal dune vegetation of southwestern Australia to killing by *Armillaria luteobubalina*. *Aust. J. Bot.* 46, 321–334.

Treseder, K. K. (2008). Nitrogen additions and microbial biomass: a meta-analysis of ecosystem studies. *Ecol. Lett.* 11, 1111–1120.

Tsiafouli, M. A., Thébault, E., Sgardelis, S. P., Ruiter, P. C., Putten, W. H., Birkhofer, K., *et al.* (2015). Intensive agriculture reduces soil biodiversity across Europe. *Glob. Chang. Biol.* 21, 973–985.

Van der Heijden, M. G. A., Bardgett, R. D. and van Straalen, N. M. (2008). The unseen majority: soil microbes as drivers of plant diversity and productivity in terrestrial ecosystems. *Ecol. Lett.* 11, 296–310.

Van der Heijden, M. G., Klironomos, J. N., Ursic, M., Moutoglis, P., Streitwolf-Engel, R., Boller, T., *et al.* (1998). Mycorrhizal fungal diversity determines plant biodiversity, ecosystem variability and productivity. *Nature* 396, 69–72.

Van der Putten, W. H., Bardgett, R. D., Bever, J. D., Bezemer, T. M., Casper, B. B., Fukami, T., *et al.* (2013). Plant–soil feedbacks: the past, the present and future challenges. *J. Ecol.* 101, 265–276.

Van der Putten, W. H., Klironomos, J. N. and Wardle, D. A. (2007). Microbial ecology of biological invasions. *ISME J.* 1, 28–37.

Van der Putten, W. H., Van Dijk, C. and Peters, B. A. M. (1993). Plant-specific soil-borne diseases contribute to succession in foredune vegetation. *Nature* 362, 53–56.

van Elsas, J. D., Chiurazzi, M., Mallon, C. A., Elhottova, D., Kristufek, V. and Salles, J. F. (2012). Microbial diversity determines the invasion of soil by a bacterial pathogen. *Proc. Natl. Acad. Sci. USA* 109, 1159–1164.

Venette, R. C. and Cohen, S. D. (2006). Potential climatic suitability for establishment of *Phytophthora ramorum* within the contiguous United States. *For. Ecol. Manage.* 231, 18–26.

Wagg, C., Bender, S. F., Widmer, F. and van der Heijden, M. G. A. (2014). Soil biodiversity and soil community composition determine ecosystem multifunctionality. *Proc. Natl. Acad. Sci. USA* 111, 5266–5270.

Walder, F., Niemann, H., Natarajan, M., Lehmann, M. F., Boller, T. and Wiemken, A. (2012). Mycorrhizal networks: common goods of plants shared under unequal terms of trade. *Plant Physiol.* 159, 789–797.

Wall, D. H., Bardgett, R. D., Behan-Pelletier, V., Herrick, J. E., Jones, H., Ritz, K., *et al.* (Eds.) (2012). *Soil Ecology and Ecosystem Services*. Oxford University Press, UK.

Wall, D. H., Nielsen, U. N. and Six, J. (2015). Soil biodiversity and human health. *Nature* 528, 69–76.

Wardle, D. A. (2002). *Communities and Ecosystems: Linking the Above-ground and Below-ground Components*, Vol. 34. Princeton University Press, New Jersey.

Wardle, D. A., Bardgett, R. D., Callaway, R. M. and Van der Putten, W. H. (2011). Terrestrial ecosystem responses to species gains and losses. *Science* 332, 1273–1277.

Wardle, D. A., Bardgett, R. D., Klironomos, J. N., Setälä, H., van der Putten, W. H. and Wall, D. H. (2004). Ecological linkages between above-ground and below-ground biota. *Science* 304, 1634–1637.

Wei, Z., Yang, T. J., Friman, V. P., Xu, Y. C., Shen, Q. R. and Jousset, A. (2015). Trophic network architecture of root-associated bacterial communities determines pathogen invasion and plant health. *Nat. Commun.* 6, 9.

Weste, G. (1981). Changes in the vegetation of schlerophyll shrubby woodland associated with invasion by *Phytophthora cinnamomi*. *Aust. J. Bot.* 29, 261–276.

Wittebolle, L., Marzorati, M., Clement, L., Balloi, A., Daffonchio, D., Heylen, K., *et al.* (2009). Initial community evenness favours functionality under selective stress. *Nature* 458, 623–626.

Xu, Q. F., Jiang, P. K., Wu, J. S., Zhou, G. M., Shen, R. F. and Fuhrmann, J. J. (2015). Bamboo invasion of native broadleaf forest modified soil microbial communities and diversity. *Biol. Invasions* 17, 433–444.

Zhou, J., Xue, K., Xie, J., Deng, Y., Wu, L., Cheng, X., *et al.* (2012). Microbial mediation of carbon-cycle feedbacks to climate warming. *Nat. Clim. Chang.* 2, 106–110.

6

Building Predictive Models for Diverse Microbial Communities in Soil

Steven D. Allison

Department of Ecology and Evolutionary Biology,
Department of Earth System Science,
University of California Irvine,
321 Steinhaus, Irvine, CA 92697, USA
allisons@uci.edu

Introduction

The key role of microbes in soil processes has been appreciated for decades. In his book *Principles of Soil Microbiology*, Waksman (1927) noted that soil organic matter "depends upon the activities of the soil microorganisms, which are in their turn influenced by environmental soil conditions including moisture, aeration, soil reaction and presence of available nitrogen and mineral nutrients" (p. 669). This statement foreshadowed a current, twenty-first-century, explosion of interest in linking microbial communities with soil functioning and global change (Bardgett *et al.*, 2008). Despite this renewed interest, our ability to make reliable, quantitative predictions of microbial functioning has advanced relatively little since the 1920s. The aim of this chapter is to provide relevant background and a vision to overcome this challenge.

As key players in all of the planet's biogeochemical cycles, microbes influence ecosystem process rates and their responses to human-caused perturbation

(Singh *et al.*, 2010). For example, microbes control soil carbon sequestration through decomposition of organic matter (a loss pathway) and formation of stable organic residues (an input pathway). Bacteria, fungi and archaea mineralise nutrients that support plant growth in unmanaged and agricultural ecosystems. Soil methanogens produce methane, a potent greenhouse gas, while methanotrophs consume methane (Gulledge and Schimel, 2000; McCalley *et al.*, 2014). Nitrifiers and denitrifiers produce nitric and nitrous oxide gases that contribute to air pollution and greenhouse warming, respectively (Firestone *et al.*, 1980; Firestone and Davidson, 1989).

Now is an opportune time to make progress on predictive models of microbial processes in soil. That microbes influence biogeochemical processes has long been known, but there is increased urgency to predict how these processes and their associated ecosystem services will respond to human-caused environmental changes. Harking back to Waksman, environmental conditions such as temperature, moisture, nutrient availability and organic matter inputs are being altered at a global scale through changes in land use, climate and nutrient inputs.

Reliable predictions require more than just recognition that microbes are important for a process. Fundamental, mechanistic understanding must be established from empirical studies. The mechanisms must be unified into a theoretical framework that provides a basis for quantitative mathematical models. The models must then be scaled, analysed and validated against independent observations. Often, the models fail to predict patterns in the data and must be refined over and over again. Perhaps not surprisingly, this challenging series of steps has not yet been accomplished at a large scale with soil microbes (Wieder *et al.*, 2015).

Still, several recent scientific developments make a focus on prediction tractable and worthwhile. The molecular revolution (genomics, metabolomics, proteomics, etc.) and the microbiome concept have increased the potential for rapid advances in fundamental understanding of diverse microbial communities (Tringe *et al.*, 2005; Sharon *et al.*, 2013). Although more work must be done to fully interpret the datasets emerging from these advances, it is now feasible to quantify the genetic content and distribution of soil microbes at high spatial, temporal and taxonomic resolution. Well-conceived models are essential for translating genetic data into predictions with relevance for critical environmental issues.

Another key discovery is that biological processes interact dynamically with the physical environment to influence ecosystem processes. Soils are complex adaptive systems (Levin, 2002) in which biological interactions may be quantitatively important relative to abiotic factors in determining function. Rather than just responding to the environment, as Waksman suggested, soil microbes alter the environment through a range of eco-evolutionary mechanisms that were not well understood a century ago. For example, social evolution theory (related to game theory) has informed our understanding of mutualistic and antagonistic interactions that play out among microbes (Allison, 2005; West *et al.*, 2006; Folse and Allison, 2012; Foster and Bell, 2012).

In the short term, microbes can acclimate to environmental changes through physiological mechanisms. Changes in gene expression may result in acclimation to temperature, substrate availability and time of day (Hurley *et al.*, 2014). Yeasts can acclimate their physiology in anticipation of future environmental changes (Mitchell *et al.*, 2009). Acclimated microbes should respond differently to environmental changes than non-acclimated microbes, meaning that predictive models need to account for physiological mechanisms (Crowther and Bradford, 2013).

Likewise, numerous studies have shown that the genetic content of microbial communities also changes with the environment (Shade *et al.*, 2012). Both evolutionary and ecological mechanisms contribute to such responses. Genetic variants with higher fitness in a given environment will increase in frequency due to natural selection. Experimental evolution studies have shown, for example, that bacterial genes associated with stress tolerance and protein stability increase rapidly in frequency under selection by high temperatures (Tenaillon *et al.*, 2012). Shifts in the environment often alter the competitive interactions in soil microbial communities, leading to changes in composition and functioning through ecological mechanisms (Allison and Martiny, 2008).

Advances in our understanding of physiological, evolutionary and ecological mechanisms in microbial communities are relevant for building predictive models. Such models must be able to predict phenomena such as hysteresis and historical contingencies. Hysteresis occurs when the relationship between a process rate, such as soil respiration, and an environmental driver, such as temperature, differs based on the previous state of the system (i.e. whether the system was recently warm or cold; Updegraff *et al.*, 1998). Historical contingencies, or legacies, arise when a current process rate depends on the history of the system in addition to current environmental conditions (Evans and Wallenstein, 2012). Both hysteresis and historical contingencies are likely driven by physiological, evolutionary and ecological responses of microbial communities.

History of Microbial Biomass Models

Models of microbial processes in soil have evolved considerably over time. The first large-scale biogeochemical models such as RothC and CENTURY included microbial biomass, but it played no direct role in organic matter decomposition

(Jenkinson and Rayner, 1977; Parton *et al.*, 1988). Instead the models assumed that microbial activity was a linear function of the substrate pool size. Factors like substrate chemistry, temperature, moisture and soil texture were included as coefficients in the linear function. Schimel (2001) defines this approach as *implicit* with the dynamics and influence of the microbial community implied indirectly by the environmental conditions. An implicit approach is convenient because linear models are easy to analyse mathematically (Xia *et al.*, 2013) and they make predictions that are consistent with empirical data from laboratory and field experiments (Powlson *et al.*, 2011).

On the other hand, models with explicit microbes represent a direct role for microbial biomass in soil processes (Schimel, 2001). The defining feature of microbial-explicit models is the mathematical coupling of dynamic microbial biomass with substrate pools (Todd-Brown *et al.*, 2012). Substrate decomposition rates depend on the quantity of microbial biomass, and microbial biomass depends on the quantity of the substrate. In some microbial-explicit models, decomposition rates depend on enzymes produced by microbes rather than the quantity of microbial biomass *per se* (Allison *et al.*, 2010). Organic matter loss rates in all microbial-explicit models are nonlinear with respect to substrate concentration because of the coupling to microbial biomass.

The concept of explicit microbes dates back at least to Waksman (1927), but microbial-explicit mathematical models did not appear until almost 50 years later. In 1979, O. L. Smith described and validated an analytical model of soil organic matter decomposition (Smith, 1979a, b). The model represented enzymatic degradation of multiple substrate pools with Michaelis–Menten kinetics and included carbon, nitrogen and phosphorus submodels along with a plant component. Yet Smith's model failed to gain a major foothold in the biogeochemical literature, possibly because it was viewed as too complex (Andrén and Paustian, 1987) or contained too many equations (Fawcett and Higginson, 2012). Not until almost 25 years later did the idea begin to take hold with models by Schimel and Weintraub (2003) and Fontaine and Barot (2005).

In the last 10 years, there has been an explosion of interest in microbial-explicit models. They have begun to include aspects of microbial functional diversity (Allison, 2005; Moorhead and Sinsabaugh, 2006) and explore responses to changing environmental conditions (Allison *et al.*, 2010; Davidson *et al.*, 2012; Sistla *et al.*, 2014). Newer models consider important microbial interactions with soil mineralogy (Wang *et al.*, 2013; Wieder *et al.*, 2014). Several models have been scaled up to simulate soil carbon pools and dynamics in the Earth system (Wieder *et al.*, 2013; Sulman *et al.*, 2014). Enough conceptual diversity has developed to allow model intercomparisons and data assimilation at the local to global scale (Lawrence *et al.*, 2009; Li *et al.*, 2014; Wang *et al.*, 2014; Hararuk *et al.*, 2015), and readers may consult recent review articles dedicated to microbial-explicit models (Todd-Brown *et al.*, 2012; Wieder *et al.*, 2015).

All of this attention has stimulated calls to incorporate microbial-explicit models into Earth system models as an alternative to conventional first-order models (Todd-Brown *et al.*, 2012; Treseder *et al.*, 2012; Wieder *et al.*, 2015). Recent analyses have revealed major shortcomings with the conventional approach for modelling soil organic matter dynamics (Todd-Brown *et al.*, 2013, 2014). Still, scaling of microbial-explicit models should be done with caution. The modelling community has a responsibility to analyse the variation, advantages and shortcomings inherent in the conceptual and mathematical diversity of current microbial models. Such an effort will help instil confidence in global predictions that arise from microbial-explicit models and ultimately appear in climate assessment reports that influence societal decisions. To that end, the following sections lay out key approaches at the heart of microbial modelling.

Microbial Modelling Approaches

Under the implicit approach of conventional biogeochemical models, microbial communities are represented through environmental response functions (Figure 6.1). For instance, a Q_{10} function is commonly used to describe how decay coefficients (*k*-values) and soil respiration increase with increasing temperature (Davidson *et al.*, 2006). This approach assumes that the rate of enzyme-driven microbial metabolism increases exponentially with increasing temperature. Most models also assume that the rate is the same for metabolism of different soil substrates and microbial communities. Response functions for other environmental variables such as soil moisture and texture have different forms, but are also assumed to apply across substrates and ecosystems.

The simplest explicit approach for representing microbial communities is to include a single biomass pool that is mathematically linked to substrate inputs and outputs (German *et al.*, 2012). In this case, the model fluxes are a function of the biomass pool size which is in turn dependent on substrate availability and environmental variables. For instance, the biochemical response to temperature is captured by modifying the kinetic parameters for enzyme catalysis. The form of this function is often very similar to the Q_{10} function employed by conventional models. However, with the microbial-explicit model, the kinetic parameters are biomass-specific, so changes in microbial biomass also influence the total rate of substrate decay by enzymes.

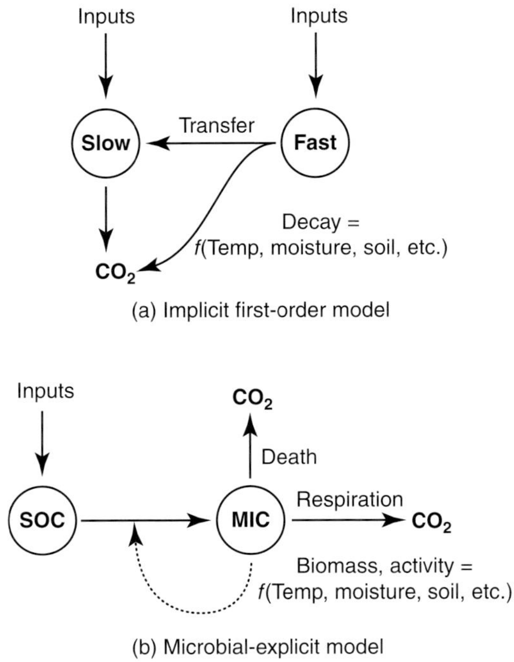

Figure 6.1 Distinction between the implicit approach to modelling microbial processes (a) and the microbial-explicit approach (b). SOC = soil organic carbon; MIC = microbial biomass carbon.

On the one hand, this change in model approach is relatively subtle. There are no new pools: most conventional models already include pools for microbial biomass. The environmental response functions are nearly identical and involve the same parameters. Both model types include kinetic parameters — *k*-values in the implicit models and Vmax values in the explicit models. Thus microbial-explicit models with a single biomass pool are not inherently more complex than implicit models.

On the other hand, differences in the mathematical structure of microbial-explicit models — how the pools interact — have a big impact on the model behaviours. In particular, the explicit models are less sensitive to changing inputs (Allison *et al.*, 2010; Wieder *et al.*, 2013). As substrate pool size increases, so does microbial biomass. Because the two pools are mathematically coupled, the increase in biomass increases substrate turnover and reduces substrate pool size.

Microbial-explicit models reflect Waksman's (paraphrased) observation — microbes eat soil carbon.

In general, microbial-explicit models differ from implicit models in their response to perturbations. The up-and-down dynamics of coupled substrate and microbial pools manifest as oscillations in response to changes in pool sizes (Wang *et al.*, 2014). For example, if microbial biomass declines due to temperature effects on growth efficiency, substrate pool sizes increase. The increase then supports more biomass and the system exhibits oscillations that dampen over time.

It remains unclear whether the behaviours of simple microbial-explicit models are realistic. There is little empirical evidence that soil substrate and microbial biomass pools oscillate in response to disturbance. However, few studies have looked for such behaviour, and there is some evidence from soil warming experiments, such as Harvard Forest, that the transient dynamics of microbial-explicit models are consistent with observational data (Allison *et al.*, 2010). The insensitivity of microbial-explicit models to substrate inputs is also debatable, as inputs are clearly required to generate soil organic matter in the first place. Still, the global-scale relationship between net primary production and soil carbon stocks is weak (Todd-Brown *et al.*, 2013), making it difficult to reject the microbial model predictions.

Another key question for microbial models is whether diverse communities should be represented as a single biomass pool. This simplifying assumption reduces the potential to incorporate many of the biological processes known to be important in microbial communities. To address this issue, a number of microbial-explicit models include aspects of functional diversity in the microbial biomass pool by defining distinct functional groups or allowing for variation in functional traits (Moorhead and Sinsabaugh, 2006; Wieder *et al.*, 2014).

Functional group approaches, common in vegetation models, separate biomass into classes with different functional parameters. For example, in vegetation models, functional groups are assigned based on leaf morphology (broadleaf or needleleaf), growth form (grass, shrub, tree), leaf longevity (deciduous vs. evergreen) and other characteristics. The parameters are static within a functional group, so grasses cannot become trees, for instance. Moorhead and Sinsabaugh (2006) used an analogous approach with litter-decomposing microbes. They defined opportunist, decomposer and miner guilds that target early, middle and late stages of litter decay, respectively. As each functional guild consumes substrate, it alters the chemical environment and shifts the competitive landscape such that guilds bloom and die off in succession as litter decays. As with plant functional groups, the functional parameters of the microbial guilds do not change — succession is driven by the environment.

The functional group approach is one example of trait-based approaches that are now being applied to model functional diversity in microbial communities. A trait can be defined as a parameter associated with the microbial biomass. For example, Moorhead and Sinsabaugh specified guilds based on traits like enzyme maximum catalytic rate (Vmax), enzyme half-saturation constant (Km), maximum substrate uptake rate, basal respiration rate and C:N ratio. Other trait-based models use a continuous approach rather than specifying guilds *a priori*. Sistla *et al.* (2014) allowed the C:N ratio of the microbial biomass to vary based on carbon substrate and soil nitrogen availability. Lower C:N ratios corresponded to bacterially dominated communities, whereas higher ratios corresponded to fungal-dominated communities.

A major challenge with trait-based approaches is that trait parameters are often uncertain, especially when specifying many functional groups. To circumvent this issue, some models use a stochastic approach to assign trait parameters and allow environmental conditions to select, or filter, the resulting functional groups. This idea was pioneered by Follows *et al.* (2007) who randomly assigned trait parameters to 78 hypothetical functional groups of phytoplankton and then observed the abundance and geographic distribution of each group after years of simulated competition in the global ocean. At the end of the simulation, the distributions of the dominant hypothetical groups matched the distributions of several known lineages of phytoplankton. There was also correspondence between the randomly assigned trait parameters of the hypothetical groups and the measured traits of the known lineages — real patterns emerged from the model dynamics.

Allison (2012) applied a similar approach to litter-decomposing microbial communities. The model was able to replicate patterns in litter decomposition and enzyme activity from a Hawaiian rainforest, but the abundance distribution of hypothetical functional groups was more difficult to validate. Unlike with phytoplankton, the geographic distributions and functional trait parameters for most litter- and soil-associated microbes are unknown.

Key Microbial Traits for Modelling

The increasing interest in trait-based modelling approaches justifies a closer look at the key trait parameters involved in microbial functioning (Table 6.1). Community ecologists have developed a useful conceptual framework that

Table 6.1 Microbial trait parameters.

Parameter	Description	Values	Units	Reference
Enzyme-specific Vmax	Quantity of product formed per unit time per unit enzyme	24–60	$g\ g^{-1}\ day^{-1}$	(Allison *et al.*, 2010; Wang *et al.*, 2013)
Biomass-specific Vmax	Quantity of product formed per unit time per unit microbial biomass	0.24	$g\ g^{-1}\ day^{-1}$	(German *et al.*, 2012; Li *et al.*, 2014)
Enzyme Km	Michaelis–Menten half-saturation constant for enzymes	0.050–0.600	$g\ cm^{-3}$	(Allison *et al.*, 2010; German *et al.*, 2012; Wang *et al.*, 2013)
Substrate Km	Reverse Michaelis–Menten half-saturation constant for substrate	0.0003	$g\ C\ g^{-1}$ soil	(Schimel and Weintraub, 2003)
Biomass-specific enzyme production	Quantity of enzyme produced per unit time per unit microbial biomass	0–0.00028	$g\ C\ g^{-1}\ C\ day^{-1}$	(Allison, 2005, 2012, 2014)
Uptake-specific enzyme production	Quantity of enzyme produced per unit resource uptake	0–0.12	$g\ C\ g^{-1}\ C$	(Allison, 2005, 2012, 2014; Kaiser *et al.*, 2014)
Enzyme-specific uptake Vmax	Rate of resource uptake per unit enzyme	14400	$g\ g^{-1}\ day^{-1}$	(Allison, 2005)
Biomass-specific uptake Vmax	Rate of resource uptake per unit microbial biomass	0.012–0.24	$g\ g^{-1}\ day^{-1}$	(Wang *et al.*, 2013; Li *et al.*, 2014)

(Continued)

Table 6.1 (*Continued*)

Parameter	Description	Values	Units	Reference
Uptake Km	Michaelis–Menten half-saturation constant for uptake proteins	1×10^{-6} to 3×10^{-4}	$g\ cm^{-3}$	(Allison, 2005; Allison *et al.*, 2010)
Enzyme specificity	Inverse of the number of substrates targeted by a given enzyme	1/3 to 1		(Allison, 2012)
Growth efficiency	Fraction of resource uptake allocated to biomass growth	0.14–0.77	$g\ g^{-1}$	(Six *et al.*, 2006; Sinsabaugh *et al.*, 2013)
Basal respiration rate	Rate of respiration per unit microbial biomass for cellular maintenance	0.064–0.216	$g\ g^{-1}\ day^{-1}$	(Allison, 2005; Kaiser *et al.*, 2014)
Enzyme respiration rate	Rate of respiration per unit enzyme produced	0.1–1	$g\ C\ g^{-1}\ C$	(Allison, 2005, 2012)
Bacterial cell size	Mass of individual bacterial cells	30–300	fg C	(Allison, 2005; Kaiser *et al.*, 2014)
Fungal cell size	Mass of individual bacterial cells	750–7500	fg C	(Allison, 2014; Kaiser *et al.*, 2014)
Cellular N quota	Minimum fraction of cellular biomass as nitrogen	0.1	$g\ g^{-1}$	(Allison, 2012)
Cellular P quota	Minimum fraction of cellular biomass as phosphorus	0.015	$g\ g^{-1}$	(Allison, 2012)

Biomass turnover rate	Fraction of microbial biomass loss per unit time	0.01–0.04	day^{-1}	(Allison, 2005, 2012)
Enzyme turnover rate	Fraction of enzyme loss per unit time	0.020–0.024	day^{-1}	(Allison, 2012; Wang *et al.*, 2013)
Osmolyte content	Fraction of microbial biomass allocated to osmolytes	0.10	$g\ C\ g^{-1}\ C$	(Schimel *et al.*, 2007)
Residue content	Fraction of microbial biomass allocated to resistant compounds	0.5–0.94	$g\ g^{-1}$	(Allison *et al.*, 2010; Kaiser *et al.*, 2014)
Vmax temperature sensitivity	Activation energy for Vmax	34–53	$kJ\ mol^{-1}$	(Allison *et al.*, 2010; Allison, 2012; Wang *et al.*, 2013)
Km temperature sensitivity	Activation energy for Km	20–30	$kJ\ mol^{-1}$	(Wang *et al.*, 2013; Allison, 2014)
Growth efficiency temperature sensitivity	Change in growth efficiency per change in temperature	–0.016 to 0	$g\ g^{-1}\ °C^{-1}$	(Allison *et al.*, 2010)
Turnover temperature sensitivity	Change in turnover rate per change in temperature	0.003–0.004	$day^{-1}\ °C^{-1}$	(Hagerty *et al.*, 2014)

distinguishes *response* and *effect* traits (Lavorel and Garnier, 2002; Webb *et al.*, 2010). Response traits govern the response of organismal physiology or abundance to environmental conditions, whereas effect traits describe how the organism affects biogeochemical processes. The categories are not mutually exclusive, as traits underlying growth, stoichiometry and turnover influence both ecosystem processes and microbial responses to resource availability.

Growth-related traits are fundamental for microbial-explicit modelling and fall into both response and effect categories. Typically, the instantaneous growth rate of microbial biomass in these models is an emergent property determined by trait interactions with environmental conditions. The key underlying traits are related to resource acquisition and metabolic losses such as respiration. For heterotrophic microbes, resource acquisition is determined by kinetic parameters for uptake such as biomass-specific Vmax and Km. Biomass-specific Vmax can be specified directly or result from multiplying enzyme-specific Vmax by the cellular investment in uptake enzymes, which allows for cellular regulation of uptake investment.

In many microbial-explicit soil models, resource acquisition by microbes also depends on extracellular enzyme traits (Allison, 2005; Wang *et al.*, 2013; Kaiser *et al.*, 2014). Analogous to uptake, there are enzyme-specific kinetic parameters for Vmax and Km. Other trait parameters may determine investment in enzyme production expressed as a fraction of microbial biomass or microbial uptake. Therefore, instantaneous resource acquisition rates for heterotrophic microbes may depend on a suite of kinetic traits that interact with the availability of high- and low-molecular-weight substrates.

At a minimum, microbial-explicit models must include trait parameters for a convex (e.g. saturating) function that relates resource acquisition rate to resource pool size (Wutzler and Reichstein, 2008). There must be a parameter that determines the maximum rate of resource acquisition, and the function must level off, otherwise microbial biomass grows indefinitely and consumes all substrates. Functions that level off with increasing substrate concentration (e.g. Michaelis–Menten kinetics), or increasing enzyme concentration (e.g. reverse Michaelis–Menten; Schimel and Weintraub, 2003) both serve this purpose.

A number of other trait parameters specify metabolic losses from microbial biomass. Most models include a growth efficiency parameter corresponding to the fraction of substrate uptake converted into microbial biomass, with the remainder being respired (Allison *et al.*, 2010; Allison, 2014). Some models also include a basal respiration parameter that specifies respiration rate as a fraction of microbial biomass (Allison, 2005; Kaiser *et al.*, 2014). If there are pathways for enzyme or other metabolite production, these may be associated with respiratory fluxes as well (Allison, 2005; Kaiser *et al.*, 2014). Together, respiratory losses, metabolite production and substrate uptake combine to determine the emergent growth rate of microbial biomass.

Cell size is also an important trait for growth (Yoshiyama and Klausmeier, 2008). Given a similar biomass-specific growth rate, larger cells divide less frequently and have lower population growth rates. Larger cells also have a lower surface area-to-volume ratio, which can reduce cell-specific resource uptake rates. In models, specifying larger minimum cell sizes or larger cell sizes required for duplication results in lower microbial population sizes and potentially lower resource acquisition rates. Smaller populations reduce the spatial extent of microbial biomass and may be at higher risk for stochastic extinction.

Aside from the cell size, trait parameters may also specify growth form. Some microbes, such as fungi and actinomycetes, display filamentous growth, whereas many bacteria and archaea grow as colonies on surfaces. Still others are free-living in aqueous phases. Filamentous growth may be specified in cellular automaton models with probabilities of movement differing for each growth direction (Boswell, 2008). Instead of new biomass spreading out from a central point at random, the growth becomes directional. Directionality rules may be combined with physiological differences associated with filamentous growth, particularly the ability to translocate carbon and nutrients along the filaments. In this way, fungal growth strategies can be represented in microbial models as large cells that grow directionally with connectivity of nutrient pools (Allison, 2014).

For models including nutrient dynamics, cellular stoichiometry is a key trait. Stoichiometry may be specified as a fixed parameter (Kaiser *et al.*, 2014) or vary within maximum and minimum limits (Allison, 2012; Sistla *et al.*, 2014). When metabolic processes result in biomass stoichiometry that deviates from the fixed value or exceeds the limits, the elements in excess are lost. For example, cells respire excess carbon as CO_2 or mineralise excess nitrogen. In some models, cellular stoichiometry may also affect resource uptake rates through a nutrient demand function (Allison, 2005). When microbes with different life-history traits and stoichiometric ratios are represented in microbial models, feedback in the microbial community can alleviate nutrient limitation (Kaiser *et al.*, 2014).

To prevent microbial biomass from growing indefinitely, models must specify parameters for biomass turnover. Processes such as predation, starvation and environmental stress may trigger microbial turnover, although the mechanism is usually not defined explicitly in models. Turnover rate parameters may reflect traits of the microbial biomass as well as external drivers, such as predation pressure. Many modelling studies have specified turnover rates as constants (Schimel and Weintraub, 2003; Moorhead and Sinsabaugh, 2006; Allison *et al.*, 2010), but microbial traits such as cell wall thickness, investment in osmolytes and antibiotic production could cause variation in turnover rates. In their MIMICS model, Wieder *et al.* (2014) assign different turnover parameters to microbial functional groups with fast vs. slow growth strategies.

Related to turnover are traits that affect the fate of dead microbial biomass. Microbial residues are thought to be important contributors to soil organic matter formation (Grandy and Neff, 2008). Most conventional and microbial-explicit models include transfer coefficients that specify the fraction of turnover entering one or more soil carbon pools, which may include dissolved or polymeric forms (Schimel and Weintraub, 2003; Allison *et al.*, 2010; German *et al.*, 2012; Kaiser *et al.*, 2014). Many studies use a constant transfer coefficient, although the MIMICS model specifies a higher transfer of residues to soil organic pools from the slow-growing microbial functional group (Wieder *et al.*, 2014), and Kaiser *et al.*'s (2014) model represents variation in the stoichiometry of residues. MIMICS assumes that microbial traits such as cell-wall chemistry differ across functional groups and alter the fraction of dead biomass that is stabilised as SOC vs. being respired quickly as CO_2 or transferred to fast-turnover pools like dissolved organic carbon (DOC).

Microbial modelling studies have started to explore the consequences of response traits related to temperature and moisture. Response trait parameters are used as coefficients in functions that specify how resource acquisition, growth and turnover traits vary with environmental conditions. For example, enzyme Vmax can be specified as a constant (Schimel and Weintraub, 2003), or a function of temperature. Several microbial-enzyme models use the Arrhenius function to describe the temperature relationship, effectively defining the activation energy as a response trait parameter (Allison *et al.*, 2010; Wang *et al.*, 2013; Wieder *et al.*, 2013).

This response function approach has been implemented in a number of microbial-explicit models with a range of trait parameters. Equilibrium soil carbon pools are very sensitive to varying the slope parameter in a linear function describing the dependence of microbial growth efficiency on temperature (Allison *et al.*, 2010; Li *et al.*, 2014). An analogous sensitivity is apparent when microbial turnover rates increase as a function of increasing temperature (Hagerty *et al.*, 2014). Moisture responses are not yet well studied, but there are efforts underway to define moisture response functions for enzyme kinetic parameters and microbial turnover rates (Allison, unpublished).

Due to physiological and evolutionary constraints, traits are often correlated. For example, high rates of resource acquisition correlate with low growth yields (Pfeiffer *et al.*, 2001), and this rate-yield trade-off is well established in microbial physiology (Frank, 2010). Allocating energy to resource uptake increases the rate of acquisition but reduces the fraction of acquired resources available for growth. There is also a positive correlation between extracellular enzyme Vmax and Km, meaning that enzymes with higher catalytic rates also require higher substrate concentrations to achieve the maximum rate (Sinsabaugh *et al.*, 2014).

Life-history strategies, or syndromes, may arise from correlations among multiple traits with different strategies residing in different regions of trait space. Among plants, the ruderal (weedy) strategy is characterised by high leaf nitrogen, high rates of photosynthesis, high specific leaf area and high vulnerability to herbivory. For microbes, Fierer *et al.* (2007) have proposed a copiotroph–oligotroph life-history continuum, whereby copiotrophs have higher resource acquisition rates, higher nutrient demands and potentially lower growth yields than oligotrophs. Different trait-based strategies represent a key organising principle for communities of microscopic phytoplankton (Litchman *et al.*, 2015).

Modelling efforts can exploit trait correlations and life-history continua to represent microbial processes. When assigning trait parameters in models, such correlations reduce the dimensionality of the parameter space because some trait parameter combinations do not occur. This approach has been used in both marine and terrestrial systems with complex microbial communities (Follows *et al.*, 2007; Allison, 2012). Members of the community may possess any strategy along the life-history continuum but cannot break the rules and fall into an unoccupied area of trait space. Such trade-offs may constrain processes such as decomposition at the ecosystem scale (Allison, 2014).

Defining strategies *a priori* is impossible if the trait correlations are unknown, but there is an alternative inverse approach. Initial trait parameters can be assigned to cover a large volume of trait space with environmental selection ruling out unfavourable trait combinations (Allison, 2012). Although computationally more intensive, this approach allows for viable strategies to emerge from the model dynamics. Both approaches can be applied simultaneously with known trade-offs parameterized *a priori* and additional refinement of strategies occurring through environmental selection.

Scaling Up Microbial Models

Microbes interact at the micron scale, whereas biogeochemical models aim to predict microbial processes at the ecosystem to global scale. Bridging these scales is an important challenge because accurate predictive models rely on integration of key mechanisms across scales. Although most current models operate on a single scale, a hierarchical approach is necessary to address the challenges of scaling microbial models.

Several models capture key mechanisms of microbial interaction at the micron scale. These agent-based models reveal how social interactions and spatial structure in microbial communities influence biogeochemical processes (Allison, 2005, 2012; Folse and Allison, 2012; Kaiser *et al.*, 2014). The upshot is that micron-scale interactions must be considered in macroscale processes. Although they represent less than 1 mm^2 of physical space, these models are too complex conceptually and computationally to run at larger scales where different sources of environmental variation contribute to biogeochemical processes.

Moving up from the micron scale, other important mechanisms operate at the scale of soil aggregates and the rhizosphere. Aggregates couple the physical influence of soil minerals with biological processes of organic matter formation and turnover (Jastrow, 1996; Sollins *et al.*, 1996; Kleber *et al.*, 2007). In the rhizosphere, root exudation and turnover affect substrate availability to the microbial biomass, leading to processes such as nutrient mineralisation and priming of soil organic matter decomposition (Talbot *et al.*, 2008; Cheng *et al.*, 2014). The importance of these processes means that microbial-explicit models should represent the spatial scale of aggregates and the rhizosphere. The CORPSE model (Sulman *et al.*, 2014) is an excellent example of how physical stabilisation and rhizosphere processes can be represented to predict microbial effects on soil carbon storage under elevated CO_2.

At the plot scale and beyond, established tools in ecosystem ecology can be applied to scale up microbial models. The soil state factor approach (Jenny, 1980) defines climate, parent material, topography, vegetation type and time as the key drivers of soil formation. Landscape variation in state factors can be used to specify the local environment for microbial processes. For example, by overlaying maps of temperature, precipitation, elevation, plant chemistry and soil type, it becomes possible to specify the environment for microbial interactions at any terrestrial point on Earth.

A hierarchical approach involves a set of nested microbial models, each capturing key mechanisms at an appropriate spatial scale. Rather than running micron-scale, agent-based models at every grid point in a global model, a hierarchical approach distils emergent patterns from small scales and applies them at larger scales. Micron-scale models can be run across environmental gradients, such as temperature and moisture, to construct response functions for macroscale processes like enzymatic decomposition and heterotrophic respiration. The parameters from these response functions can then be used in larger-scale models with greater confidence because they emerge from microscale processes. Likewise, aggregate or rhizosphere models incorporating the emergent microscale functions can be run across state factor gradients to scale up processes to the ecosystem or global level.

There are statistical and computational techniques available to reduce model complexity in a mechanistically informed way. Although not yet applied to soil microbial biomass models, emulation approaches have been used in other fields to distil the complexity of numerical models (Castruccio *et al.*, 2014). Emulation fits complex model output to a statistical model. Parameters from the statistical model are then used to emulate the behaviour of the more complex model. Castruccio *et al.* (2014) used a statistical emulator to represent the relationship between radiative forcing and temperature in climate model outputs. A similar approach could be used to construct functions for the response of microbial processes, such as heterotrophic respiration, to environmental variables based on outputs from microscale agent-based models. The functional parameters could then be used in larger scale models.

An alternative but less mechanistic approach to scaling involves the application of community-averaged trait values. Rather than capturing mechanisms operating at the micron or aggregate scale that shape community composition and the associated distribution of traits, empirical approaches can provide information about trait values averaged across the community. These values can then be applied in ecosystem- to global-scale models. In particular, metagenomic techniques allow for quantification of gene frequencies in a sampled community (Berlemont *et al.*, 2014). If these frequencies are assumed to correspond to traits, they can be summed or averaged to estimate community-wide trait values (Fierer *et al.*, 2014). Alternatively, community-averaged traits may be measured directly; for example, most empirical assays of extracellular enzyme potential and microbial growth efficiency represent community averages, and these have been applied in ecosystem models (German *et al.*, 2012; Hagerty *et al.*, 2014).

Challenges and Future Directions

Ideally, recent advances in microbial ecology could inform revised global models that make more accurate predictions of soil biogeochemistry. This objective is within reach, but there are important challenges to overcome, including model parameterisation, validation and scaling approaches. A recent review by Wieder *et al.* (2015) described specific recommendations to address these challenges.

One of Wieder *et al.*'s (2015) major recommendations focused on coordination between modellers and empiricists regarding microbial parameters. Some parameters such as growth efficiencies and temperature response coefficients are represented in many microbial-explicit models, yet relatively few studies measure them. Conversely, few microbial-explicit models have considered moisture responses, whereas many empirical studies have focused on this topic (Lennon *et al.*, 2012; Placella *et al.*, 2012; Meisner *et al.*, 2013).

In light of the recent focus on microbial trait-based models, there should also be greater attention to assembling model-relevant trait data (Litchman and Klausmeier, 2008). Numerous datasets exist, but they are often dispersed across the literature, and new efforts to synthesise trait information in databases would benefit modelling efforts. Trait databases are available for plants (Kattge *et al.*, 2011) and are being assembled for phytoplankton (Litchman *et al.*, 2015), bacteria (Martiny *et al.*, 2013) and fungi (Treseder and Lennon, 2015). Still, these databases should be expanded and leveraged to engage modellers who aim to represent microbial traits.

Microbial-explicit models should be thoroughly tested against empirical data and compared amongst themselves to identify strengths and weaknesses. Adequate testing is essential to ensure confidence in larger-scale predictions based on microbial-explicit models. Bayesian inference is particularly useful for model-data integration because empirical data can be assimilated to both test models and estimate parameters along with uncertainties (Hararuk *et al.*, 2015). These approaches are most useful when multiple independent empirical datasets are available. If validation datasets are limited, it becomes impossible to identify unique parameter values in microbial-explicit models (Sierra *et al.*, 2015).

In addition to trait parameter information, future microbial model development would benefit from databases of empirically measured output variables. For example, datasets on soil carbon and nutrient stocks, microbial biomass pools, litter decay rates and rates of soil respiration over time are essential for model validation and data assimilation (Wieder *et al.*, 2015). Measurements of these variables in response to disturbances such as temperature change, resource pulses, moisture pulses and nutrient addition can provide additional constraints on model predictions.

New scaling approaches are needed to assemble the hierarchy of models that will allow for better global-scale predictions. Model emulators represent one promising approach but have not yet been applied to microbial systems. Monte Carlo simulation approaches are also promising for analysing emergent behaviours of microscale models. One issue with microbial-explicit models is that they assume homogeneity of biomass and soil carbon pools. When these homogeneous pools are mathematically coupled, they oscillate in response to perturbations (Li *et al.*, 2014; Hararuk *et al.*, 2015), a behaviour for which there is little empirical support, especially at large scales.

Still, we know that microbes and soil organic matter interact at the micron scale where there is high variation in biomass density and substrate availability (Sierra and Müller, 2015). When this heterogeneity is represented through repeated random sampling of microbial trait parameters and substrate input rates, pool size oscillations occur at different frequencies. If the pool dynamics of many random simulations are aggregated, the oscillatory behaviour dampens and disappears (Figure 6.2). These results suggest that microbial mechanisms operating in heterogeneous microsites are consistent with empirical data when aggregated at larger scales. By coupling Monte Carlo-type simulations with statistical

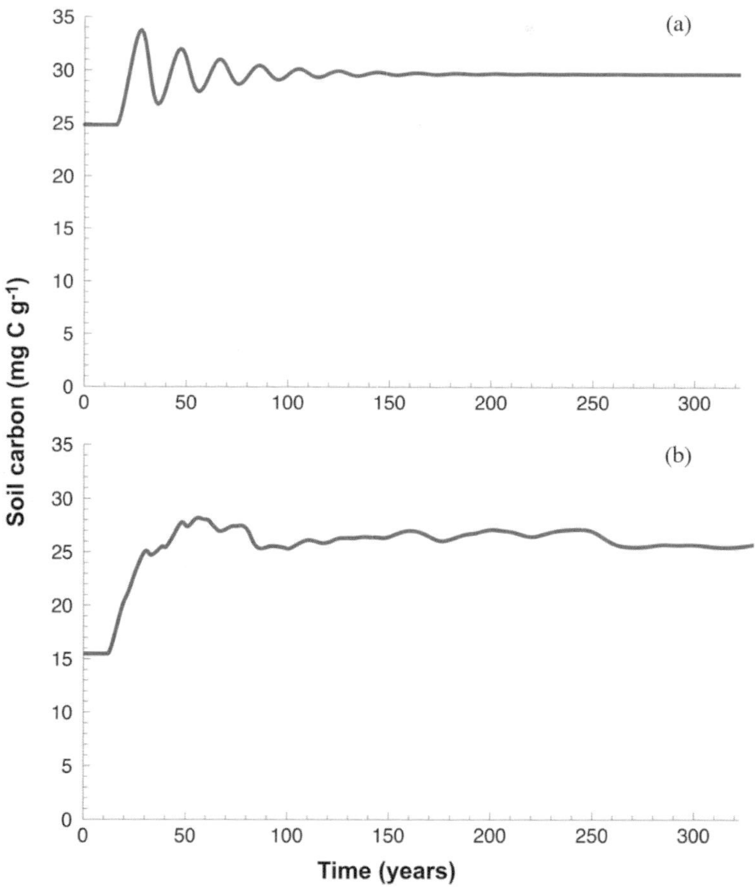

Figure 6.2 Aggregation of micro-scale processes to generate macroscale emergent properties. (a) When a two-pool microbial-explicit model (German *et al.*, 2012) is perturbed by a 5°C temperature increase, the soil carbon pool oscillates asymptotically to a higher equilibrium value based on parameters from Li *et al.* (2014). (b) When heterogeneous microsites are averaged, the oscillatory pattern diminishes (W. Sherman, unpublished). The line represents the average of 68 microsites, each defined by unique parameter values for Vmax, Km, biomass turnover and substrate input drawn at random from plausible distributions.

emulation, we may be able to incorporate microbial mechanisms into models while accurately representing emergent behaviours at large scales.

An additional need for scaling microbial models centres on state factor relationships. It is not yet clear how emergent microbial properties and responses vary with state factor gradients. There have been some studies of microbial trait parameters across environmental gradients (German *et al.*, 2012; Whitaker *et al.*, 2014), but more studies with additional environmental drivers are needed to scale up to the global level. Additional gridded data products on state factor variables at the global scale represent another key need for scaling up these models.

Conclusions

Now is an opportune time to make progress on modelling and scaling microbial processes to improve biogeochemical predictions. New empirical tools, particularly sequencing technology, are making unprecedented volumes of data available with the potential to inform microbial models. Realizing this potential will require new theory to translate sequences into ecologically relevant traits. Coupled with increased efforts to assemble data on microbial traits and processes, these advances will provide a basis for parameterizing new trait-based models. Increases in computational power to support Monte Carlo approaches and richer data sources from remote sensing and ecological experiments will facilitate the upscaling of microbial models.

The recent blossoming of microbial modelling approaches presents both an opportunity and a challenge. There is great potential for improving Earth system models by representing microbial processes with greater fidelity and elegance. The diversity of modelling perspectives is stimulating healthy debate about the best way to account for microbes in the Earth system. At the same time, all microbial-explicit models are relatively unproved. More rigorous validation is necessary to avoid scaling up theoretical approaches with fundamental flaws. Still, this challenge can be overcome with sufficient intellectual investment to yield Earth system predictions with much greater certainty and confidence than is currently possible.

Acknowledgements

I thank the US National Science Foundation program in Advancing Theory in Biology and the Office of Science (BER), US Department of Energy program in Microbial Communities and Carbon Cycling for financial support.

References

Allison, S. D. (2005). Cheaters, diffusion, and nutrients constrain decomposition by microbial enzymes in spatially structured environments. *Ecol. Lett.* 8, 626–635.

Allison, S. D. (2012). A trait-based approach for modelling microbial litter decomposition. *Ecol. Lett.* 15, 1058–1070.

Allison, S. D. (2014). Modeling adaptation of carbon use efficiency in microbial communities. *Front. Microbiol.* 5, 571.

Allison, S. D. and Martiny, J. B. H. (2008). Resistance, resilience, and redundancy in microbial communities. *Proc. Natl. Acad. Sci. USA* 105, 11512–11519.

Allison, S. D., Wallenstein, M. D. and Bradford, M. A. (2010). Soil-carbon response to warming dependent on microbial physiology. *Nat. Geosci.* 3, 336–340.

Andrén, O. and Paustian, K. (1987). Barley straw decomposition in the field: a comparison of models. *Ecology* 68, 1190–1200.

Bardgett, R. D., Freeman, C. and Ostle, N. J. (2008). Microbial contributions to climate change through carbon cycle feedbacks. *ISME J.* 2, 805–814.

Berlemont, R., Allison, S. D., Weihe, C., Lu, L., Brodie, E. L., Martiny, J. B. H., *et al.* (2014). Cellulolytic potential under environmental changes in microbial communities from grassland litter. *Front. Microbiol.* 5, 639.

Boswell, G. P. (2008). Modelling mycelial networks in structured environments. *Mycol. Res.* 112, 1015–1025.

Castruccio, S., McInerney, D. J., Stein, M. L., Liu Crouch, F., Jacob, R. L. and Moyer, E. J. (2014). Statistical emulation of climate model projections based on precomputed GCM runs. *J. Clim.* 27, 1829–1844.

Cheng, W., Parton, W. J., Gonzalez-Meler, M. A., Phillips, R., Asao, S., Mcnickle, G. G., *et al.* (2014). Synthesis and modeling perspectives of rhizosphere priming. *New Phytol.* 201, 31–44.

Crowther, T. W. and Bradford, M. A. (2013). Thermal acclimation in widespread heterotrophic soil microbes. *Ecol. Lett.* 16, 469–477.

Davidson, E. A., Janssens, I. A. and Luo, Y. (2006). On the variability of respiration in terrestrial ecosystems: moving beyond Q_{10}. *Glob. Chang. Biol.* 12, 154–164.

Davidson, E. A., Samanta, S., Caramori, S. S. and Savage, K. E. (2012). The dual Arrhenius and Michaelis–Menten (DAMM) kinetics model for decomposition of soil organic matter at hourly to seasonal time scales. *Glob. Chang. Biol.* 18, 371–384.

Evans, S. E. and Wallenstein, M. D. (2012). Soil microbial community response to drying and rewetting stress: does historical precipitation regime matter? *Biogeochemistry* 109, 101–116.

Fawcett, T. W. and Higginson, A. D. (2012). Heavy use of equations impedes communication among biologists. *Proc. Natl. Acad. Sci. USA* 109, 11735–11739.

Fierer, N., Barberan, A. and Laughlin, D. C. (2014). Seeing the forest for the genes: using metagenomics to infer the aggregated traits of microbial communities. *Front. Microbiol.* 5, 614.

Fierer, N., Bradford, M. A. and Jackson, R. B. (2007). Towards an ecological classification of soil bacteria. *Ecology* 88, 1354–1364.

Firestone, M. K. and Davidson, E. A. (1989). Microbiological basis of NO and N_2O production and consumption in soil. In Andreae, M. O. and Schimel, D. S. (eds.), *Exchange of Trace Gases Between Terrestrial Ecosystems and the Atmosphere.*. Wiley, New York, pp. 7–21.

Firestone, M. K., Firestone, R. B. and Tiedje, J. M. (1980). Nitrous oxide from soil denitrification: factors controlling its biological production. *Science* 208, 749–751.

Follows, M. J., Dutkiewicz, S., Grant, S. and Chisholm, S. W. (2007). Emergent biogeography of microbial communities in a model ocean. *Science* 315, 1843–1846.

Folse, H. J. and Allison, S. D. (2012). Cooperation, competition, and coalitions in enzyme-producing microbes: social evolution and nutrient depolymerization rates. *Front. Microbiol.* 3, 338.

Fontaine, S. and Barot, S. (2005). Size and functional diversity of microbe populations control plant persistence and long-term soil carbon accumulation. *Ecol. Lett.* 8, 1075–1087.

Foster, K. R. and Bell, T. (2012). Competition, not cooperation, dominates interactions among culturable microbial species. *Curr. Biol.* 22, 1845–1850.

Frank, S. A. (2010). The trade-off between rate and yield in the design of microbial metabolism. *J. Evol. Biol.* 23, 609–613.

German, D. P., Marcelo, K. R. B., Stone, M. M. and Allison, S. D. (2012). The Michaelis–Menten kinetics of soil extracellular enzymes in response to temperature: a cross-latitudinal study. *Glob. Chang. Biol.* 18, 1468–1479.

Grandy, A. S. and Neff, J. C. (2008). Molecular C dynamics downstream: the biochemical decomposition sequence and its impact on soil organic matter structure and function. *Sci. Total Environ.* 404, 297–307.

Gulledge, J. and Schimel, J. P. (2000). Controls on soil carbon dioxide and methane fluxes in a variety of taiga forest stands in interior Alaska. *Ecosystems* 3, 269–282.

Hagerty, S. B., van Groenigen, K. J., Allison, S. D., Hungate, B. A., Schwartz, E., Koch, G. W., *et al.* (2014). Accelerated microbial turnover but constant growth efficiency with warming in soil. *Nat. Clim. Chang.* 4, 903–906.

Hararuk, O., Smith, M. J. and Luo, Y. (2015). Microbial models with data-driven parameters predict stronger soil carbon responses to climate change. *Glob. Chang. Biol.* 21, 2439–2453.

Hurley, J. M., Dasgupta, A., Emerson, J. M., Zhou, X., Ringelberg, C. S., Knabe, N., *et al.* (2014). Analysis of clock-regulated genes in *Neurospora* reveals widespread posttranscriptional control of metabolic potential. *Proc. Natl. Acad. Sci. USA* 111, 16995–17002.

Jastrow, J. D. (1996). Soil aggregate formation and the accrual of particulate and mineral-associated organic matter. *Soil Biol. Biochem.* 28, 665–676.

Jenkinson, D. S. and Rayner, J. H. (1977). The turnover of soil organic matter in some of the Rothamsted classical experiments. *Soil Sci.* 123, 298–305.

Jenny, H. (1980). *The Soil Resource: Origin and Behavior. Ecol. Stud. Ser.* Springer, New York.

Kaiser, C., Franklin, O., Dieckmann, U. and Richter, A. (2014). Microbial community dynamics alleviate stoichiometric constraints during litter decay. *Ecol. Lett.* 17, 680–690.

Kattge, J., Díaz, S., Lavorel, S., Prentice, I. C., Leadley, P., Bönisch, G., *et al.* (2011). TRY — a global database of plant traits. *Glob. Chang. Biol.* 17, 2905–2935.

Kleber, M., Sollins, P. and Sutton, R. (2007). A conceptual model of organo-mineral interactions in soils: self-assembly of organic molecular fragments into zonal structures on mineral surfaces. *Biogeochemistry* 85, 9–24.

Lavorel, S. and Garnier, E. (2002). Predicting changes in community composition and ecosystem functioning from plant traits: revisiting the Holy Grail. *Funct. Ecol.* 16, 545–556.

Lawrence, C. R., Neff, J. C. and Schimel, J. P. (2009). Does adding microbial mechanisms of decomposition improve soil organic matter models? A comparison of four models using data from a pulsed rewetting experiment. *Soil Biol. Biochem.* 41, 1923–1934.

Lennon, J. T., Aanderud, Z. T., Lehmkuhl, B. K. and Schoolmaster, D. R. (2012). Mapping the niche space of soil microorganisms using taxonomy and traits. *Ecology* 93, 1867–1879.

Levin, S. A. (2002). Complex adaptive systems: exploring the known, the unknown, and the unknowable. *Bull. Am. Math. Soc.* 40, 3–19.

Li, J., Wang, G., Allison, S. D., Mayes, M. A. and Luo, Y. (2014). Soil carbon sensitivity to temperature and carbon use efficiency compared across microbial-ecosystem models of varying complexity. *Biogeochemistry* 119, 67–84.

Litchman, E., de Tezanos Pinto, P., Edwards, K. F., Klausmeier, C. A., Kremer, C. T. and Thomas, M. K. (2015). Global biogeochemical impacts of phytoplankton: a trait-based perspective. *J. Ecol.* 103, 1384–1396.

Litchman, E. and Klausmeier, C. A. (2008). Trait-based community ecology of phytoplankton. *Annu. Rev. Ecol. Evol. Syst.* 39, 615–639.

Martiny, A. C., Treseder, K. and Pusch, G. (2013). Phylogenetic conservatism of functional traits in microorganisms. *ISME J.* 7, 830–838.

McCalley, C. K., Woodcroft, B. J., Hodgkins, S. B., Wehr, R. A., Kim, E.-H., Mondav, R., *et al.* (2014). Methane dynamics regulated by microbial community response to permafrost thaw. *Nature* 514, 478–481.

Meisner, A., Bååth, E. and Rousk, J. (2013). Microbial growth responses upon rewetting soil dried for four days or one year. *Soil Biol. Biochem.* 66, 188–192.

Mitchell, A., Romano, G. H., Groisman, B., Yona, A., Dekel, E., Kupiec, M., *et al.* (2009). Adaptive prediction of environmental changes by microorganisms. *Nature* 460, 220–224.

Moorhead, D. L. and Sinsabaugh, R. L. (2006). A theoretical model of litter decay and microbial interaction. *Ecol. Monogr.* 76, 151–174.

Parton, W. J., Stewart, J. W. B. and Cole, C. V. (1988). Dynamics of C, N, P, and S in grassland soils — a model. *Biogeochemistry* 5, 109–131.

Pfeiffer, T., Schuster, S. and Bonhoeffer, S. (2001). Cooperation and competition in the evolution of ATP-producing pathways. *Science* 292, 504–507.

Placella, S. A., Brodie, E. L. and Firestone, M. K. (2012). Rainfall-induced carbon dioxide pulses result from sequential resuscitation of phylogenetically clustered microbial groups. *Proc. Natl. Acad. Sci. USA* 109, 10931–10936.

Powlson, D. S., Whitmore, A. P. and Goulding, K. W. T. (2011). Soil carbon sequestration to mitigate climate change: a critical re-examination to identify the true and the false. *Eur. J. Soil Sci.* 62, 42–55.

Schimel, J. (2001). Biogeochemical models: implicit versus explicit microbiology. In Schulze, E. D., Harrison, S. P., Heimann, M., Holland, E. A., Lloyd, J. J., Prentice, I. C., *et al.* (eds.), *Global Biogeochemical Cycles in the Climate System*. Academic Press, San Diego, CA, pp. 177–183.

Schimel, J., Balser, T. C. and Wallenstein, M. (2007). Microbial stress-response physiology and its implications for ecosystem function. *Ecology* 88, 1386–1394.

Schimel, J. P. and Weintraub, M. N. (2003). The implications of exoenzyme activity on microbial carbon and nitrogen limitation in soil: a theoretical model. *Soil Biol. Biochem.* 35, 549–563.

Shade, A., Peter, H., Allison, S. D., Baho, D. L., Berga, M., Bürgmann, H., *et al.* (2012). Fundamentals of microbial community resistance and resilience. *Front. Microbiol.* 3, 417.

Sharon, I., Morowitz, M. J., Thomas, B. C., Costello, E. K., Relman, D. A. and Banfield, J. F. (2013). Time series community genomics analysis reveals rapid shifts in bacterial species, strains, and phage during infant gut colonization. *Genome Res.* 23, 111–120.

Sierra, C. A., Malghani, S. and Müller, M. (2015). Model structure and parameter identification of soil organic matter models. *Soil Biol. Biochem.* 90, 197–203.

Sierra, C. A. and Müller, M. (2015). A general mathematical framework for representing soil organic matter dynamics. *Ecol. Monogr.* 85, 505–524.

Singh, B. K., Bardgett, R. D., Smith, P. and Reay, D. S. (2010). Microorganisms and climate change: terrestrial feedbacks and mitigation options. *Nat. Rev. Microbiol.* 8, 779–790.

Sinsabaugh, R. L., Belnap, J., Findlay, S. G., Shah, J. J. F., Hill, B. H., Kuehn, K. a., *et al.* (2014). Extracellular enzyme kinetics scale with resource availability. *Biogeochemistry* 121, 287–304.

Sinsabaugh, R. L., Manzoni, S., Moorhead, D. L. and Richter, A. (2013). Carbon use efficiency of microbial communities: stoichiometry, methodology and modelling. *Ecol. Lett.* 16, 930–939.

Sistla, S. A., Rastetter, E. B. and Schimel, J. P. (2014). Responses of a tundra system to warming using SCAMPS: a stoichiometrically coupled, acclimating microbe-plant-soil model. *Ecol. Monogr.* 84, 151–170.

Six, J., Frey, S. D., Thiet, R. K. and Batten, K. M. (2006). Bacterial and fungal contributions to carbon sequestration in agroecosystems. *Soil Sci. Soc. Am. J.* 70, 555–569.

Smith, O. L. (1979a). Application of a model of the decomposition of soil organic matter. *Soil Biol. Biochem.* 11, 607–618.

Smith, O. L. (1979b). An analytical model of the decomposition of soil organic matter. *Soil Biol. Biochem.* 11, 585–606.

Sollins, P., Homann, P. and Caldwell, B. A. (1996). Stabilization and destabilization of soil organic matter: mechanisms and controls. *Geoderma* 74, 65–105.

Sulman, B., Phillips, R. P., Oishi, A. C., Shevliakova, E. and Pacala, S. W. (2014). Microbe-driven turnover offsets mineral-mediated storage of soil carbon under elevated CO_2. *Nat. Clim. Chang.* 4, 1099–1102.

Talbot, J. M., Allison, S. D. and Treseder, K. K. (2008). Decomposers in disguise: mycorrhizal fungi as regulators of soil C dynamics in ecosystems under global change. *Funct. Ecol.* 22, 955–963.

Tenaillon, O., Rodriguez-Verdugo, A., Gaut, R. L., McDonald, P., Bennett, A. F., Long, A. D., *et al.* (2012). The molecular diversity of adaptive convergence. *Science* 335, 457–461.

Todd-Brown, K. E. O., Hoffman, F. M., Post, W. M., Randerson, J. T. and Allison, S. D. (2013). Causes of variation in soil carbon simulations from CMIP5 Earth system models and comparisons with observations. *Biogeosciences* 10, 1717–1736.

Todd-Brown, K. E. O., Hopkins, F. M., Kivlin, S. N., Talbot, J. M. and Allison, S. D. (2012). A framework for representing microbial decomposition in coupled climate models. *Biogeochemistry* 109, 19–33.

Todd-Brown, K. E. O., Randerson, J. T., Hopkins, F., Arora, V., Hajima, T., Jones, C., *et al.* (2014). Changes in soil organic carbon storage predicted by Earth system models during the 21st century. *Biogeosciences* 11, 2341–2356.

Treseder, K. K., Balser, T. C., Bradford, M. A., Brodie, E. L., Dubinsky, E. A., Eviner, V. T., *et al.* (2012). Integrating microbial ecology into ecosystem models. *Biogeochemistry* 109, 7–18.

Treseder, K. K. and Lennon, J. T. (2015). Fungal traits that drive ecosystem dynamics on land. *Microbiol. Mol. Biol. Rev.* 79, 243–262.

Tringe, S. G., von Mering, C., Kobayashi, A., Salamov, A. A., Chen, K., Chang, H. W., *et al.* (2005). Comparative metagenomics of microbial communities. *Science* 308, 554–557.

Updegraff, K., Bridgham, S. D., Pastor, J. and Weishampel, P. (1998). Hysteresis in the temperature response of carbon dioxide and methane production in peat soils. *Biogeochemistry* 43, 253–272.

Waksman, S. A. (1927). *Principles of Soil Microbiology*. The Williams and Wilkins Company, Baltimore, MD.

Wang, Y. P., Chen, B. C., Wieder, W. R., Leite, M., Medlyn, B. E., Rasmussen, M., *et al.* (2014). Oscillatory behavior of two nonlinear microbial models of soil carbon decomposition. *Biogeosciences* 11, 1817–1831.

Wang, G., Post, W. and Mayes, M. (2013). Development of microbial-enzyme-mediated decomposition model parameters through steady-state and dynamic analyses. *Ecol. Appl.* 23, 255–272.

Webb, C. T., Hoeting, J. A., Ames, G. M., Pyne, M. I. and Poff, N. L. (2010). A structured and dynamic framework to advance traits-based theory and prediction in ecology. *Ecol. Lett.* 13, 267–283.

West, S. A., Griffin, A. S., Gardner, A. and Diggle, S. P. (2006). Social evolution theory for microorganisms. *Nat. Rev. Microbiol.* 4, 597–607.

Whitaker, J., Ostle, N., Nottingham, A. T., Ccahuana, A., Salinas, N., Bardgett, R. D., *et al.* (2014). Microbial community composition explains soil respiration responses to changing carbon inputs along an Andes-to-Amazon elevation gradient. *J. Ecol.* 102, 1058–1071.

Wieder, W. R., Allison, S. D., Davidson, E. A., Georgiou, K., Hararuk, O., He, Y., *et al.* (2015). Explicitly representing soil microbial processes in Earth system models. *Global Biogeochem. Cycles* 29, 1782–1800.

Wieder, W. R., Bonan, G. B. and Allison, S. D. (2013). Global soil carbon projections are improved by modelling microbial processes. *Nat. Clim. Chang.* 3, 909–912.

Wieder, W. R., Grandy, A. S., Kallenbach, C. M. and Bonan, G. B. (2014). Integrating microbial physiology and physio-chemical principles in soils with the MIcrobial-MIneral Carbon Stabilization (MIMICS) model. *Biogeosciences* 11, 3899–3917.

Wutzler, T. and Reichstein, M. (2008). Colimitation of decomposition by substrate and decomposers — a comparison of model formulations. *Biogeosciences* 5, 749–759.

Xia, J., Luo, Y., Wang, Y.-P. and Hararuk, O. (2013). Traceable components of terrestrial carbon storage capacity in biogeochemical models. *Glob. Chang. Biol.* 19, 2104–2116.

Yoshiyama, K. and Klausmeier, C. A. (2008). Optimal cell size for resource uptake in fluids: a new facet of resource competition. *Am. Nat.* 171, 59–70.

7

Dynamic Compound-Specific Stable Isotope Probing of the Soil Microbial Biomass

Richard P. Evershed*, Alice F. Charteris, Peter J. Maxfield†, Ian D. Bull and Timothy D. J. Knowles

Organic Geochemistry Unit, School of Chemistry, University of Bristol, Cantock's Close, Bristol BS8 1TS, UK
**R.P.Evershed@bristol.ac.uk*

Introduction

The concept of following the assimilation of stable isotopically labelled substrates into newly synthesised soil microbial biomass biochemicals (= biomarkers) was first demonstrated nearly 20 years ago in studies of unculturable soil methanotrophs (Radajewski *et al.*, 2000; Bull *et al.*, 2000). These landmark papers laid the foundations for an approach that has come to be generally known as stable isotope probing (SIP). The major advantage of SIP in studying the soil microbial biomass lies in its effectiveness for investigating unculturable soil microorganisms, that is, the vast majority of the soil bacteria, actinomycetes, fungi, algae and protozoa. Historically, SIP has developed along two distinct paths, namely those

† Present address: Biological, Biomedical and Analytical Sciences, University of the West of England, Frenchay Campus, Coldharbour Lane, Bristol BS16 1QY, UK.

applications focussing on genomic material (DNA and RNA) and those targeting small molecules, such as PLFAs.

The basic soil SIP experiment involves presenting the microbial community with a stable isotopically labelled substrate, most commonly ^{13}C-labelled. After a period of incubation, soils are sampled and specific cellular biochemicals extracted. The increase in mass associated with uptake of the stable isotope label is either utilised to separate the labelled components gravimetrically (e.g. DNA and RNA) or to analyse them directly by isotope ratio mass spectrometry (IRMS). The original experiment performed by Boschker and co-workers traced ^{13}C-labelled acetate utilisation in aquatic sediments (Boschker *et al.*, 1998), with the resulting phospholipids being solvent extracted, hydrolysed and the phospholipid fatty acids (PLFAs) analysed by gas chromatography-combustion-IRMS (GC-C-IRMS) to determine the pattern of uptake of the ^{13}C-label into different components — the distribution of the stable isotopically labelled PLFAs was used as a taxonomic fingerprint for the microbes assimilating different substrates. In the original experiment, ^{13}C from acetate was recovered mainly from PLFAs with even numbers of carbon atoms (16:1ω7c, 16:0, 18:1ω7c), whereas fatty acids with odd numbers of carbon atoms (a15:0, 15:0, 17:1ω6, 17:0) were primarily labelled upon incubation with propionate. Comparison of the ^{13}C-PLFA compositions with those of known microorganisms showed that the two substrates were predominantly consumed by different specialized groups of sulphate-reducing bacteria (SRB). The PLFA ^{13}C-labelling pattern for the acetate consumers was similar to *Desulfotomaculum acetoxidans* and *Desulfofrigus* spp., two acetate-consuming SRB, while those of the propionate consumers did not resemble any known strain. Despite the obvious effectiveness of this experiment, one of the criticisms of ^{13}C-PLFA-SIP has been the lack of specificity in its ability to identify specific microorganisms taxonomically. We will show in this contribution, however, that such criticism overlooks how the use of specific ^{13}C or ^{15}N-labelled tracers together with GC-C-IRMS analysis of different classes of small molecules can be extremely effective in exploring critical aspects of soil biomass characteristics, which extend well beyond taxonomic assignments.

DNA- and RNA-SIP are, without question, the preferred techniques for identifying active microorganisms assimilating particular carbon substrates and nutrients into cellular biomass. The initial demonstration of this approach immediately showed its value as a cultivation-independent technique for linking metabolic function to specific members of microbial communities inhabiting complex environmental matrices, such as soils. As with PLFA-SIP, soils are incubated with stable isotopically labelled substrates, but the technique departs from the PLFA approach since extracted nucleic acids are subjected to density gradient ultracentrifugation to separate nucleic acids of differing masses due to ^{13}C-labelling. Purification of DNA from caesium chloride retrieves ^{13}C-labelled and unlabelled DNA (differentiated by IRMS), followed by polymerase chain reaction amplification and molecular characterisation (e.g. fingerprinting, microarrays, clone libraries, metagenomics). The approach has proved to be extremely powerful in a wide variety of environmental settings. For the procedure to succeed, however, a very high level of ^{13}C-labelling is required to ensure that physical separation is achieved during the centrifugation step. Even higher levels of labelling are required for ^{15}N-DNA/RNA SIP due to the lower molar ratios of N/C.

We do not dwell further in this contribution on the use of DNA/RNA-SIP approaches to soil biomass studies. Instead, we focus on exploring an under-appreciated dimension of small-molecule SIP, which we now term *dynamic-SIP*. This is an approach we have come to use routinely in our laboratory over the last 10 years. We will focus initially on dynamic ^{13}C-SIP and then move on to ^{15}N- and dual ^{13}C,^{15}N-SIP. Overall, we will demonstrate how the small-molecule SIP technique has progressed from the original and widely used ^{13}C-PLFA taxonomic approach to a significantly more powerful tool for exploring soil biomass dynamics in relation to: (i) revealing the assimilation of different substrates; (ii) the turnover rates of different biomass C pools; (iii) the connectivity between C and N cycles and (iv) trophic interactions. This contribution can be viewed as an update to our previous reviews of ^{13}C-PLFA SIP applied to soils (Evershed *et al.*, 2006, 2007; Maxfield and Evershed, 2011), but with an emphasis on the dynamic-SIP approach. Our aim here is to demonstrate how small-molecule dynamic-SIP is much more than a taxonomic technique, yielding information generally inaccessible to genomic and bulk stable isotope approaches.

General Considerations for SIP in the Soil System

SIP in various forms has been applied in investigations of microbial communities in a diversity of environments ranging from ocean sediments to the human gut, but with soil SIP arguably being the most challenging of all the systems studied (Murrell and Whiteley, 2011). ^{13}C-SIP is far more commonly used than any other isotope system, with substrates used in soil SIP studies including $^{13}CH_4$, $^{13}CO_2$, $^{13}CH_3OH$, ^{13}C-acetate, ^{13}C-propionate, ^{13}C-butyrate, ^{13}C-pyruvate, ^{13}C-mono- and polysaccharides, ^{13}C-alkyl halides, ^{13}C-aromatics (including phenolic and PAHs) and complex substrates such as ^{13}C-labelled plants or cultured microbes. After incubation, the vast majority of studies target either DNA/RNA or PLFAs. The major advantages of ^{13}C-PLFA SIP are that the preparation of PLFAs from soils

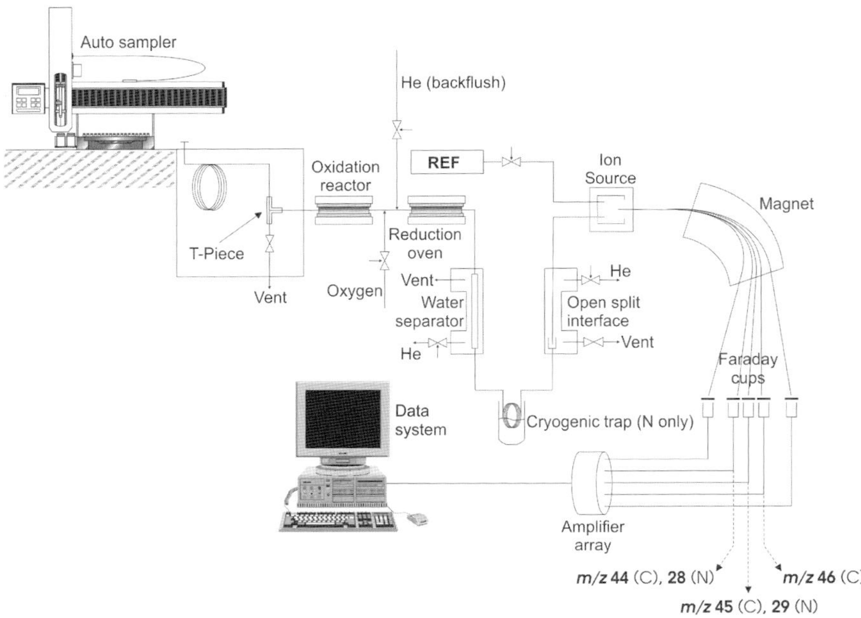

Figure 7.1 The instrument that made small-molecule SIP possible (Merritt and Hayes, 1994). A GC-C-IRMS configured for either compound-specific carbon or nitrogen isotope analyses. For $\delta^{13}C$ value determinations ions at *m/z* 44 ($^{13}C^{16}O^{16}O$), 45 ($^{13}C^{16}O^{16}O$+$^{12}C^{16}O^{17}O$) and 46 ($^{13}C^{16}O^{17}O$+$^{12}C^{16}O^{18}O$+$^{12}C^{17}O^{17}O$) are continuously measured; monitoring *m/z* 46 enables the ^{17}O (via determination of ^{18}O content) contribution to *m/z* 45 to be accounted for. For $\delta^{15}N$ determinations, ions at *m/z* 28 ($^{14}N^{14}N$) and 29 ($^{14}N^{15}N$) are monitored. The exact derivatisation, standardisation and calibration, GC columns and temperature programmes and data analysis protocols are given at the end of this chapter for all the compound classes referred to in the following sections. Further detailed discussions of the technical aspects of the instrumentation in the context of ecological research are given in Evershed *et al.* (2007).

is entirely routine (see "Experimental Methods" at the end of this chapter) but is combined with the exceptional sensitivity of the final analysis (Mottram and Evershed, 2003) and rests upon the use of GC-C-IRMS (Figure 7.1).

Indeed, it is the sensitivity aspect, combined with the selectivity conferred through the choice of ^{13}C-labelled substrate that makes the ^{13}C-PLFA SIP such a powerful tool in studying the soil microbial biomass. The high sensitivity of the technique has a number of important practical advantages. The first was economics — stable isotopically labelled substrates are amongst the most expensive chemicals used in ecological research. The high sensitivity of the

GC-C-IRMS means that less isotope (^{13}C, ^{15}N) is required per treatment which has obvious cost savings. This has critical advantages in terms of improved experimental design as scale can be increased, that is, increased replication or more frequent sampling in time-course experiments, which amounts to enhanced rigour and ultimately greater insights into biomass function. The second advantage involves appropriate application rates — this is critical to avoid perturbing the activities of microbial biomass away from the norm, with advantages coming from targeting of specific compounds providing insights into the wider fate of the isotopically labelled substrate that can never be achieved through bulk stable isotope analysis of soil except by applying substrates at supra-normal rates. As we will show repeatedly below, the enhanced sensitivity of compound-specific approaches offers considerable, and in some cases hitherto unexplored, scope in both ^{13}C- and ^{15}N-SIP studies.

One final important point is that the most successful SIP experiments are predicated on the use of highly enriched stable isotopically labelled substrates in "pulse-chase" or continuous incubation experiments. Such experiments have grown in popularity since they circumvent the sensitivity limitations experienced when using natural abundance-labelled substrates, for example, C_4 plant litters or animal dung, while such substrates might have appeal on grounds of greatly reduced cost and availability in large amounts, except in some special circumstances (e.g. Dungait *et al.*, 2008, 2009), their applications will be significantly more limited than when isotopically enriched substrates are employed.

Dynamic-SIP

Here we define *dynamic-SIP* as those experiments which involve regular sampling over a time-course throughout the incubation period. We actually regard such experiments as fundamental in optimising experimental designs in order to: (i) establish optimal incubation times, (ii) ensure that equilibrium conditions have been reached, (iii) detect changes in microbial populations with time, (iv) detect cross-feeding between microbes in long-term incubations and (v) provide numerical estimates for nutrient fluxes and pool sizes. Given the hundreds of reported SIP studies, it is surprising that dynamic-SIP experiments are performed so infrequently. Table 7.1 lists examples of dynamic ^{13}C-SIP experiments performed during the last decade.

As can be seen from Table 7.1, these studies include both laboratory and field experiments and a wide range of ^{13}C-labelled substrates. Dynamic-SIP approaches can be divided into two broad categories, namely those applying: (i) complex

Table 7.1 Examples of dynamic ^{13}C-SIP studies of the soil microbial biomass arranged by substrate applied.

Substrate	Environment	Biomarkers	Observation	Reference
Chemically defined				
$^{13}CH_4$	Rice field/forest soils	PLFAs	Methanotrophs affected by N fertilisation	(Mohanty *et al.*, 2006)
$^{13}CH_4$	Upland grassland	PLFAs	New approach to estimating methanotroph population characteristics	(Maxfield *et al.*, 2006)
$^{13}CH_4$	Upland grassland	PLFAs	Methanotrophs affected by inorganic fertilisation	(Maxfield *et al.*, 2008)
$^{13}CH_4$	Volcanic soils	PLFAs	High populations of methanotrophs in Andisols	(Maxfield *et al.*, 2009)
$^{13}CH_4$	Rothamsted Classical Experiment	PLFAs	Methanotrophs affected by inorganic fertilisation	(Maxfield *et al.*, 2011)
$^{13}CH_4$	Landfill cap soil	Various cellular biochemicals	New technique for elucidating total C dynamics (incorporation, turnover and decay) at the molecular level	(Maxfield *et al.*, 2012)
$^{13}CO_2$	Eutric cambisol	Neutral sugars	Root derived glucose-^{13}C converted to microbial sugars with time	(Darien *et al.*, 2004)
$^{13}CO_2$	Douglas-fir forest soil	PLFAs	Differential utilisation of substrates; also affected by plant inputs	(Brant *et al.*, 2006)

$^{13}CO_2$	Desert soil	PLFAs	Fungal-to-bacterial PLFA-C increased under elevated CO_2 compared to ambient conditions	(Jin and Evans, 2010)
$^{13}CO_2$	Upland grassland	PLFAs, DNA, RNA	Complex C dynamics and pathways of photosynthate entering soil	(Leake *et al.*, 2006)
$^{13}CO_2$	Boreal pine forest soil	PLFAs	Photosynthate found mainly in fungi and Collembola with long-term C allocation affected by N additions	(Högberg *et al.*, 2010)
$^{13}CO_2$	Sandy loam agricultural soil	PLFAs	Plant species influences magnitude and distribution of plant C to soil microorganisms	(Ladygina and Hedland, 2010)
$^{13}CO_2$	Sandy peat agricultural soil	PLFAs, DNA, RNA	C flow of photosynthate to soil varies between GM plant and parental cultivar	(Hannula *et al.*, 2012)
$^{13}CO_2$	Paddy soil (Hydragric Anthrosol)	PLFAs	Differing microbial communities established by manipulating water status	(Yao *et al.*, 2012)
$^{13}CO_2$	Fore-dune soil	PLFAs, RNA	Elevated CO_2-induced changes in rhizosphere C flow and dynamics	(Drigo *et al.*, 2013)
$^{13}CO_2$	Drained boreal organic peat soil	PLFAs	Flow of ^{13}C maximised at 3 days in fungi but after 1 year ^{13}C increased in bacteria	(Tavi *et al.*, 2013)

(Continued)

Table 7.1 (*Continued*)

Substrate	Environment	Biomarkers	Observation	Reference
$^{13}CO_2$	Mountain meadow dystric cambisol	PLFAs	Shading produced no reduction ^{13}C in fungi and bacteria but increased residence time	(Bahn *et al.*, 2013)
$^{13}CO_2$	Sandy grassland soil	PLFAs	Faster turnover of C in mycorrhizal vs. saprotrophic fungi vs. bacteria	(Balasooriya *et al.*, 2014)
$^{13}CO_2$	Jena Biodiversity Experiment	PLFAs, RNA	C turnover is higher in RNA than PLFAs	(Malik *et al.*, 2015)
^{13}C-glucose, glutamate, oxalate, phenol	Douglas-fir forest soil	PLFA	Differential utilisation of substrates; also affected by plant inputs	(Brant *et al.*, 2006)
^{13}C-glucose	Contaminated soil	PLFA	Changing contributions of fungi in response to contaminants	(Bartling *et al.*, 2009)
^{13}C-phenanthrene	Haplic Chernozem	PLFA	Rape oil application changed assimilation by microbes	(Mellendorf *et al.*, 2010)
^{13}C-Cu-*N*-cyclo-hexyldiazenium dioxide	Soil–wood–sawdust mixture	PLFA and RNA	Defined bacterial community stimulated	(Jakobs-Schonwandt *et al.*, 2010)
^{13}C-aspartic acid	Tree-rhizosphere	PLFA	Stem injection of substrates into large trees can be used to trace below-ground rhizodeposition	(Churchland *et al.*, 2012)

^{13}C-alanine and glutamic acid	Loamy luvisol	PLFA	Position-specific incorporation of ^{13}C atoms from AAs provide insights into mechanisms and kinetics of assimilation by microbial groups	(Apostel *et al.*, 2013)
^{13}C-glucose	Permanent grassland	PLFA	Differences in assimilation of glucose by bacteria and fungi over 52-week incubation. NPK decreased biomass	(Lemanski and Scheu, 2014)
^{13}C-glucose	Woburn Grassland and Broadbalk soils	PLFA	Different responses of SMB to different glucose concentrations in different soils	(Dungait *et al.*, 2011)
^{13}C-lignin and ^{13}C-cellulose	Abanilla and Santomera semi-arid soils	PLFA	Minor part of the microbial community assimilated C from substrates in 120 days	(Torres *et al.*, 2014)
Plant litter				
^{13}C-,^{15}N-litter	Sand–clay growth medium	PLFAs	Transition from fungal to bacterial PLFAs as decomposition progressed	(Herman *et al.*, 2012)
^{13}C-aspen litter	Forest soil	PLFAs	Chronic N deposition decreased ^{13}C mobilisation into microbial biomass	(Gan *et al.*, 2013)

(Continued)

Table 7.1 (*Continued*)

Substrate	Environment	Biomarkers	Observation	Reference
^{13}C-barley residues	Humid and semi-arid Cambisol and Haplic Kastanozem	PLFAs	Overall plant residue turnover slower under semi-arid conditions	(Helgason *et al.*, 2014)
^{13}C-eucalyptus leaf litter	Remnant native woodland soil	PLFAs, DNA, RNA	Distinct microbial communities alter turnover and fate of OM in response to temperature	(Creamer *et al.*, 2015)
^{13}C-wheat grains, leaves, roots	Conventionally tilled Luvisol	PLFAs	Double exponential model described PLFA synthesis-degradation kinetics	(Bai *et al.*, 2016)
Microbial cells				
^{13}C-*E.coli*	Agricultural reference soil	PLFAs, DNA	Biomass C retained in soil in soil food web and non-living SOM	(Kindler *et al.*, 2006)
^{13}C-*E.coli*	Agricultural reference soil	PLFAs, DNA	Biomass C recycled down to 10% in living organisms with 33% stabilised in non-living SOM pool	(Kindler *et al.*, 2009)
^{13}C-methanotrophs	Rice field soil	PLFAs	Eukaryotes, probably protozoa, preferentially graze on methanotrophs	(Murase *et al.*, 2011)

substrates, such as ^{13}C-labelled plant litter (Williams *et al.*, 2006; Moore-Kucera and Dick, 2008; Herman *et al.*, 2012; Gan *et al.*, 2013; Helgason *et al.*, 2014; Creamer *et al.*, 2015; Bai *et al.*, 2016) or ^{13}C-labelled microbial cultures (Kindler *et al.*, 2006, 2009; Murase *et al.*, 2011); or (ii) pure ^{13}C-labelled compounds (Brant *et al.*, 2006; Bartling *et al.*, 2009; Mellendorf *et al.*, 2010; Jakobs-Schonwandt *et al.*, 2010; Dungait *et al.*, 2011; Churchland *et al.*, 2012; Apostel *et al.*, 2013; Lemanski and Scheu, 2014). $^{13}CO_2$ pulse-chase experiments are amongst the most widely applied, designed to deliver ^{13}C-labelled root exudate derived from newly formed ^{13}C-photosynthate to explore C allocation from plants to the below-ground biomass (Leake *et al.*, 2006; Högberg *et al.*, 2010; Jin and Evans, 2010; Ladygina and Hedland, 2010; Hannula *et al.*, 2012; Yao *et al.*, 2012; Bahn *et al.*, 2013; Dias *et al.*, 2013; Drigo *et al.*, 2013; Tavi *et al.*, 2013; Balasooriya *et al.*, 2014; Malik *et al.*, 2015). In both categories, ^{13}C-PLFAs are the most commonly investigated biomarkers with time courses varying widely (days to years) to provide temporal and quantitative insights into the processing of ^{13}C-labelled substrates by the soil microbial community, often in response to changes in particular environmental variables (e.g. Creamer *et al.*, 2015). A small number of studies have built in DNA and RNA-SIP investigations to provide deeper taxonomic assessments (e.g. Leake *et al.*, 2006).

Dynamic $^{13}CH_4$ SIP

Dynamic $^{13}CH_4$ SIP has been extensively employed in studies of methanotrophic communities in soils (Table 7.1). The approach has proved especially effective in the case of the unculturable high-affinity methanotrophic bacteria where an active methanotrophic population assimilating $^{13}CH_4$ can be confirmed by ^{13}C-enriched PLFAs detected by GC-C-IRMS. Indeed, as noted above, one of the first applications of ^{13}C-PLFA to soil microbes was in the study of methanotrophs (Bull *et al.*, 2000). We have extended this technique to characterise methanotrophic bacteria based on a combination of ^{13}C-labelled hopanoids and PLFAs (Crossman *et al.*, 2001, 2005), recently incorporating this approach into a new type of $^{13}CH_4$ dynamic-SIP experiment, termed stable isotope switching (SIS), which we will discuss in detail below (Maxfield *et al.*, 2012).

A major development of this approach has been our implementation of $^{13}CH_4$ time-series ^{13}C-PLFA labelling which facilitated, for the first time, enumeration of high-affinity methanotroph populations (Maxfield *et al.*, 2006). This approach involves performing laboratory incubations of soils at ambient CH_4 concentrations using synthetic air containing 2 ppmv of $^{13}CH_4$. Flow-through chambers are employed to maintain a stable $^{13}CH_4$ concentration throughout the 11-week incubations (see Maxfield and Evershed, 2011, for chamber design). When using this approach, soils are sampled and analysed at weekly intervals by GC, GC-MS and GC-C-IRMS to identify and quantify individual PLFAs and trace the incorporation of ^{13}C-label into the microbial biomass. Incorporation of the ^{13}C-label is generally seen throughout the experiment, with the rate of incorporation decreasing in the later stages (9–10 weeks) as the methanotroph population becomes fully ^{13}C-labelled. The $\delta^{13}C$ values of individual PLFAs showed that ^{13}C was incorporated into different components selectively and at various rates, reflecting the diversity of PLFA sources. Quantitative assessments of ^{13}C-PLFAs have shown that high-affinity methanotrophic populations are usually of constant structure throughout the experiment. In the original experiment, performed using soils from Bronydd-Mawr (Wales), the dominant ^{13}C-labelled PLFA was 18:1ω7c, with 16:1ω5 present at lower abundance, suggesting the presence of novel type II methanotrophs. We used the concentration of PLFAs per bacterial cell determined by Frostegård and Bååth (1996) to determine the biomass of CH_4-oxidising bacteria; at optimum labelling this was estimated to be about 7.2×10^6 cells g^{-1} of soil (dry weight), remarkably, only one ten-thousandth of the total bacterial population! Cross-feeding or recycling of biomass C is a recognised phenomenon that may thwart some types of SIP investigation, resulting in a loss of specificity. However, due to the low concentration and high specificity of the $^{13}CH_4$ labelled substrate, two factors supported this approach with high-affinity methanotrophs: (i) recycling of ^{13}C-label from the methanotrophic biomass is a slower process than initial $^{13}CH_4$ incorporation, with only about 5–10% of ^{13}C-PLFAs reflecting this process, and (ii) the relatively low cell numbers of high-affinity methanotrophs compared with the total microbial biomass (10^6 vs. 10^{10} cells, respectively) means that any leakage of ^{13}C-label into other microbial pools will effectively constitute a loss of label by dilution into the high background C pools, such that the ^{13}C-PLFA profile of the high-affinity methanotroph population will dominate. Thus, ^{13}C-labelled PLFA distributions determined at any time point during $^{13}CH_4$ incubation can be used for chemotaxonomic assessments; however, extended incubations are required to achieve maximum ^{13}C-labelling for methanotrophic biomass determinations.

We have applied this technique in different environments to study variations in low- and high-affinity methanotrophic populations in relation to a range of environmental variables. For example, as shown in Figure 7.2, we have linked a reduction of CH_4 sink capacity to inhibition of the high-affinity methanotroph populations resulting from long-term inorganic fertiliser applications to upland grassland soils (Maxfield *et al.*, 2008). While it is well known that agricultural practices, including tillage (Hütsch, 1998) and fertiliser addition (Steudler *et al.*,

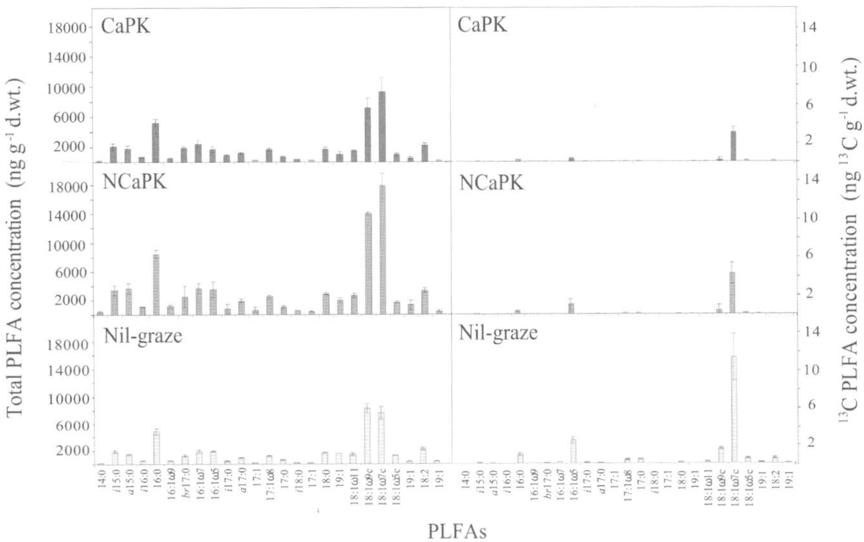

Figure 7.2 Concentrations of the extracted total PLFAs compared with ^{13}C-PLFA concentration following 17–18 weeks of incubation under 2 ppmv $^{13}CH_4$ for Bronydd-Mawr NCaPK, CaPK and Nil-graze soils. Error bars represent ±1 SD.

1989; Mosier *et al.*, 1991; Adamsen and King, 1993; Castro *et al.*, 1995; Gulledge *et al.*, 1997; Gulledge and Schimel, 1998), inhibit CH_4 oxidation in soils, much debate surrounds the proposed mechanism(s).

Our results show that inhibition of high-affinity methanotrophic bacteria in the Bronydd Mawr long-term low-input fertilised soil is caused by a general fertilisation-induced salt effect rather than a nitrogen-specific inhibitory mechanism. Addition of NH_4^+ to soils has been reported to suppress CH_4 uptake rates (Mosier *et al.*, 1991; Adamsen and King, 1993; Castro *et al.*, 1995; Gulledge *et al.*, 1997; Gulledge and Schimel, 1998). Inhibition also has been observed in short-term studies (Gulledge *et al.*, 1997; Wang and Ineson, 2003; Crossman *et al.*, 2006) for a range of ions including K^+, Cl^- and SO_4^{2-}, confirming that general salt effects rather than by-products of NH_4^+ oxidation can inhibit CH_4 oxidation. Reay and Nedwell (2004) observed a clear relationship between inhibition of CH_4 oxidation and NO_3^- rather than NH_4^+; however, the mechanism of CH_4 oxidation inhibition by NO_3^- rather than NO_3^- and NH_4^+ remains unclear. Gulledge and Schimel (1998) demonstrated the importance of non-ammoniacal salts on CH_4 oxidation by treating soils with K_2SO_4, Na_2SO_4, KCl, $(NH_4)_2SO_4$ and NH_4Cl. Methane oxidation was inhibited, to some extent, by all the salts irrespective of NH_4^+ levels, which were naturally low in the soils. It was suggested that inhibition was not a result of desorption of NH_4^+ from the soils but rather a direct effect of SO_4^{2-} or Cl^- ions on CH_4 oxidation.

With the exception of two recent laboratory studies, all previous investigations have used changes in CH_4 oxidation rates to detect patterns of inhibition, thereby making it impossible until now to explain the underlying cause of inhibition. Our results confirm that loss of CH_4 oxidation capacity is the result of a shift in microbial populations resulting in a significant decrease in the numbers of methanotrophic bacteria. Our results also confirmed why problems have been encountered in studying the mechanism(s) of fertiliser addition impacts on high-affinity methanotrophic bacteria. The low relative abundance (<0.04%) of methanotrophs in total microbial biomass precludes use of other culture-independent microbiological methods, such as earlier attempts using gene probing (Reay *et al.*, 2001; Seghers *et al.*, 2003) and most probable number techniques (Hütsch, 2001). Frustratingly, the exact mechanism(s) underlying the loss of high-affinity methanotrophic activity remains unknown but it is clear that the loss of CH_4 sink capacity as judged by flux measurement and methanotroph population estimates can be as much as 80%.

Similar effects of fertiliser additions were originally observed through flux measurements on the arable soils of Rothamsted by Powlson and co-workers (Hütsch *et al.*, 1993; Willison *et al.*, 1995). Our more recent work with dynamic $^{13}CH_4$ ^{13}C-PLFA-SIP using flow-through chambers has confirmed the very small methanotroph populations in these heavily fertilised and extensively tilled soils (Maxfield *et al.*, 2011). Overall, methanotrophic biomass was smaller in the Rothamsted soils compared with the other mineral soils studied to date, as indicated by poor ^{13}C incorporation. High-affinity methanotrophs similar to known type II methanotrophs were most abundant in all of the soils studied, except in two plots receiving farmyard manure (FYM and FYM + N). Manuring resulted in a shift to a population similar to known type I methanotrophs. Methanotrophic biomass was elevated in soils that had received the largest input of N fertiliser, though without detectable differences in CH_4 oxidation rates, indicating the potential mediation of atmospheric CH_4 oxidation by non-methanotrophs, which are most likely to be nitrifying bacteria.

These results clearly demonstrate the impact that inorganic fertilisers have on the capacity of soil to function as a sink for tropospheric CH_4 and provide a strong link to changes in high-affinity methanotroph community structure. Placing these findings in the wider context of greenhouse gas emissions and mitigation strategies requires consideration of national and international inventories for CH_4 sources and sinks. Significantly, the UK has the highest anthropogenic CH_4 emissions in Europe after France (Bergamaschi *et al.*, 2005), with the largest source of CH_4 emissions in the UK being agriculture (NETCEN, 2003). The lack of a detailed UK inventory of CH_4 sources and sinks precludes calculation of an accurate estimation of the proportion of the UK CH_4 sink inhibited by inorganic fertiliser (N + non-N). Our conservative estimate is that low-input fertiliser inhibition of high-affinity methanotrophic bacteria in the UK is as much as 5% of the EU-15 soil sink. Considering that the surface area of the UK (244×10^3 km^2) is less than 10% of the total EU-15 surface area (3154×10^3 km^2) then inhibition due to low-input fertilisation may be >50% of the potential UK soil CH_4 sink. Thus, the regional and indeed global impacts of inorganic fertiliser treatments on soil microbial diversity as an integral component of system C dynamics are worthy of continued detailed assessment.

Indeed, we have extended this work to other soil types. Andisols, soils derived from volcanic ash, have the capacity to support high levels of atmospheric CH_4 oxidation (Singh *et al.*, 2007). The unique properties of Andisols, specifically their low bulk density, high C content and high water holding capacity (WHC), would provide a favourable environment for the development of methanotroph populations. The low bulk density, in particular, would promote gas diffusion, enhancing the supply of CH_4 and O_2 while maintaining a relatively high moisture content. Furthermore, high-affinity methanotrophs thrive in soils that have good substrate and moisture supplies. We characterised methanotrophic bacteria in Andisols via the dynamic $^{13}CH_4$ SIP time-series ^{13}C-PLFA labelling approach discussed above. Three Andisols were incubated under 2 ppmv CH_4 for up to 18 weeks, enabling high-affinity methanotrophs to be selectively characterised and quantified. PLFA profiles from all soils were broadly similar, but the magnitude of the high-affinity methanotrophic populations determined through ^{13}C-PLFA-SIP displayed sizeable differences. Substantial incorporation of ^{13}C indicated very large high-affinity methanotrophic populations in two of the soils. Such high values are far in excess (10×) of those observed for the aforementioned mineral soils incubated under similar conditions (Maxfield *et al.*, 2008, 2011). Two of the three Andisols studied also displayed high but variable CH_4 oxidation rates ranging from 0.03 to 1.58 nmol CH_4 g^{-1} dw t h^{-1}. These findings suggest that Andisols may oxidize significant amounts of atmospheric CH_4 despite their low area of coverage globally. Andisols cover only about 0.7% of the global land surface (Brady and Weil, 2002), which equates to a land area of ca. 0.91 million km^2. Total uptake of atmospheric CH_4 globally by Andisols may be greater than previously thought if methanotophic activity is similar to the rates observed in the Hapludand and Ultic Fulvudand soils we investigated. The much lower levels of methanotrophic activity displayed a greater net CH_4 flux, suggesting the possible role of low-affinity methanotrophs at the site. It should also be noted that

Andisols are extensively cultivated globally, in particular, in Japan and Italy because of their intrinsically high C content and favourable physical properties. As discussed above, it is well established that certain agricultural practices, for example, the application of inorganic N-fertilisers inhibit the activity of methanotrophic bacteria in mineral soils (e.g. Mosier *et al.*, 1991; Gulledge *et al.*, 1997; Boeckx *et al.*, 1998; Hütsch, 1998). A similar relative degree of inhibition of methanotrophy in Andisols may represent a much greater loss of atmospheric CH_4 oxidation capacity.

Stable Isotope Switching

Recent advances in SIP have enabled a direct linkage between microbial population structure and function; however investigations of the redistribution of ^{13}C-label in C pools other than PLFAs and DNA/RNA are rare. To address this we introduced a new SIP approach, stable isotope switching (SIS), which enables the simultaneous assessment of C uptake, turnover and decay, and the elucidation of soil food webs within complex soils or other environmental matrices (Maxfield *et al.*, 2012). SIS utilises a stable isotope labelling approach whereby the ^{13}C-labelled substrate is switched part way through the incubation to a natural abundance substrate, which allows for the reallocation of C following fixation by methanotrophs into 2° and 3° non-methanotrophic consumers. Figure 7.3 shows a schematic of the experimental design and potential uptake and redistribution trajectories. Regular sampling throughout the experiment is a prerequisite of the approach.

The development work used a landfill cover soil from Odcombe (Somerset, UK) which previous work had shown contained a high abundance of low affinity methanotrophs (Crossman *et al.*, 2004). During incubation lasting 153 days, C assimilation and dissimilation processes were monitored through bulk IRMS and GC-C-IRMS targeting a wide range of biomolecular components including: lipids, proteins and carbohydrates (Figure 7.4, LHS). Carbon assimilation by primary consumers (methanotrophs; Figure 7.4a) and sequential assimilation into secondary (Gram-negative and -positive bacteria; Figure 7.4b) and tertiary consumers (Eukaryotes; Figure 7.4c) was observed (the full experimental details for sample preparation and data analysis are given in Experimental Methods at the end of this chapter, and in Maxfield *et al.*, 2012). Mass balance calculations showed that up to 45% of the bacterial membrane lipid C was directly derived from CH_4 and at the conclusion of the experiment ca.50% of the bulk soil C derived directly from CH_4 was retained within the soil. At the time this work was reported, it appeared that this was the first estimate of soil organic C allocation

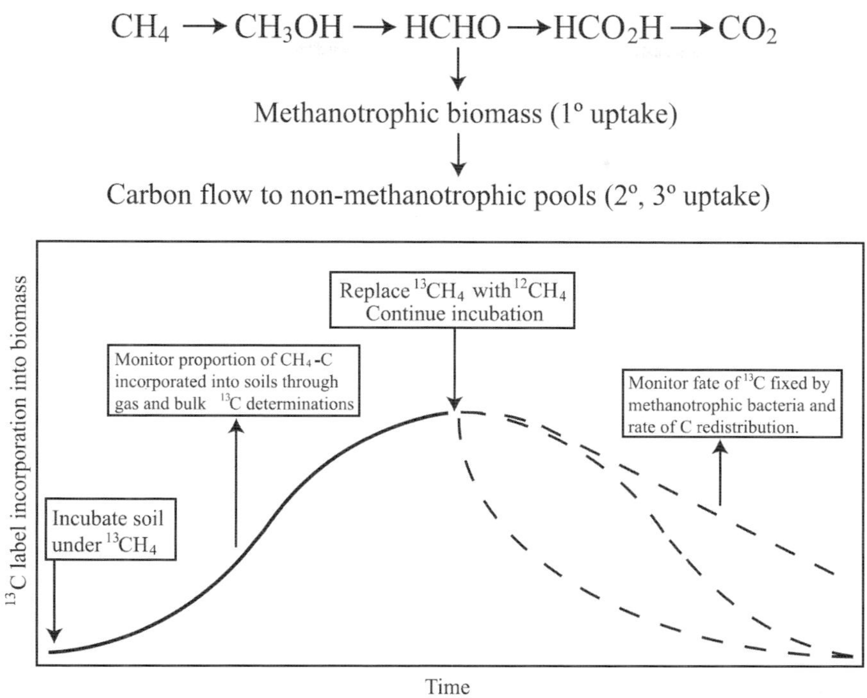

Figure 7.3 Schematic representation of an SIS experiment based on previously observed (solid line) and theoretical (dashed lines) curves indicating how the ^{13}C-label from $^{13}CH_4$ is incorporated by methanotrophs and turned over within soil ecosystems. The time scale to achieve full labelling of the microbial population will vary depending on the labelled substrate, concentration, delivery method and the nature of the environmental matrix studied. (Adapted from Maxfield *et al.*, 2012.)

from methanotrophy. The high concentration of C retained is comparable with levels observed in lakes that have high levels of benthic methanogenesis. SIS opens the way for a new generation of SIP studies aimed at elucidating total C dynamics (incorporation, turnover and decay) at the molecular level in a wide range of complex environmental and biological matrices.

Amino Acid Stable Isotope Probing (AA SIP)

As demonstrated in the SIS experiment described above and in the work summarised in Table 7.1, small-molecule SIP using GC-C-IRMS has been extended beyond PLFAs to a wider range of biomarkers/analytes (e.g. sugars and AAs). As N-containing molecules, AAs also raise the possibility for

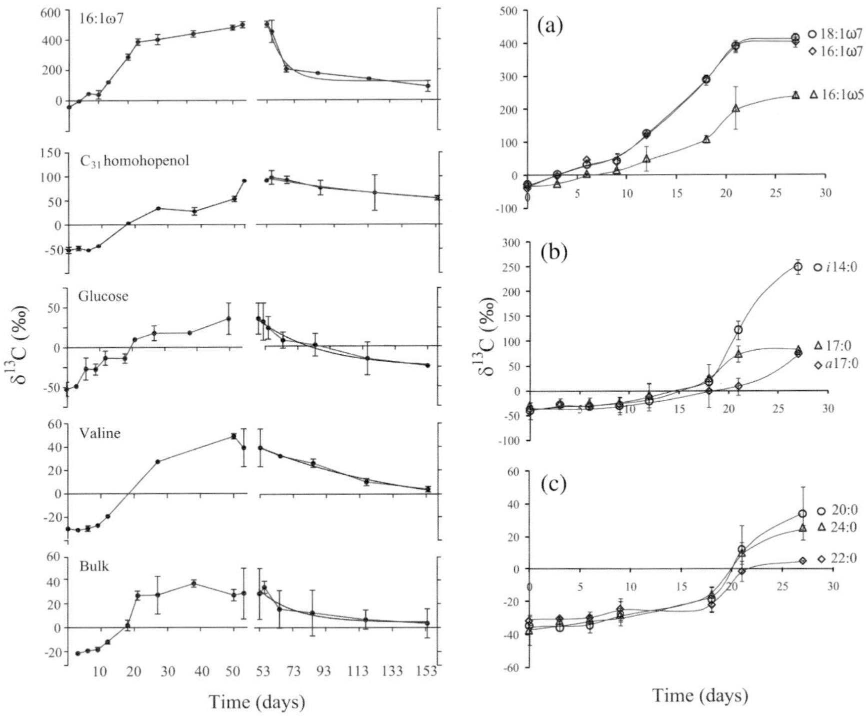

Figure 7.4 Graphs (LHS) showing the differential rates and extents of uptake of the ^{13}C-label into the bulk soil C pool and different biochemical components of the microbial community of a landfill cap soil after 50 days of incubation with only 1 at.% $^{13}CH_4$ (1% CH_4 balance synthetic air) in a flow-through chamber. The different time points derive from the analysis of replicate soils sampled at intervals throughout the experiment. The part of the graphs from 53 days onwards shows the decay of the label in the different pools after switching to $^{12}CH_4$ on day 50. The data obtained clearly illustrate the new levels of C pool and kinetic data available through our new approach. The graphs on the RHS show ^{13}C-label incorporation by selected PLFAs: (a) primary consumers (methanotrophs), hence rapid uptake of ^{13}C-label, (b) secondary consumers and (c) tertiary consumers, and hence, increasingly slow uptake of ^{13}C-label into the latter PLFAs. Note the different *y*-axis scales for plots (a)–(c). Error bars represent 1 standard deviation from the mean ($n = 3$).

^{15}N-SIP incubation experiments. We initially used combined ^{13}C and ^{15}N-SIP of AAs to trace the fates of N and C within metabolites central to soil organic N cycling (Knowles *et al.*, 2010) and then, during work focusing on the transformations and fate of fertiliser-N compounds in soils, we further developed our conceptual use of the approach to provide a powerful tool to assess microbial assimilation and newly synthesised soil protein (Charteris *et al.*, 2016).

Knowles *et al.* (2010) traced the fate of ^{13}C and ^{15}N from dual-labelled U-^{13}C,^{15}N-glutamate and U-^{13}C,^{15}N-glycine into *de novo* biosynthesised AAs in laboratory time-course experiments. Glutamate was selected as a substrate because of its centrality to cellular AA biochemistry, while glycine was chosen because of its more peripheral position in AA biosynthetic pathways. The AAs (2 μg) were applied to a grassland soil in laboratory mesocosms (10 g) and the fate of the ^{13}C and ^{15}N followed by sampling at $t = 3$, 6, 12, 24, 48, 96, 192, 384 and 768 h. In a preliminary experiment, sampling was began after 1 day, but it was clear from the analyses that the initial phase of processing had been missed, so the experiment was repeated with an earlier sampling time being included to capture the full trajectory of assimilation. Hydrolysable soil AAs were extracted, derivatised and their $\delta^{13}C$ and $\delta^{15}N$ values determined using GC-C-IRMS as described in the Experimental Methods section at the end of the chapter.

^{15}N- and ^{13}C-stable isotopic analyses of soil AAs demonstrated that both ^{15}N and ^{13}C from glutamate and glycine were incorporated into other, endogenous AAs through *de novo* biosynthesis using N and C from these labelled substrates. The total proportions of ^{15}N and ^{13}C used in the *de novo* biosynthesis of endogenous AAs and remaining within glutamate and glycine were compared. As anticipated, it was found that glutamate and glycine functioned differently as biochemical substrates in the biosynthesis of new AAs, with clear differences existing in the degree of N and C utilisation. The higher peak values of 30% glutamate N and 22% glutamate C being used in the synthesis of other AAs vs. 8% glycine N and 2.8% glycine C reflects their different positions in AA biosynthetic pathways. The total amount of label incorporated into other AAs began to decrease after $t = 8$ days probably as a result of N and C mineralisation and their use in the biosynthesis of new cellular biochemicals. By applying linear and nonlinear regressions, several important parameters were derived, namely rate constants, magnitudes of fluxes and measures of biosynthetic proximity, which describe the rate and magnitude of N and C flux through primary metabolic processes. The significant differences in N and C processing demonstrate a decoupling of the N and C cycles at the molecular level (Figure 7.5), that is, after 32 days the magnitude of N flux into newly biosynthesised AAs was double that of C from both substrates. We anticipate that the parameters derived will have potential for use in developing detailed models of soil organic N and C processing at the molecular level (Knowles *et al.*, 2010).

As noted above, application of this approach to investigations of the transformations and fate of fertiliser-N compounds in grassland soils prompted its development into a more widely applicable conceptual technique to trace the fate of applied N in soils and indeed, other complex ecosystems (Charteris *et al.*, 2016).

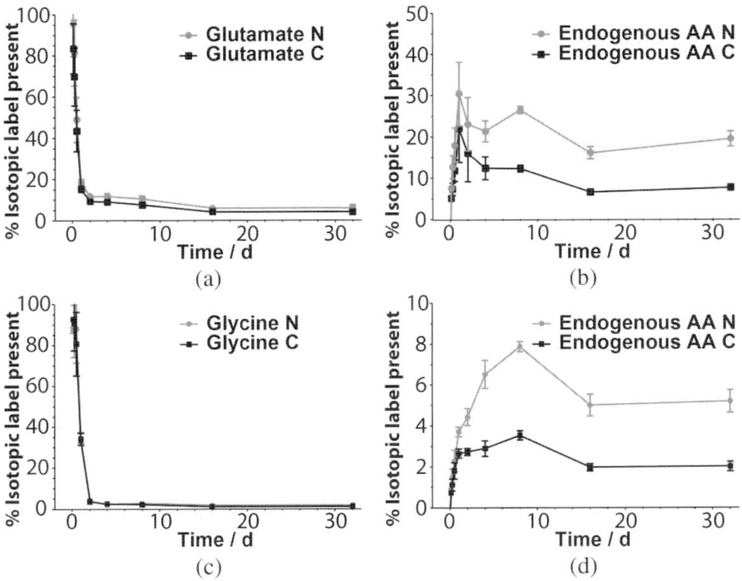

Figure 7.5 Variation (%) substrate label within the applied AA and endogenous soil AAs over time. Panels show the % of glutamate-derived N and C over time within (a) soil glutamate and (b) endogenous soil AAs in U-^{13}C,^{15}N-glutamate-amended soils; and the % of glycine-derived N and C over time within (c) soil glycine and (d) endogenous soil AAs in U-^{13}C,^{15}N-glycine-amended soils. Note different scales on *y*-axis; error bars represent ±SEM ($n = 3$).

Compound-specific AA ^{15}N-SIP using GC-C-IRMS offers a number of insights, inaccessible via existing techniques, providing:

(i) A sensitive and relatively selective means of assessing microbial assimilation of ^{15}N-labelled substrates/amendments is applied at environmentally relevant concentrations and appropriately low ^{15}N enrichments to minimise perturbations and ^{15}N discrimination/isotopic fractionation, respectively (bulk N isotope analysis cannot provide such insights and GC/MS ^{15}N-SIP studies require highly ^{15}N-labelled substrates for sufficient product AA ^{15}N enrichment, making such studies more susceptible to ^{15}N discrimination/isotopic fractionation effects and are much more expensive).

(ii) Valuable insights into microbial biochemical assimilation pathways can be gained and differences are readily revealed in the microbial processing of N-containing substrates of differing chemical/biochemical natures.

(iii) Detailed quantitative insights can be gained into the dynamics of N cycling from an applied substrate through the soil protein pool.

(iv) Estimates are provided for newly synthesised soil protein, which are inaccessible based on currently available methods.

Our compound-specific AA ^{15}N-SIP method is particularly valuable in determining the fate of N-fertilisers in relation to the soil organic N (mainly biomass) pool as organic N greatly exceeds inorganic N in soils and few other techniques are able to provide direct quantitative estimates of this partitioning (perhaps with the exception of microbial biomass N extraction; Charteris *et al.*, 2016). Accordingly, laboratory time-course incubation experiments were undertaken with the common inorganic N fertiliser compounds $^{15}NH_4^+$ and $^{15}NO_3^-$ and the high sensitivity and selectivity GC-C-IRMS was again exploited to determine the assimilation of the applied ^{15}N-labelled substrate by the soil microbial biomass and provide estimates of newly synthesised soil protein (see protocols at the end of the chapter). As with the glutamate and glycine incubations summarised above, the assimilation pathways of $^{15}NH_4^+$ and $^{15}NO_3^-$ were revealed via patterns in AA $\delta^{15}N$ values with time, reflecting known biosynthetic pathways (e.g. ammonium uptake occurs first via glutamate). Comparing these data to those for organic substrates such as glutamate, clear differences can be seen in the rates and fluxes of the applied N substrates into the soil protein pool (glutamate > ammonium > glycine > nitrate). This work constitutes a significant advance as the cycling of agriculturally relevant N additions is undetectable via bulk soil N content and $\delta^{15}N$ measurements and AA concentrations (Figure 7.6).

Although it does not include a time-course, the work of Redmile-Gordon *et al.* (2015) provides an example of how this conceptual compound-specific ^{15}N SIP approach might be extended or refined by considering different soil protein fractions. Finally, the method could potentially be adapted to investigate N cycling into other N-containing biochemical pools, for example, amino sugars, as appropriate extraction, derivatisation and GC-C-IRMS protocols are developed.

Looking to the Future

Dynamic-SIP is a proven approach in soil research with applications involving a wide range of cellular biochemical products. The survey of recently published examples show that there has been a focus on DNA/RNA- and PLFA-SIP, mainly as taxonomic tools, to trace the fate of the ^{13}C-label from applied substrates in the soil microbial biomass. In some cases, the magnitude of ^{13}C-labelling is used as an indication of different taxonomic groups receiving the label and their relative importance, both spatially and temporally (e.g. within the rhizosphere). From our preliminary work with the SIS technique, however, it is clear that there

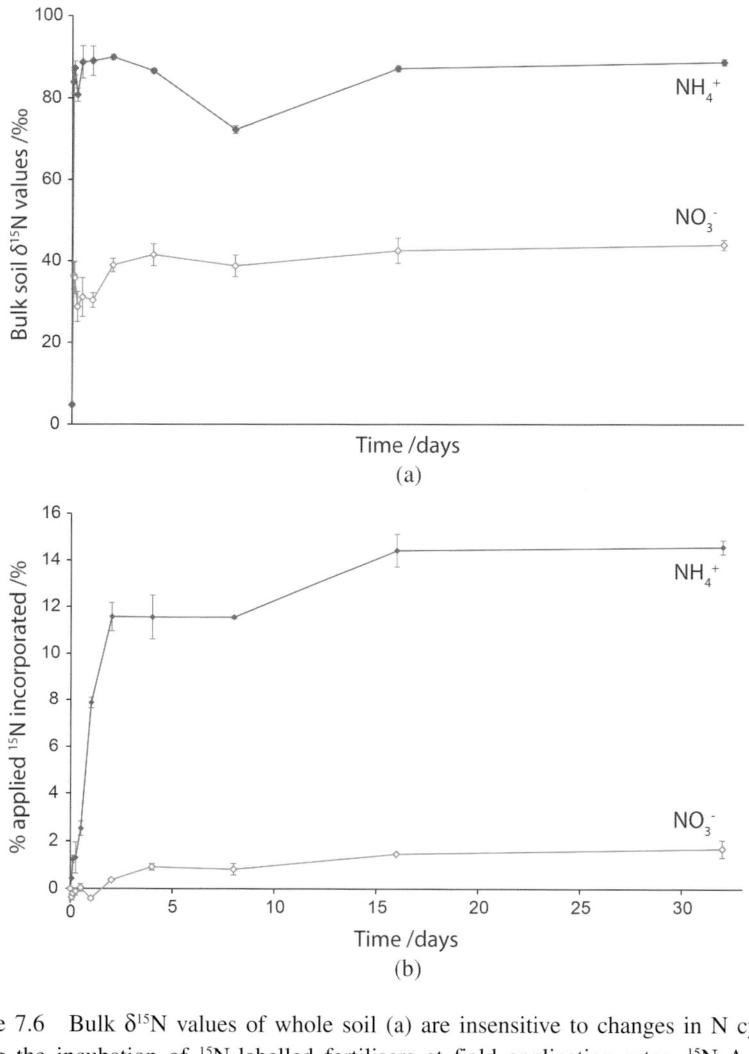

Figure 7.6 Bulk $\delta^{15}N$ values of whole soil (a) are insensitive to changes in N cycling during the incubation of ^{15}N-labelled fertilisers at field application rates. ^{15}N AA SIP allows for the percentage of applied $^{15}NH_4^+$ and $^{15}NO_3^-$ incorporated (b) into the total hydrolysable AA pool or soil protein pool to be determined. Error bars are ±SE ($n = 3$). Calculations for $^{15}NH_4^+$ and $^{15}NO_3^-$ are straightforward summations of the percentage of the applied ^{15}N incorporated into each AA, indicating a considerable difference in the use of the applied ^{15}N in *de novo* AA biosynthesis.

are additional benefits that accrue, when considering this methodology and its variants, beyond its use just as a taxonomic technique. We showed in our SIS studies that the stable isotope label from applied substrates can be traced into the full range of cellular biochemicals produced by soil biota. This opens up the SIP technique as a highly selective and sensitive technique for investigating the cycling of a wide range of soil natural substrates as well as agricultural inputs and amendments. Where dynamic-SIP has been used to determine the rates of uptake and turnovers of different biochemical pools, it is clear that these can vary widely even within a compound class, for example, AAs, as a result of the specific demands of different organisms and the associated biochemical pathways. Such information has the potential to feed into a new generation of molecular-based SOM cycling models.

The emergence of SIP at the turn of the millennium was, in large part, driven by a combination of technological advances, notably the commercial availability of GC-C-IRMS, which enabled compound-specific stable C isotope analyses to be performed on soil-derived PLFAs. As we have shown, a wide range of other cellular biochemicals can be made amenable to GC-C-IRMS using the appropriate protocols, for example, hydrolysis and derivatisation for biopolymers, such as proteins and carbohydrates. There is also scope for using high-temperature GC-C-IRMS, employing short columns with thin stationary phase coatings to determine stable isotope compositions of high-molecular-weight lipids. However, new opportunities also exist to investigate the isotopic composition of polar and high-molecular-weight compounds using high-performance liquid chromatography-IRMS or direct infusion IRMS. This instrument uses a solution rather gas-phase oxidation to generate CO_2 from organic compounds for C isotope analysis and thereby offers as yet untapped potential for its incorporation into soil SIP studies. Such an instrument would have value in determining $\delta^{13}C$ values of genomic material isolated from density centrifugation gradients.

Additionally, numerous untapped opportunities exist to extend the applications of SIP with GC-C-IRMS to study other isotope systems, including $^{15}N/^{14}N$, $^{2}H/^{1}H$, $^{18}O/^{16}O$. We have shown, through our work above, the advantages of using compound-specific ^{15}N- and $^{15}N,^{13}C$-AA SIP to track the fate of nitrogenous substrates into newly biosynthesised biomass protein. The sensitivity and selectivity of GC-C-IRMS will enable a new generation of naturalistic experiments, to probe the functional behaviour of the soil microbial biomass, to be performed. Such studies have hitherto been impossible to conduct using conventional GC-MS or bulk isotope approaches.

Finally, we cannot overemphasise that *dynamic*-SIP should be the norm in experimental design in this area. SIP experiments performed without regular sampling over a proper time-course cannot reveal the true functional dynamics of the soil microbial biomass with respect to a given substrate. Without this information, we will never be able to properly predict the fate of substrates or biomass response to a given set of environmental variables. We accept that such experiments are considerably more demanding in terms of infrastructure resource and manpower, but not performing *dynamic*-SIP experiments will result in potentially vital aspects of soil biomass behaviour remaining undetected.

Experimental Methods

Dynamic $^{13}CH_4$-SIP

This methodology follows that of Maxfield *et al.* (2006). Soil from a depth of 5 cm was selected for the $^{13}CH_4$ incubation due to its high CH_4 oxidation rate (0.04 nM CH_4 g^{-1} dw h^{-1}). Soil samples for $^{13}CH_4$ incubation were weighed (10 g) into 5 cm diameter Petri dishes and placed into an incubation chamber (63 L; Perspex®) sealed at the base via immersion in a water-filled tray to prevent gas leakage. The water in the base of the chamber maintained atmospheric moisture saturation and prevented samples from drying. Double-distilled water (DDW) was added periodically (ca. 2-week intervals) to the soils during incubation to maintain moisture levels. $^{13}CH_4$ (2 ppmv) pre-mixed in synthetic air was passed through the chamber at a rate of 44 mL min^{-1} in order to renew the headspace every 24 h. Each soil sample was incubated in triplicate with incubation times ranging from 2 to 11 weeks. A second, identical set of "control" soils were incubated in a separate building under ambient CH_4 (2 ppmv) levels. All incubations were carried out in temperature-controlled rooms at 20°C. Throughout the experiment, selected soil samples were removed for isotope analysis, and headspace was monitored by GC analysis to verify that CH_4 oxidation activity was being maintained.

SIS CH_4 incubation

SIS is the method developed by Maxfield *et al.* (2012). Soils containing low affinity methanotrophs were incubated in the same flow-through incubation chamber described above. Synthetic air premixed with CH_4 (1.33%) was constantly flowed through the chamber to maintain a steady CH_4 mixing ratio close to that observed at the sampling site. For the initial 50 days of the incubation, 1% of the CH_4 was

$^{13}CH_4$ (mixed from >99% ^{13}C; CK Gas Products Ltd., Hook, UK; equivalent to 133 ppmv $^{13}CH_4$). Following 50 days of the incubation, the $^{13}CH_4$ was switched and pre-mixed natural abundance CH_4 (13,300 ppmv) was flowed through the chamber. The flow-rate through the chamber was maintained at 44 mL min^{-1} throughout the entire incubation period in order to flush the entire headspace every 24 h. Soil samples were removed, in triplicate, at regular intervals throughout the incubation (t = 0, 3, 6, 9, 12, 18, 21, 27, 38, 50, 53, 56, 65, 85, 117, 154 days) and initial soil moisture contents were maintained through the regular addition of DDW to each sample (determined gravimetrically). All of the soil samples were stored at -20°C until required for analysis. Soil samples were freeze-dried and ground prior to extraction and analysed for PLFAs, total AAs, total monosaccharides, glycolipids (GLFAs), free fatty acids (FFAs), hopanoids, sterols, *n*-alkanes and *n*-alkanols, as described below.

^{15}N- and dual ^{13}C,^{15}N-AA SIP

The soil was air-dried sufficiently to allow sieving (2 mm) without smearing it, which would result in the destruction of the microstructure and adjusted to 50% water holding capacity by addition of DDW through an atomizer. Incubation experiments were carried out in glass incubation tubes (10 cm × 2 cm i.d.), to which 10.0 g soil (6.06 g dry wt. equivalent) were added and gently compacted to a density of ca. 1.1 g cm^{-3} (field density) by tapping the tubes on a solid surface. Incubations were carried out under aerobic conditions and soils were maintained at 50% water holding capacity. A study by Jones and Shannon (1999) showed that sieving soil reduced the AA mineralisation rate for the first 24 h, but subsequent incubation (up to 40 days) revealed no difference in the AA mineralisation rate. To ensure this period was exceeded, a pre-incubation period of 4 days was employed before soils received either ^{15}N-labelled ammonium chloride ($^{15}NH_4Cl$, 10 at.%, 400 μg in 200 μL DDW; Sigma-Aldrich, St. Louis, USA), ^{15}N-labelled potassium nitrate ($K^{15}NO_3$, 10 at.%, 400 μg in 200 μL DDW; Sigma-Aldrich, St. Louis, USA), ^{15}N,^{13}C-labelled glutamic acid and glycine (^{15}N,^{13}C-Glu, 98 at.%, 2 mg in 200 μL 0.1 M hydrochloric acid; HCl; ^{15}N,^{13}C-Glu from Spectra Stable Isotopes, Columbia, USA and HCl was reagent grade from Fisher Scientific, Loughborough, UK) or for the control samples, DDW (200 μL). Substrates were introduced by injection, and the needle was drawn up through the soil as the plunger was depressed in order to achieve a good distribution. The incubation experiments were halted by immersion in liquid nitrogen (N_2) to instantly cease all biological activity after periods of 1.5, 3, 6 and 12 h and 1, 2, 4, 8, 16 and 32 days in the dark and stored at -20°C until freeze-drying. All incubations were carried out in triplicate, so there were three tubes for each time point of each treatment (see also Knowles *et al.*, 2010). The 10 at.% ^{15}N enrichment of $^{15}NH_4^+$ and $^{15}NO_3^-$ was chosen based on research highlighting changes in ^{15}N discrimination and isotopic fractionation in biological mechanisms at very high enrichments (Mathieu *et al.*, 2007; Tang and Maggi, 2012). Nitrogen-15 enrichments of 10 at.% were considered low enough for these effects to be negligible. The ^{15}N-Glu incubation experiments were carried out earlier, hence the undesirably high ^{15}N enrichment of the applied Glu.

PLFA analysis

Following incubation, all soil samples were freeze-dried, homogenized by grinding and extracted using a modified Bligh-Dyer monophasic solvent system (Maxfield *et al.*, 2006). Briefly, soil (2 g, dw) was extracted using Bligh-Dyer solvent containing buffered water (0.05 M KH_2PO_4; pH 7.2)/chloroform/methanol in a ratio of 4:5:10 (v/v/v). After addition of solvent (3 mL), the mixture was sonicated (15 min) and centrifuged (3200 rpm, 5 min). The supernatant was decanted and the residue extracted as above (×3). The extracts were combined and evaporated under N_2 to yield the total lipid extract (TLE). The TLE was separated into three fractions based on polarity using a 3-mL STRATA aminopropyl-bonded silica cartridge (Phenomenex). Fractions were eluted sequentially with 2:1 (v/v) dichloromethane/isopropanol (8 mL; yielding neutrals), 2% acetic acid in diethyl ether (12 mL; FFAs) and methanol (8 mL; polar fraction including phospholipids). The PLFA components of the polar fraction were released by saponification (1 mL 0.5 M methanolic NaOH, 1 h, 80°C) and an internal standard (I.S.) added (10 μL of 0.1 mg mL^{-1} solution of C_{19} *n*-alkane in hexane). The PLFAs were then methylated using a 14% (v/v) BF_3-methanol solution (100 μL; Aldrich, UK, 1 h, 80°C) and extracted with chloroform (3 × 2 mL). The phospholipid ester-linked fatty acid methyl esters were dissolved in hexane for analysis by GC, GC-MS and GC-C-IRMS (see below).

Dimethyl disulphide (DMDS) derivatives of mono-unsaturated fatty acids were prepared to facilitate determination of double-bond positions and geometries in monounsaturated fatty acids. A 100-μL aliquot of the methyl esters in dichloromethane (DCM) was added to 0.25 M I_2/diethyl ether (100 μL) and DMDS (1 mL) and then heated (60°C, 24 h in the dark). Excess I_2 was removed by addition of aqueous $Na_2S_2O_3$ (5%, 2 mL) and the DMDS derivatives extracted with hexane (2 × 2 mL) for GC-MS analysis (see below and Maxfield and Evershed, 2011, for mass spectral interpretations).

Lipid extraction and fractionation

Lipid class separation was achieved using a modified silicic acid fractionation protocol (Dickson *et al.*, 2009) to yield three fractions: simple lipids, glycolipids and phospholipids. Simple lipids were further fractioned using silicic acid columns into four fractions: hydrocarbons (eluted with hexane, 4 mL), ketones and wax esters (DCM, 6 mL), alcohols (DCM/methanol (MeOH), 1:1, v/v, 4 mL) and polar poly-functionalised compounds (MeOH, 4 mL). Alcohols and hydrocarbons were subjected to a further urea adduction procedure to separate cyclic from acyclic components (Nwadinigwe and Nwobodo, 1994).

Trimethylsilylation

The PLFA, GLFA and FFA fractions were methylated as above. As with standard GC analysis, all neutral polar lipids were derivatised with *N*,*O*-bis(trimethylsilyl) trifluoroacetamide (BSTFA) + 1% trimethylchlorosilane (TMCS) before analysis by GC, GC-MS and GC-C-IRMS (Evershed, 1994).

Extraction and derivatisation of carbohydrates

The method of Blakeney *et al.* (1983), modified by Docherty *et al.* (2001), was employed to prepare alditol acetate derivatives of total monosaccharides. Soils were hydrolysed with H_2SO_4 and the released monosaccharides reduced with $NaBH_4$ to their corresponding alditols; excess $NaBH_4$ was removed by the addition of glacial acetic acid. The alditols were acetylated by reaction with *N*-methylimidazole and acetic anhydride. A standard mix of rhamnose, fucose, galactose, mannose, xylose, arabinose glucose, inositol and pentaerythritol was prepared according to Docherty *et al.* (2001). DDW (400 μL) was added to the dry monosaccharide standards, and the solutions derivatised as above. A 20 $\mu g\ \mu L^{-1}$ I.S. was made up by dissolving 0.04 g pentaerythritol in 2 mL 1 M NH_3 solution. Corrections for added derivative C and propagation of errors followed Docherty *et al.* (2001).

Extraction, isolation and derivatisation of hydrolysable amino acids

Finely ground, freeze-dried incubation soil samples (100 mg) were weighed into culture tubes and 100 μL of norleucine (Nle; 400 $\mu g\ mL^{-1}$ in 0.1 M HCl; Sigma-Aldrich, St. Louis, USA) was added as an I.S. Hydrolysis with 5 mL 6 M HCl was carried out at 100°C for 24 h under an atmosphere of N_2. Acid hydrolysis extracts both free and proteinaceous AAs as well as catalysing the breakdown of living microbial biomass (Roberts and Jones, 2008). The relatively harsh conditions are necessary for the cleavage of peptide bonds between hydrophobic residues (e.g. isoleucine; Ile, leucine; Leu and valine; Val), but also result in the deamination of asparagines (Asn) to aspartate (Asp) and glutamine (Gln) to Glu, and the complete destruction of cysteine (Cys) and tryptophan (Trp; Fountoulakis and Lahm, 1998; Roberts and Jones, 2008). The technique may also partially destroy serine (Ser; ca. 10% loss), threonine (Thr; ca. 5% loss) and tyrosine (Tyr; loss depends on level of trace impurities in hydrolysis agent; Roberts and Jones, 2008) and has the potential to hydrolyse AA chains from non-proteinaceous sources, such as peptidoglycan, resulting in an overestimation of some AAs, mostly alanine (Ala), Glu, lysine (Lys) and glycine (Gly; Roberts and Jones, 2008). The technique is, however, considered the most reliable method for determining the total protein content of soils (Roberts and Jones, 2008) and as such, we equate total hydrolysable AA concentrations to the size of the soil protein pool. The hydrolysis is performed under N_2 as the presence of oxygen (O_2) can induce the thermal breakdown of hydroxyl- and sulphur-containing AAs (e.g. methionine; Met, Ser, Thr and Tyr; Roberts and Jones, 2008). Hydrolysates were collected by centrifugation, dried under a stream of N_2 at 60°C and stored at 20°C under 1 mL 0.1 M HCl. AAs were isolated from hydrolysates by cation exchange column chromatography using acidified Dowex 50WX8 200–400 mesh ion exchange resin (Acros Organics, New Jersey, USA; Metges and Petzke, 1997). This was followed by conversion to their *N*-acetyl, *O*-isopropyl derivatives for analysis (Corr *et al.*, 2007; Knowles *et al.*, 2010). Derivatising agents were supplied by Sigma-Aldrich (Steinheim, Germany): acetyl chloride was puriss. p.a. grade; trimethylamine had ≥99.50% purity; and acetic anhydride was ReagentPlus® grade. All solvents were of HPLC grade and were supplied by Rathburn Chemicals Ltd. (Walkerburn, Scotland). DDW was produced using a Bibby Aquatron still. Where not applied to living soil, DDW was extracted with DCM prior to use in order to remove dissolved organic compounds.

Instrumental analyses

GC analyses were performed using a Hewlett-Packard Series 5890 Series II gas chromatograph (Agilent Technologies UK Ltd., Edinburgh, UK) with a flame ionisation detector and H_2 carrier gas (10 psi). Non-polar fractions were analysed using a Chrompack CPSil-5CB (50 m × 0.32 mm i.d. × 0.12 μm film thickness). The temperature programme was 50–200°C at 10°C min^{-1}, to 300°C at 3°C min^{-1} (held for 20 min). Polar compounds were analysed using a Varian VF23ms (Varian BV,

Middelburg, The Netherlands) 50% cyanopropyl equivalent fused-silica column (60 m × 0.32 mm i.d. × 0.25 μm film thickness). The temperature programme for fatty acid derivatives was 50°C (2 min) to 100°C at 15°C min^{-1}, to 240°C at 4°C min^{-1} (held for 20 min). The temperature programme for monosaccharides was 50°C (1 min) to 200°C at 20°C min^{-1}, to 230°C at 4°C min^{-1} (held for 22 min). The temperature programme for amino acids was 40°C (1 min) to 120°C at 15°C min^{-1}, to 190°C at 3°C min^{-1}, to 250°C at 5°C min^{-1} (held for 20 min).

GC-MS analyses were performed using a Thermo Finnigan Trace GC-MS (Thermo Fisher Scientific, Hemel Hempstead, UK). All the GC conditions were the same with the exception of helium being used as carrier gas. The interface was held at the maximum oven temperature, the ion source was held at 200°C and the quadrupole mass analyser operated in EI mode, scanning over the range *m/z* 50–650 at 1.7 scans s^{-1}. The emission current was maintained at 300 μA and electron energy was 70 eV. The data were acquired and analysed using the XCalibur software Version 1.2 (ThermoFisher Scientific, Hemel Hempstead, UK).

GC-C-IRMS isotope analyses of the various compound classes were carried out using a variety of instruments produced by ThermoFisher.

For C isotope analyses use was made of an instrument incorporating a Varian 3500 GC (Varian BV, Middelburg, The Netherlands) coupled to a Finnigan MAT DELTA S isotope ratio mass spectrometer (Finnigan MAT, Hemel Hempstead, UK). Analytes in He were combusted to CO_2 using a modified Type I Finnigan MAT combustion interface with a CuO/Pt combustion reactor set to a temperature of 850°C prior to entry into the mass spectrometer source via an open split. The ionisation source electron energy was 100 eV with an electron current of 1 mA. Detection was via three Faraday cup collectors set at *m/z* 44, 45 and 46. The chromatographic conditions were the same as those described previously. All sample analyses were conducted in duplicate to ensure reliable $\delta^{13}C$ values. Samples were calibrated against reference CO_2 of known isotopic composition which was introduced directly into the source three times at the beginning and end of every run. Compound-specific IRMS performance was determined using a suite of externally calibrated reference fatty acid methyl esters. Samples were analysed with a precision <0.5‰ (±1 standard deviation, $n = 5$).

The $\delta^{15}N$ values of individual AAs were determined as their *N*-acetyl, *O*-isopropyl derivatives and were determined using a ThermoFinnigan Trace 2000 gas chromatograph coupled with a ThermoFinnigan DeltaPlus XP isotope ratio mass spectrometer via a ThermoFinnigan Combustion III Interface (Thermo Electron Corporation, Waltham, USA). Samples were introduced using a CTC Analytics GC Pal autosampler (Zwingen, Switzerland) and via a programmable temperature vaporisation inlet (Thermo Electron Corporation, Waltham, USA). The carrier gas was helium (He; The BOC Group plc., Guildford, UK) at a flow rate of 1.4 mL min^{-1} and the gas chromatograph was fitted with a DB-35 column (30 m × 0.32 mm i.d. × 0.5 μm stationary phase thickness; Agilent Technologies, Santa Clara, USA). The temperature programme utilised was: 40°C (5 min) to 120°C at 15°C min^{-1}, to 180°C at 3°C min^{-1}, then to 210°C at 1.5°C min^{-1} and finally to 270°C at 5°C min^{-1}. The oxidation reactor comprised Cu, Ni and Pt wires (high purity from OEA Laboratories Ltd., Callington, UK) maintained at 980°C; the reduction reactor comprised Cu wires maintained at 650°C. AA $\delta^{15}N$ values were determined relative to that of a monitoring gas of known (previously determined using in-house AA standards) N-isotopic composition introduced directly into the ion source via an open split in four pulses at the beginning and end of each run. The $\delta^{15}N$ values of the in-house AA standards were determined off-line by elemental analysis-isotope ratio mass spectrometry (EA-IRMS) by ThermoFisher Scientific (Bremen, Germany) and by NERC CEH (Merlewood/ Lancaster, UK) using primary reference materials (NIST 8547 IAEA-N-1 ammonium sulphate; $\delta^{15}N$ +0.4‰). In order to adhere to the identical treatment principle and ensure the GC-C-IRMS system was functioning properly, each sample was bracketed by the in-house AA standard mixture of known $\delta^{15}N$ values and sample AA $\delta^{15}N$ values accepted only when at least 75% of the AAs in the standard mixture run either side of the sample were within ±1‰ and the others were within ±1.5‰, and when this was also true on average over the course of the run. Data were acquired and analysed using Isodat NT 3.0 (ThermoFinnigan, Thermo Electron Corporation, Waltham, USA).

Samples in which the substrate AAs (or others) were significantly more enriched in ^{15}N or ^{13}C than endogenous AAs were analysed twice: once where the more enriched peaks were vented before entering the oxidation and reduction reactors (Figure 2b in Knowles *et al.*, 2010), in order to avoid possible reactor memory effects that may affect the δ values of subsequent eluting peaks, and once where all compounds were allowed to pass through the reactors and enter the mass spectrometer in order to determine δ values for the more enriched compounds (Figure 2a in Knowles *et al.*, 2010).

Calculations for ^{15}N-SIP of AA

If the total hydrolysable AA pool is taken to be representative of the soil protein pool, then any ^{15}N enrichment (E) in hydrolysable AAs can be summed to represent newly synthesised soil protein in the soil at that time:

$$\text{Newly synthesised soil protein} \approx \sum E \text{ in hydrolysable AAs} \qquad (7.1)$$

The E of an AA may be expressed as the number of moles of ^{15}N derived from the applied substrate that are present in that AA in the soil:

$$E = n_N \times \text{AFE}, \tag{7.2}$$

where n_N is the number of moles of N in the AA (i.e. if the molecular structure of the AA contains only one N atom, n_N is the same as the number of moles of the AA in the soil, but twice this if the AA structure consists of two N atoms and so on) and AFE is the atom fraction excess of the AA after incubation compared with the control:

$$\text{AFE} = \text{AF}_{\text{Sample}} - \text{AF}_{\text{Control}}. \tag{7.3}$$

AF is the atom fraction of ^{15}N in the AA, that is,

$$\frac{\text{Number }^{15}\text{N atoms}}{(^{14}\text{N} + {}^{15}\text{N})\text{ atoms}}. \tag{7.4}$$

This can be calculated from the AA's $\delta^{15}N$ value as in Knowles *et al.* (2010):

$$\text{AF} = \frac{R_{\text{Std}}(\delta^{15}\text{N}/1000+1)}{1+(\delta^{15}\text{N}/1000+1)}, \tag{7.5}$$

where R_{Std} is the $^{15}N/^{14}N$ ratio of AIR, the international isotopic standard for N. The data may also be expressed in terms of the percentage of the applied ^{15}N incorporated into each AA, as in Knowles *et al.* (2010):

$$\%\ \text{Incorporation} = \left(\frac{E}{N}\right) \times 100, \tag{7.6}$$

where N is the number of moles of excess ^{15}N applied (above natural abundance). Percentage incorporations reflect both the concentration and ^{15}N enrichment ($\delta^{15}N$ value) of the AA (i.e. how much was incorporated if the AA at x concentration was ^{15}N-enriched by x‰) and the percentage of applied ^{15}N incorporated into newly synthesised soil protein is determined by summing these results for individual AAs. Note that these percentage incorporation data will be affected by the conservation of applied ^{15}N in the system, thus if ^{15}N is lost from the system (e.g. over time), AA percentage incorporations may become skewed as there is less ^{15}N available for incorporation than expected. The incubation experimental design described aims to limit any ^{15}N losses from the system in order to obviate this issue, but it is equally possible to calculate percentage incorporations at time, t, based on the moles of applied ^{15}N retained (N_R; above natural abundance/control soil values) in the system at time, t (Eq. (7.7)), if bulk recovery of the applied ^{15}N is low or decreases with time.

$$N_R = \text{AFE}\left(\frac{\%\ \text{TN}}{1400}\right), \tag{7.7}$$

where AFE is the atom fraction excess of ^{15}N in the bulk soil (calculated from bulk and control soil $\delta^{15}N$ values using Eqs. (7.3) and (7.5)) and % TN is the percentage total N content of the soil. Percentage retention of applied ^{15}N was calculated as follows:

$$\%\ \text{Retention}\ ^{15}\text{N} = \left(\frac{N_R}{N}\right) \times 100 \tag{7.8}$$

It can also be argued, however, that loss from the system is just another process competing against AA biosynthesis for N, so should not be discounted in this way. All of these approaches are valid and the most appropriate one will depend on the specifics of the experimental design and desired outcomes.

Statistical analyses for SIS

Two approaches were taken to analyse the ^{13}C-label incorporation curves. Linear and nonlinear regressions were performed using GraphPad Prism version 5.02 for Windows (GraphPad Software, San Diego, CA, USA), R 2.8.1 (R foundation) and Microsoft Excel (Microsoft, Redmond, WA, USA). Linear regression was applied to %-incorporation data from the first 9 days of the incubation to obtain initial zero-order rate constants, whereas separate nonlinear regressions were applied to incorporation data from the first 50 days of the incubation before the switchover to unlabelled CH_4, and the C turnover data from 50 to 150 days. The equations of the fitted lines were of the form:

Incorporation: $$F = F^0 + (P - Y^0) * (1 - \exp(-K * t)). \tag{7.9}$$

Decay: $$F = (F^0 - P) * \exp(-K * t) + P, \tag{7.10}$$

where F is the fraction of excess ^{13}C incorporated into organic material, t is the incubation time in days, F^0 is the initial F value at initial t, P is the plateau at maximum ^{13}C incorporation (incorporation) or maximum ^{13}C loss (decay) and K is the first-order rate constant expressed in inverse days.

Isotopic corrections for added derivative C

All $\delta^{13}C$ values were corrected for derivatisation using a mass balance equation (7.11). The $\delta^{13}C$ value of each batch of BF_3-MeOH was determined by bulk IRMS.

$$n_{cd}\delta^{13}C_{cd} = n_c\delta^{13}C_c + n_d\delta^{13}C_d. \tag{7.11}$$

The fractional abundance (F) of ^{13}C in the control (F_c) and enriched (F_e) PLFAs was used to calculate the concentration of ^{13}C incorporated into PLFAs from the total PLFA concentration. The fractional abundance expresses the amount of ^{13}C as a proportion of the total amount of carbon in the PLFA (Eq. (7.12)).

$$\text{Incorporation} = (F_e - F_c)[\text{PLFA}]_e$$

$$F = \frac{^{13}C}{^{13}C + {}^{12}C}. \tag{7.12}$$

Statistical analysis

Bacterial populations were compared by cluster analysis using SYSTAT Version 7.0. The distribution of PLFAs was used as a measure of similarity or distance. The analysis was performed on Euclidean distances between standardised data using averages.

Acknowledgements

None of the work presented in this contribution would have been possible without the continued support of the UK Natural Environment Research Council: PhD studentships (P.M. and A.F.C.), a research grant (R.P.E.) and partial funding of the mass spectrometry facilities (Contract No. R8/H10/63; (http://www.lsmsf.co.uk)). The UK Biochemistry and Biotechnology Research Council is also thanked for a PhD studentship (T.D.J.K.).

References

Adamsen, A. P. S. and King, G. M. (1993). Methane consumption in temperate and sub-Arctic forest soils — rates, vertical zonation, and responses to water and nitrogen. *Appl. Environ. Microbiol.* 59, 485–490.

Apostel, C., Dippold, M., Glaser, B. and Kuzyakov, Y. (2013). Biochemical pathways of amino acids in soil: assessment by position-specific labeling and ^{13}C-PLFA analysis. *Soil Biol. Biochem.* 67, 31–40.

Bahn, M., Lattanzi, F. A., Hasibeder, R., Wild, B., Koranda, M., Danese, V., *et al.* (2013). Responses of below-ground carbon allocation dynamics to extended shading in mountain grassland. *New Phytol.* 198, 116–126.

Bai, Z., Liang, C., Bode, S., Huygens, D. and Boeckx, P. (2016). Phospholipid ^{13}C stable probing during decomposition of wheat residues. *Appl. Soil Ecol.* 98, 65–74.

Balasooriya, W. K., Denef, K., Huygens, D. and Boeckx, P. (2014). Translocation and turnover of rhizodeposit carbon within soil microbial communities of an extensive grassland ecosystem. *Plant Soil* 376, 61–73.

Bartling, J., Kotzerke, A., Mai, M., Esperschütz, J., Buegger, F. and Schloter, M. (2009). Microbial community structure and function during abnormal curve development of substrate-induced respiration measurements. *Chemosphere* 77, 1488–1494.

Bergamaschi, P., Krol, M., Dentener, F., Vermeulen, A., Meinhardt, F. and Graul, R. (2005). Inverse modelling of national and European CH_4 emissions using the atmospheric zoom model TM5. *Atmos. Chem. Phys.* 5, 2431–2460.

Blakeney, A. B., Harris, P. J., Henry, R. J. and Stone, B. A. (1983). A simple and rapid preparation of alditol acetates for monosaccharide analysis. *Carbohydr. Res.* 13, 291–299.

Boeckx, P., Van Cleemput, O. and Meyer, T. (1998). The influence of land use and pesticides on methane oxidation in some Belgian soils. *Biol. Fert. Soils* 27, 293–298.

Boschker, H. T. S., Nold, S. C., Wellsbury, P., Bos, D., de Graaf, W., Pel, R., *et al.* (1998). Direct linking of microbial populations to specific biogeochemical processes by ^{13}C-labelling of biomarkers. *Nature* 392, 801–805.

Brady, N. C. and Weil, R. R. (2002). *The Nature and Properties of Soils*, 13th edn. Prentice-Hall, Upper Saddle River, NJ.

Brant, J. B., Sulzman, E. W. and Myrold, D. D. (2006). Microbial community utilization of added carbon substrates in response to long-term carbon input manipulation. *Soil Biol. Biochem.* 38, 2219–2232.

Bull, I. D., Parekh, N., Hall, G., Ineson, P. and Evershed, R. P. (2000). Detection and classification of atmospheric methane oxidizing bacteria in soil. *Nature* 405, 175–178.

Castro, M. S., Steudler, P. A. and Melillo, J. M. (1995). Factors controlling atmospheric methane consumption by temperate forest soils. *Global Biogeo. Cycl.* 9, 1–10.

Charteris, A. F., Knowles, T. D., Michaelides, J. K. and Evershed, R. P. (2016). Compound-specific amino acid ^{15}N stable isotope probing of nitrogen assimilation by the soil microbial biomass using gas chromatography/combustion/isotope ratio mass spectrometry. *Rapid Commun. Mass Spectrom.* 30(16), 1846–1856.

Churchland, C., Weatherall, A., Briones, M. J. I. and Grayston, S. J. (2012). Stable-isotope labeling and probing of recent photosynthates into respired CO_2, soil microbes and soil mesofauna using a xylem and phloem stem-injection technique on Sitka spruce (*Picea sitchensis*). *Rapid Commun. Mass Spectrom.* 26, 2493–2501.

Corr, L. T., Berstan, R. and Evershed, R. P. (2007). Optimisation of derivatisation procedures for the determination of $\delta^{13}C$ values of amino acids by gas chromatography/combustion/isotope ratio mass spectrometry. *Rapid Commun. Mass Spectrom.* 21, 3759–3771.

Creamer, C. C., de Menezes, A. B., Krull, E. S., Sanderman, J., Newton-Walters, R. and Farrell, M. (2015). Microbial community structure mediates response of soil C decomposition to litter addition and warming. *Soil Biol. Biochem.* 80, 175–188.

Crossman, Z. M., Abraham, F. and Evershed, R. P. (2004). Stable isotope pulse-chasing and compound specific stable carbon isotope analysis of phospholipid fatty acids to assess methane oxidizing bacterial populations in landfill cover soils. *Environ. Sci. Technol.* 38, 1359–1367.

Crossman, Z. M., Ineson, P. and Evershed, R. P. (2005). The use of ^{13}C labelling of bacterial lipids in the characterisation of ambient methane-oxidising bacteria in soils. *Org. Geochem.* 36, 769–778.

Crossman, Z. M., McNamara, N., Parekh, N., Ineson, P. and Evershed, R. P. (2001). A new method for identifying the origins of simple and complex hopanoids in sedimentary materials using stable isotope labelling with $^{13}CH_4$ and compound specific stable isotope analyses. *Org. Geochem.* 32, 359–364.

Crossman, Z. M., Wang, Z.-P., Ineson, P. and Evershed, R. P. (2006). Investigation of the effect of ammonium sulfate on populations of ambient methane oxidising bacteria by ^{13}C-labelling and GC/C/IRMS analysis of phospholipid fatty acids. *Soil Biol. Biochem.* 38, 983–990.

Darien, D., Marol, C. and Balesdent, J. (2004) The dynamics of neutral sugars in the rhizosphere of wheat: An approach by ^{13}C pulse-labelling and GC/C/IRMS. *Plant Soil.* 267, 243–253.

Dias, A. C. F., Dini-Andreote, F., Hannula, S. E., Andreote, F. D., Silva, M. S. P., Salles, J. F., *et al.* (2013). Different selective effects on rhizosphere bacteria exerted by genetically modified versus conventional potato lines. *PLoS ONE* 8, 1–12.

Dickson, L., Bull, I. D., Gates, P. J. and Evershed, R. P. (2009). A simple modification of a silicic acid lipid fractionation protocol to eliminate free fatty acids from glycolipid and phospholipid fractions. *J. Microbiol. Methods* 78, 249–254.

Docherty, G., Jones, V. and Evershed, R. P. (2001). Practical and theoretical considerations in the gas chromatography/combustion/isotope ratio mass spectrometry $\delta^{13}C$ analysis of small polyfunctional compounds. *Rapid Commun. Mass Spectrom.* 15, 730–738.

Drigo, B., Kowalchuk, G. A., Knapps, B. R., Pijl, A. S., Boschker, H. T. S. and van Veen, J. A. (2013). Impacts of 3 years of elevated atmospheric CO_2 on rhizosphere carbon flow and microbial community dynamics. *Global Change Biol.* 19, 621–636.

Dungait, J. A. J., Bol, R., Bull, I. D. and Evershed, R. P. (2009). Tracking the fate of dung-derived carbohydrates in a temperate grassland soil using compound-specific stable isotope analysis. *Org. Geochem.* 40, 1210–1218.

Dungait, J. A. J., Kemmitt, S. J., Michallon, L., Guob, S., Wen, Q., Brookes, P. C. and Evershed, R. P. (2011). Variable responses of the soil microbial biomass to trace concentrations of ^{13}C-labelled glucose, using ^{13}C-PLFA analysis. *Eur. J. Soil. Sci.* 62, 117–126.

Dungait, J. A. J., Stear, N. A., van Dongen, B. E., Bol, R. and Evershed, R. P. (2008). Off-line pyrolysis and compound-specific stable carbon isotope analysis of lignin moieties: a new method for determining the fate of lignin residues in soil. *Rapid Commun. Mass Spectrom.* 22, 1631–1639.

Evershed, R. P. (1994). Handbook of derivatives for chromatography. Blau, K. and McK. Halket, J. (eds.), *Advances in Silylation*, Chapter 2. Wiley, Chichester, pp. 52–107.

Evershed, R. P., Bull, I. D., Corr, L. T., Crossman, Z. M., van Dongen, B. E., Evans, C., *et al.* (2007). Stable isotopes in ecology and environmental science. Michener, R. and Latjha, K. (eds.), *Compound-specific Stable Isotope Analysis in Ecology and Palaeoecology*, Chapter 14. Blackwell Publishing, MA, USA, pp. 480–540.

Evershed, R. P., Crossman, Z. M., Bull, I. D., Mottram, H., Dungait, J. A. J., Maxfield, P. J. and Brennand, E. L. (2006). ^{13}C-Labelling of lipids to investigate microbial communities in the environment. *Curr. Opin. Biotechnol.* 17, 72–82.

Fountoulakis, M. and Lahm, H. W. (1998). Hydrolysis and amino acid composition analysis of proteins. *J. Chromatogr. A* 826, 109–134.

Frostegård, Å. and E. Bååth. (1996). The use of phospholipid fatty acid analysis to estimate bacterial and fungal biomass in soil. *Biol. Fert. Soils* 22, 59–65.

Gan, H., Zak, D. R. and Hunter, M. D. (2013). Chronic nitrogen deposition alters the structure and function of detrital food webs in a northern hardwood ecosystem. *Ecol. Appl.* 23, 1311–1321.

Gulledge, J., Doyle, A. P. and Schimel, J. P. (1997). Different NH_4^+-inhibition patterns of soil CH_4 consumption: a result of distinct CH_4-oxidizer populations across sites? *Soil Biol. Biochem.* 29, 13–21.

Gulledge, J. and Schimel, J. P. (1998). Low-concentration kinetics of atmospheric CH_4 oxidation in soil and mechanism of NH_4^+ inhibition. *Appl. Environ. Microbiol.* 64, 4291–4298.

Hannula, S. E., Boschker, H. T. S., de Boer, W. and van Veen, J. A. (2012). ^{13}C pulse-labeling assessment of the community structure of active fungi in the rhizosphere of a genetically starch-modified potato (*Solanum tuberosum*) cultivar and its parental isoline. *New Phytol.* 194, 784–799.

Helgason, B. L., Gregorich, E. G., Janzen, H. H., Ellert, B. H., Lorenz, N. and Dick, R. P. (2014). Long-term microbial retention of residue C is site-specific and depends on residue placement. *Soil Biol. Biochem.* 68, 231–240.

Herman, D. J., Furestone, M. K., Nuccio, E. and Hodge, A. (2012). Interactions between an arbuscular mycorrhizal fungus and a soil community mediating litter decomposition. *FEMS Microb. Ecol.* 80, 236–247.

Högberg, M. N., Briones, M. J. I., Keel, S. G., Metcalfe, D. B., Campbell, C., Midwood, A. J., *et al.* (2010). Quantification of effects of season and nitrogen supply on tree below-ground carbon transfer to ectomycorrhizal fungi and other soil organisms in a boreal pine forest. *New Phytol.* 187, 485–493.

Hütsch, B. W. (1998). Tillage and land use effects on methane oxidation rates and their vertical profiles in soil. *Biol. Fert. Soils* 27, 284–292.

Hütsch, B. W. (2001). Methane oxidation, nitrification, and counts of methanotrophic bacteria in soils from a long-term fertilization experiment ('Ewiger Roggenbau' at Halle). *J. Plant Nutr.* 164, 21–28.

Hütsch, B. W., Webster, C. P. and Powlson, D. S. (1993). Long-term effects of nitrogen fertilisation on methane oxidation in soil of the Broadbalk wheat experiment. *Soil Biol. Biochem.* 25, 1307–1315.

Jakobs-Schonwandt, D., Mathies, H., Abram, W.-R., Pritzkow, W., Stephan, I. and Noll, M. (2010). Biodegradation of a biocide (Cu-*N*-cyclohexyldiazenium dioxide) component of a wood preservative by a defined soil bacterial community. *Appl. Environ. Microbiol.* 76, 8076–8083.

Jin, V. L. and Evans, R. D. (2010). Microbial ^{13}C utilisation patterns via stable isotope probing of phospholipid biomarkers in Mojave Desert soils exposed to ambient and elevated CO_2. *Global Change Biol.* 16, 2334–2344.

Jones, D. L. and Shannon, D. (1999). Mineralization of amino acids applied to soils: impact of soil sieving, storage, and inorganic nitrogen additions. *Soil Sci. Soc. Am. J.* 63, 1199–1206.

Kindler, R., Miltner, A., Richnow, H-.H. and Kastner, M. (2006). Fate of Gram-negative bacterial biomass in soil — mineralization and contribution to SOM. *Soil Biol. Biochem.* 38, 2860–2870.

Kindler, R., Miltner, A., Thullner, M., Richnow, H-.H. and Kastner, M. (2009). Fate of bacterial biomass derived fatty acids in soil and their contribution to soil organic matter. *Soil Biol. Biochem.* 40, 29–37.

Knowles, T. D. J., Chadwick, D. R., Bol, R. and Evershed, R.P. (2010). Tracing the rate and extent of N and C flow from ^{13}C,^{15}N-glycine and glutamate into individual de novo synthesised soil amino acids. *Org. Geochem.* 41, 1259–1268.

Ladygina, N. and Hedlund, K. (2010). Plant species diversity and carbon allocation in the rhizosphere. *Soil Biol. Biochem.* 42, 162–168.

Leake, J. R., Ostle, N. J., Rangel-Castro, J. I. and Johnson, D. (2006). Carbon fluxes from plants through soil organisms determined by field $^{13}CO_2$ pulse-labelling in an upland grassland. *Appl. Soil Ecol.* 33, 152–175.

Lemanski, K. and Scheu, S. (2014). Incorporation of ^{13}C labelled glucose into soil microorganisms of grasslands: effect of fertilizer addition and plant functional group. *Soil Biol. Biochem.* 69, 38–45.

Malik, A. A., Dannert, H., Griffiths, R., Thomson, B. and Gleixner, G. (2015). Rhizosphere bacterial carbon turnover is higher in nucleic acids than membrane lipids: implications for understanding soil carbon cycling. *Front. Microbiol.* 6, 1–9.

Mathieu, O., Lévêque, J., Hénault, C., Ambus, P., Milloux, M. and Andreux, F. (2007). Influence of ^{15}N enrichment on the net isotopic fractionation factor during the reduction of nitrate to nitrous oxide in soil. *Rapid Commun. Mass Spectrom.* 21, 1447–1451.

Maxfield, P. J., Brennand, E. L., Powlson, D. S. and Evershed, R. P. (2011). Impact of land management practices on high-affinity methanotrophic bacterial populations: evidence from long-term sites at Rothamsted. *Eur. J. Soil Sci.* 62, 56–68.

Maxfield, P. J., Dildar, N., Hornibrook, E. R. C., Stott, A. W. and Evershed, R. P. (2012). Stable isotope switching (SIS): a new stable isotope probing (SIP) approach to determine carbon flow in the soil food web and dynamics in organic matter pools. *Rapid Commun. Mass Spectrom.* 26, 997–1004.

Maxfield, P. J. and Evershed, R. P. (2011). Stable isotope probing and related technologies. Murrell, C. and Whiteley, A.S. (eds.), *Phospholipid Fatty Acid Stable Isotope Probing Techniques in Microbial Ecology*, Chapter 3. ASM Press, Washington, DC, pp. 37–71.

Maxfield, P. J., Hornibrook, E. R. and Evershed, R. P. (2006). Estimating high-affinity methanotrophic bacterial biomass, growth, and turnover in soil by phospholipid fatty acid ^{13}C labelling. *Appl. Environ. Microbiol.* 72, 3901–3907.

Maxfield, P. J., Hornibrook, E. R. and Evershed, R. P. (2008). Acute impact of agriculture on high-affinity methanotrophic bacterial populations. *Environ. Microbiol.* 10, 1917–1924.

Maxfield, P. J., Hornibrook, E. R. and Evershed, R. P. (2009). Substantial high-affinity methanotroph populations in Andisols effect high rates of atmospheric methane oxidation. *Environ. Microbiol. Rep.* 1, 450–456.

Mellendorf, M., Soja, G., Gerzabek, M. H. and Watzinger, A. (2010). Soil microbial community dynamics and phenanthrene degradation as affected by rape oil application. *Appl. Soil Ecol.* 46, 329–334.

Merritt, D. A. and Hayes, J. M. (1994). Nitrogen isotopic analyses by isotope-ratio-monitoring gas chromatography/mass spectrometry. *J. Am. Soc. Mass Spectrom.* 5, 387–439.

Metges, C. C. and Petzke, K. J. (1997). Measurement of $^{15}N/^{14}N$ isotopic composition in individual plasma free amino acids of human adults at natural abundance by gas chromatography–combustion isotope ratio mass spectrometry. *Anal. Biochem.* 247, 158–164.

Mohanty, S. R., Bodelier, P. L. E., Floris, V. and Conrad, R. (2006). Differential effects of nitrogenous fertilizers on methane-consuming microbes in rice field and forest soils. *Appl. Environ. Microbiol.* 72, 1346–1354.

Moore-Kucera, J. and Dick, R. P. (2008). Application of ^{13}C-labeled litter and root materials for in situ decomposition studies using phospholipid fatty acids. *Soil Biol. Biochem.* 40, 2485–2493.

Mosier, A. R., Schimel, D. S., Valentine, D. W., Bronson, K. F. and Parton, W. J. (1991). Methane and nitrous oxide fluxes in native, fertilized, and cultivated grasslands. *Nature* 350, 330–332.

Mottram, H. R. and Evershed, R. P. (2003). Practical considerations in the gas chromatography/combustion/isotope ratio monitoring mass spectrometry of ^{13}C-enriched compounds: detection limits and carryover effects. *Rapid Commun. Mass Spectrom.* 17, 2669–2674.

Murase, J., Hordijk, K., Tayasu, I. and Bodelier, P. L. E. (2011). Strain-specific incorporation of methanotrophic biomass into eukaryotic grazers in a rice field soil revealed by PLFA-SIP. *FEMS Microbiol. Ecol.* 75, 284–290.

Murrell, J. C. and Whiteley, A. S. (Eds.) (2011). *Stable Isotopes and Related Technologies.* ASM press, Washington, DC.

NETCEN (2003). *UK Emissions of Air Pollutants 1970–2003.* 17th NAEI annual report.

Nwadinigwe, C. A. and Nwobodo, I. O. (1994). Analysis of *n*-paraffins in light crudes: molecular sieve and urea adduction techniques revisited. *Fuel* 73, 779.

Radajewski, S., Ineson, P., Parekh N. R. and Murrell, J. C. (2000). Stable-isotope probing as a tool in microbial ecology. *Nature* 403, 646–649.

Reay, D. S. and Nedwell, D. B. (2004). Methane oxidation in temperate soils: effects of inorganic N. *Soil Biol. Biochem.* 36, 2059–2065.

Reay, D. S., Radajewski, S., Murrell, J. C., McNamara, N. and Nedwell, D. B. (2001). Effects of land-use on the activity and diversity of methane oxidizing bacteria in forest soils. *Soil Biol. Biochem.* 33, 1613–1623.

Redmile-Gordon, M. A., Evershed, R. P., Hirsch, P. R., White, R. P. and Goulding, K. W. T. (2015). Soil organic matter and the extracellular microbial matrix show contrasting responses to C and N availability. *Soil Biol. Biochem.* 88, 257–267.

Roberts, P. and Jones, D. L. (2008). Critical evaluation of methods for determining total protein in soil solution. *Soil Biol. Biochem.* 40, 1485–1495.

Seghers, D., Top, E. M., Reheul, D., Bulcke, R., Boeckx, P. and Verstraete, W. (2003). Long-term effects of mineral versus organic fertilizers on activity and structure of the methanotrophic community in agricultural soils. *Environ. Microbiol.* 5, 867–877.

Singh, B. K., Tate, K. R., Kolipaka, G., Hedley, C. B., Macdonald, C. A., Millard, P. and Murrell, J. C. (2007). Effect of afforestation and reforestation of pastures on the activity and population dynamics of methanotrophic bacteria. *Appl. Environ. Microbiol.* 73, 5153–5161.

Steudler, P. A., Bowden, R. D., Melillo, J. M. and Aber, J. D. (1989). Influence of nitrogen fertilization on methane uptake in temperate forest soils. *Nature* 341, 314–316.

Tang, F. H. M. and Maggi, F. (2012). The effect of ^{15}N to ^{14}N ratio on nitrification, denitrification and dissimilatory nitrate reduction. *Rapid Commun. Mass Spectrom.* 26, 430–442.

Tavi, N. M., Martikainen, P. J., Lokko, K., Kontro, M., Wild, B., Richter, A. and Biasi, C. (2013). Linking microbial community structure and allocation of plant-derived carbon in an organic agricultural soil using $^{13}CO_2$ pulse-chase labelling combined with ^{13}C-PLFA profiling. *Soil Biol. Biochem.* 58, 207–215.

Torres, I. F., Bastida, F., Hernándes, T., Bombach, P., Richnow, H. H. and García, C. (2014). The role of lignin and cellulose in the carbon cycling of degraded soils under semiarid climate and their relation to microbial biomass. *Soil Biol. Biochem.* 75, 152–160.

Wang, Z. P. and Ineson, P. (2003). Methane oxidation in a temperate coniferous forest soil: effects of inorganic N. *Soil Biol. Biochem.* 35, 427–433.

Williams, M. A., Myrold, D. D. and Bottomley, P. J. (2006). Carbon flow from ^{13}C-labeled straw and root residues into the phospholipid fatty acids of a soil microbial community under field conditions. *Soil Biol. Biochem.* 38, 759–768.

Willison, T. W., Webster, C. P., Goulding, K. W. T. and Powlson, D. S. (1995). Methane oxidation in temperate soils — effects of land-use and the chemical form of nitrogen-fertilizer. *Chemosphere* 30, 539–546.

Yao, H., Thornton, B. and Paterson, E. (2012). Incorporation of ^{13}C-labelled rice rhizodeposition carbon into soil microbial communities under different water status. *Soil Biol. Biochem.* 53, 72–77.

8

Emerging Culture-Independent Tools to Enhance Our Understanding of Soil Microbial Ecology

Hongwei Liu*, Hooman Mirzaee*, Vivian Rincon-Florez*, Richard Moyle*, Marie Spohn†, Lilia C. Carvalhais† and Peer M. Schenk*,‡

**Plant-Microbe Interactions Laboratory, School of Agriculture and Food Sciences, The University of Queensland, Brisbane, Queensland 4072, Australia*

†Department of Soil Ecology, University of Bayreuth, 95548 Bayreuth, Germany

‡p.schenk@uq.edu.au

Introduction

Soil bacteria, fungi, protists and archaea constitute the most diverse living communities on Earth and provide essential ecosystem services to life on this planet. However, the genetic information that they encode remained largely unexplored until recently. The main reason for this is that many soil microorganisms cannot be cultured using standard techniques (Clardy *et al.*, 2006; Ritz, 2007) and methods that enable culture-independent exploration of their genomes have only recently become available (Jones *et al.*, 2009; Lauber *et al.*, 2009; Scholz *et al.*, 2012). While these new approaches gain a lot of attention, it appears that culture-independent techniques can also be very well complemented with classical microbiological culture-dependent techniques, and in some cases impressive numbers of soil microbes have been isolated from cultures. For example, a novel high-throughput *in situ* cultivation platform has been developed to cultivate and isolate hitherto uncultivated microbial species from a variety of environments, including soil. In an "isolation chip" (ichip), several hundred miniature diffusion chambers can be colonised in a single environmental cell, enabling the investigation of a large and diverse array of previously inaccessible microorganisms (Kaeberlein *et al.*, 2002; Nichols *et al.*, 2010). In this new method, significantly more species were grown in comparison with standard media. As a one-step process, "unculturable" microorganisms inside the ichip generated pure colonies under the conditions that they had been collected from in the environment (Nichols *et al.*, 2010). Furthermore, a recent study that simply used a variety of different cultivation media shows that up to 70% of soil microbes associated with *Arabidopsis* plant roots can be cultured and match the data from culture-independent next-generation 16 ribosomal RNA (rRNA) gene amplicon sequencing (Bai *et al.*, 2015). This chapter provides cases of emerging culture-independent techniques that will enhance our understanding of soil microbial ecology. Soil ecosystem function, biodiscovery from soil microbes and plant–microbe interactions are just some of the areas that will benefit from these new approaches.

It can be expected that many important functions and compounds of soil microbiomes will be discovered in the near future. Cultured microbial strains have already been the source of numerous natural products, but the cryptic uncultivated majority is believed to produce an immense and vastly unexplored source of bioactive molecules (Hibbing *et al.*, 2010), in particular, by members of the actinomycetes (Watve *et al.*, 2001). Similarly, peptide bacteriocins (Class I and II) are abundant in bacterial microbial soil ecosystems, including the soil surrounding plant roots (rhizosphere), and genome searches reveal the presence of potential bacteriocin genes in most bacteria (Dirix *et al.*, 2004; Nes and Johnsborg, 2004; Montesinos, 2007; Holtsmark *et al.*, 2008).

Soil microbiomes have been influenced by humans in many ways since the advent of agriculture. These include (1) soil amendments (additives to soil, such as fertilisers, pesticides, charcoal, manure or other organic matter sources), (2) soil cultivation (e.g. tillage) and (3) crop rotation or mixed crop systems (e.g. legumes biofertilise the soil or the use of allelopathic plants). While soil-borne pathogens, such as *Fusarium*, *Pythium* and *Phytophthora* ssp. often receive a lot of attention, the vast majority of soil microbes can be considered neutral (commensals) or beneficial to plants. Our experiments have shown that plants grown in axenic soil may only produce half the biomass than in the presence of soil microbes (Carvalhais *et al.*, 2013a). Beneficial soil microorganisms play a major role in plants for nutrient acquisition (e.g. by N fixation or P solubilisation) and disease suppression (e.g. by production of siderophores, antimicrobial compounds or antifungal chitinases). For these reasons, plants release large amounts of organic carbon (sugars and organic acids) into the rhizosphere to recruit soil microbes that provide benefits to the plants. The direct addition of certain or mixed microorganisms (e.g. *Bacillus subtilis* or *Trichoderma*) to soil has also been practised with varying results to improve plant nutrition and/or disease resistance (Cao *et al.*, 2011; Kavoo-Mwangi *et al.*, 2013). There is mounting evidence that plants can selectively attract and maintain rhizosphere microbes by root exudates to gain benefits, but the chemical language and services from these types of soil microbiome manipulations are often still poorly understood or unknown, especially for commercial crop plants. The well-studied legume–rhizobia interactions provide a good example for a chemical language where specific compounds attract specific rhizobacteria (Cooper, 2007; Fierer *et al.*, 2007). It can be expected that similar common principles between compounds, microbes attracted by them and their benefits for plants can be established for other parts of the rhizosphere microbiome, although some of these interactions will be less specific.

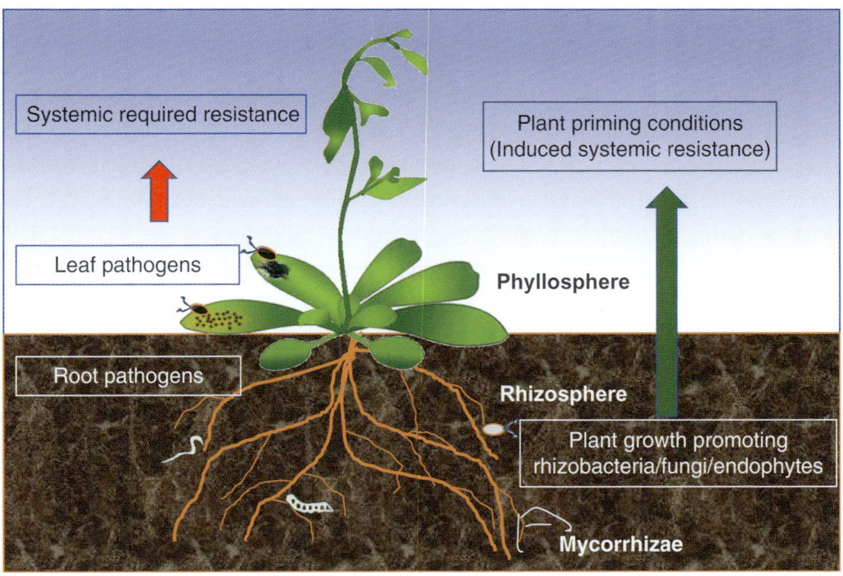

Figure 8.1 Overview of plant–microbe interactions in the rhizosphere and phyllosphere. Plant growth-promoting microbes are attracted to the rhizosphere and phyllosphere by root and leaf exudates, respectively. In return, they make nutrients available to plants, control pathogens or influence plant development, disease resistance and stress resilience by manipulating plant hormone signalling.

Figure 8.1 provides an overview of beneficial and parasitic plant–microbe interactions in the rhizosphere and phyllosphere. The following paragraphs provide examples of emerging technologies that will enhance our understanding of soil microbiomes in this and other areas.

Characterisation of Soil Microbial Genes and Enzymes

Quantitative PCR

Real-time quantitative PCR (Q-PCR) and reverse transcriptase Q-PCR (qRT-PCR) are culture-independent methods to quantify the abundance of genes in the environment (Smith and Osborn, 2009). Gene abundance measurements can be used for taxonomic and functional analysis of microbial communities colonizing environmental samples (Torsvik and Øvreås, 2002; Fierer *et al.*, 2005). Q-PCR is a highly sensitive, accurate and quick method that allows for the analysis of several samples with different genes in the same Q-PCR run (Heid *et al.*, 1996). However, this method can only be used for the quantification of known sequences and requires prior knowledge for primer design. Q-PCR/qRT-PCR is based on fluorescence chemistries using either intercalating fluorescent probes of TaqMan or SYBR Green to detect the accumulation of amplicons during PCR cycles (Van Guilder *et al.*, 2008). TaqMan chemistry requires specific hybridisation between probe and target to produce a fluorescent signal. Therefore, Q-PCR using TaqMan chemistry has higher specificity than using SYBR green. TaqMan also allows for parallel amplifications of two different sequences in one reaction by using distinguishable reporter dyes of probes. In contrast, the SYBR Green dye binds to any double-stranded DNA and does not require a probe to produce a fluorescence signal, and this reduces running costs for assays (Van Guilder *et al.*, 2008). One drawback of using SYBR Green dye is that false-positive signals may be generated during PCR due to non-specific amplifications or DNA contamination (Smith and Osborn, 2009). To achieve accurate quantification in a Q-PCR assay, optimisation for template concentration and the annealing temperature of gene-specific primers prior to Q-PCR is necessary.

Q-PCR that targets the 16S rRNA genes of genomic DNA has been proposed to evaluate relative abundances of some taxonomic bacterial groups in microbial ecology (Fierer *et al.*, 2005; De Gregoris *et al.*, 2011). Using this method, Liu *et al.* (2016a, b) evaluated the impacts of one-time strategic tillage on the abundance of five bacterial taxa (Actinobacteria, α and γ Proteobacteria, Firmicutes and Bacteriodetes) of microbial communities in Australian long-term no-till soils. Another important application of Q-PCR/qRT-PCR is for the profiling of functional genes (e.g. associated with nitrification and denitrification) in environmental samples. The understanding of nitrogen (N) and carbon (C) cycles is an important theme for microbial ecological research (Paul, 2014). For instance, using Q-PCR, Harter *et al.* (2014) quantified the abundance and expression (using reverse transcription) of soil microbial N-fixation (*nifH*), nitrification (*amoA*) and denitrification (*nirK*, *nirS* and *nosZ*) after biochar addition in soil and found N_2O emissions were reduced and N_2-fixing microorganisms were increased in abundance. This finding contributes to a better understanding of the impact of biochar on the N-cycling of soil microbial communities. Unfortunately, qRT-PCR can be challenging as it requires enough good-quality mRNA from environmental samples, for example, soil samples. This often involves the use of special protocols to remove phenolics, carbohydrates and humic acids from soil, all of which can inhibit reverse transcriptases and/or polymerases.

Soil zymography

An interesting emerging imaging technique for localising and quantifying enzyme activities in soil is zymography (Spohn *et al.*, 2013; Spohn and Kuzyakov, 2013). The advantages of soil zymography are that it is an *in situ* method and therefore no sample preparation such as soil sieving or enzyme extraction is necessary. The method is non-destructive, which means that the same soil can be measured several times (Spohn and Kuzyakov, 2014; Hofmann *et al.*, 2016), which is very useful for the determination of changes in enzyme activity over time. Moreover, it allows for the distribution of enzyme activity to be determined at a small scale since it has a resolution of about 2 mm. Soil zymography is based on the incubation of the soil attached to an agarose gel that is covered by a nylon membrane coated with a substrate that becomes fluorescent once it gets hydrolysed. After the incubation, which typically takes between 20 and 40 min, the fluorescence is made visible by excitation with UV light, and a photograph of the fluorescent membrane is taken. Calibration is performed with nylon membranes coated with solutions of standards (Spohn and Kuyzakov, 2014). Soil zymography works for a large range of hydrolases (Spohn *et al.*, 2013; Spohn and Kuzyakov, 2014). The method has been specifically useful for determining the distribution of enzyme activity in the rhizosphere (Spohn and Kuzyakov, 2013, 2014), and it has already been combined with other imaging methods, such as ^{14}C imaging (Spohn and Kuzyakov, 2013) and fluorescence *in situ* hybridisation (Spohn *et al.*, 2015) to gain insights into rhizosphere processes (Figure 8.2).

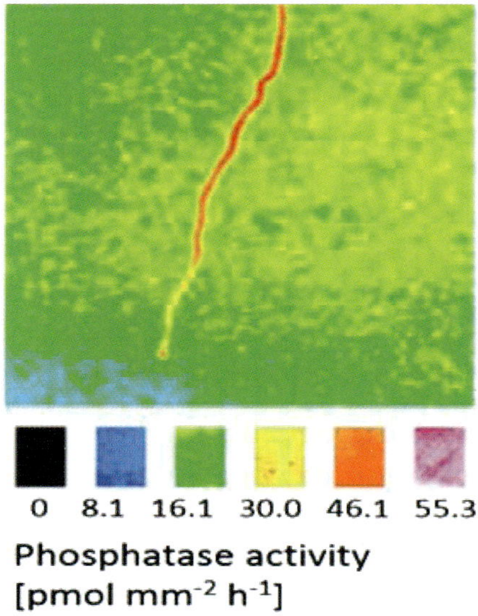

Figure 8.2 An example of using soil zymography to map and quantify phosphatase activity in the rhizosphere along a wheat root (*Triticum aestivum*) without P fertiliser addition to the soil. The comparison to calibration gels shown below the false colour image enables quantification of enzymatic activity.

Microbial Community Profiling

Fluorescence *in situ* hybridisation

Fluorescence *in situ* hybridisation (FISH) as a cytogenetic technique has been used for more than 30 years (Wilkinson, 1998; Levsky and Singer, 2003). It has become one of the most useful techniques in microbial ecology to visualise archaeal and bacterial cells in biospheres and does not require prior cultivation (Amann *et al.*, 2001). The traditional FISH technique relied on 16S rRNA as a probe target for phylogenetic identification of microbes as 16S rRNA exists in all prokaryotic cells with high copy numbers. Typical FISH entails the procedures of cell fixation and permeabilisation, hybridisation, washing steps and the detection of fluorescence signals (Pernthaler *et al.*, 2002). Microbial cells are fixed and their DNA and RNA are preserved. Those microbial cells with permeabilised conditions allow for the nucleic acid probes access and hybridisation to the target site. The detection of cells of interest in environmental samples is often achieved using epifluorescence or laser scanning microscopy or flow cytometry for spatial distribution, quantification and further studies. FISH has several advantages over other fingerprinting techniques, such as for terminal reaction fragment length polymorphism (T-RFLP), 16S rRNA sequencing or PLFA methyl esters: (1) FISH allows for the detection of both culturable and non-culturable microorganisms; (2) it enables phylogenetic identification of individual cells *in situ* in their natural environment and may assist in discovering their functions in ecosystems. FISH is well established and widely used in microbial ecology to identify microorganisms (Moter and Göbel, 2000). As a very useful tool for environmental microbiological studies, FISH has spawned numerous applications. Over the past decade, more and more FISH techniques have been applied to complex environmental samples (e.g. marine environments, plant and soil environments, biosphere; Amann *et al.*, 2001). For instance, biofilms contain a diverse range of bacterial species. Performing FISH allows for the visualisation of specific bacterial taxa in biofilms by using bacteria-specific FISH DNA probes. CARD-FISH has been recently used to confirm the presence of Actinobacteria on whole root segments of *Arabidopsis* plants. It was found that on the root surface of *Arabidopsis*, Actinobacteria were visually detected in higher numbers while the abundance of Bradyrhizobiaceae was lower using the probe HGC69a (Lundberg *et al.*, 2012). Remus-Emsermann *et al.* (2014) established a FISH protocol for measurement of the spatial distribution of different main phylogenetic lineages of bacterial communities on the leaf surface of *Arabidopsis*. They found that most of the bacterial biomass was composed of taxa affiliated to Betaproteobacteria, Alphaproteobacteria and Actinobacteria. About one-third of the bacteria found in the phylloplane exhibited infrared autofluorescence, which could be from aerobic anoxygenic phototrophs (Remus-Emsermann *et al.*, 2014).

Phylogenetic marker gene sequencing

High-throughput deep sequencing, or next-generation sequencing is widely being used for the evaluation of microbial diversity in the environment. Compared to other culture-independent methods, for example, T-RFLP, phylogenetic gene sequencing allows for the biodiversity and abundance of environmental samples to be explored with reliable taxonomic units up to the genus level. It presents a descriptive picture for microbial ecology studies and abundance fluctuations of microbes with known functions to enable functional analyses of microbial communities (Bartram *et al.*, 2011; Simon and Daniel, 2011; Carvalhais *et al.*, 2012).

16S rRNA deep sequencing has been extensively used to detect the composition and diversity of archaeal and bacterial genera in soil samples. For instance,

16S rRNA gene amplicon pyrosequencing has been used for bacterial community profiling in different compartments of *Arabidopsis* including rhizosphere soil and in roots (endophytic) (Lundberg *et al.*, 2012; Cao *et al.*, 2011). Endophytic bacteria displayed an over-representation of Actinobacteria and Proteobacteria for two different soil types tested. Our previous studies also found that an elevated level of jasmonate signalling pathway in *Arabidopsis* altered the composition but not the α diversity in the associated rhizosphere bacterial communities (Carvalhais *et al.*, 2013b). More recently, 18S rRNA and ITS amplicon deep sequencing have been used for the profiling of soil eukaryotic microbial communities to cover protists and fungi (Bates *et al.*, 2013; Hugerth *et al.*, 2014). However, the nuclear ribosomal internal transcribed spacer (ITS) region of eukaryotes is considered to be more suitable for phytogenetic profiling of eukaryotes as it shows a higher variability (Schoch *et al.*, 2012; Adl *et al.*, 2014). ITS primers are now more commonly used for amplicon sequencing to reveal fungal and protistic diversity in soils and this area of research is expected to grow.

High-Throughput (Next-Generation) Sequencing Technologies

First-generation sequencing technology, better known as Sanger sequencing, was used to complete the first bacterial genome sequence in 1995 (Fleischmann *et al.*, 1995). 454 Life Sciences launched a sequencing-by-synthesis second-generation sequencing platform in 2005. This advance in high-throughput, massively parallel sequence analysis permitted sequencing of bacterial genomes in a matter of days.

Today, entire microbiomes are analysed using next-generation sequencing technologies which possess a combination of longer read lengths, higher throughput per run and higher depth of coverage. A number of next-generation sequencing platforms exist, each with their own strengths and weaknesses. The 454 GS FLX+ platform is particularly useful at generating long reads in under a day (up to 1000 bp in length), but has relatively low throughput of ~700 Mb per run. It can also have relatively high error rates (particularly at homopolymeric tracts of sequence) and is relatively expensive compared to other platforms. The Illumina HiSeq2500 platform is higher throughput with up to 600 Gb per run and has a lower reagent cost, but takes over a week per run and generates shorter read lengths of up to 150 bp. The solid platform is also high throughput, generating ~320 Gb per run with high accuracy, but generates shorter read lengths of up to 75 bp. Thus, the appropriate selection of next-generation sequencing platform depends upon the particular aims of each experiment. In some cases, combinations of platforms are chosen, for example, pairing the 454 platform with the

Illumina platform provides both long reads for sequence scaffold assembly and high depth of coverage, respectively.

Metagenomics

Metagenomics involves simultaneously investigating multiple genomes present in the DNA of cohabiting microbial populations, recovered directly from the environment (Thomas *et al.*, 2012) Currently this is mainly achieved by direct random shotgun sequencing (Tyson *et al.*, 2004). The decline in the cost of high-throughput sequencing technologies is the main reason why, recently, metagenome datasets have massively expanded. One of the applications of metagenome analyses is to assess taxonomic information of the sample being investigated. Marker genes which are well known to provide phylogenetic information are usually targeted in the analysis. The other application is to understand the potential ongoing ecological processes in the sample. Compared to cultivation-based methods, metagenomics approaches are more likely to reveal unbiased insights into microbial community composition and function given that many of the microbes cannot be cultivated in standard laboratory culture media (Hugenholtz and Tyson, 2008). In a single environmental sample, organisms from distinct domains typically co-exist, including viruses, bacteria, archaea and microbial eukaryotes. However, the intricate nature of the ecological information contained in most ecosystems favours the study of individual domains, most often combining bacteria and archaea. The main constraint involved in including eukaryotes in metagenomics analyses is that their genomes contain a significant portion of non-coding genetic material. Another limitation stems from the fact that eukaryotic genomes are much more complex because they usually have two (or more) sets of chromosomes (Kunin *et al.*, 2008). Metagenomics provides information on the phylogenetic types, gene functions and interactions between different organisms co-inhabiting the environmental sample under evaluation. Comparisons between metagenomes have shed light into adaptive microbial strategies that allow them to thrive in environments with different chemical and physical properties and into the abundance of distinct gene families in such environments (Tringe *et al.*, 2005; Dinsdale *et al.*, 2008; Delmont *et al.*, 2011). To properly interpret the data obtained in metagenomics studies, detailed data on the chemical, biological and physical attributes need to be put into context, as well as a suitable experimental design, and statistics (Knight *et al.*, 2012).

Sequencing of environmental DNA for metagenomics analyses can be performed either deeply or shallowly, depending on the purpose of the investigation. Deep sequencing is especially required to detect rare taxa. Rare microbes can be essential for the functioning of soil ecosystems, given that these organisms are pivotal for processes like N-fixation and methanogenesis (Falkowski *et al.*, 2008; Thauer *et al.*, 2008). However, microbial groups that are rare in one environment may be abundant in another. For example, N-fixing organisms are abundant in the root nodule of a leguminous plant and may be rare in a soil sample. Shallow sequencing indicates the most abundant microbial groups present at the time of the sampling. Given that the most dominant microbial populations are possibly the most functionally relevant under standard conditions, in studies which aim at assessing the most representative organisms and processes, shallow sequencing should suffice (Knight *et al.*, 2012). Nevertheless, information on the dynamics of microbial communities can only be gained if multiple samples are taken on a time course and across a spatial gradient (Knight *et al.*, 2012).

A series of bioinformatics methods is then used to treat and analyse the data after the sequencing step has been concluded. In the case of environments with highly complex microbial communities, such as most soil samples, the sequencing depth is often not enough to perform a metagenome assembly and therefore a gene-centric approach is used for the analysis. Bioinformatics methods include quality control, clustering (for gene-centric approaches) or assembly (in case a reference genome is available), binning and taxonomic classification, functional annotation and metabolic reconstruction. A wide range of high-quality reviews on metagenomics analysis tools is available (for more detailed information, please refer to Kim *et al.*, 2013; Bragg and Tyson, 2014; Lindgreen *et al.*, 2016).

Metatranscriptomics

Metatranscriptomic approaches (also referred to as RNASeq of microbiomes) reveal a functional snapshot of the ongoing ecological processes and active taxa in the environmental sample at the transcriptional level. Apart from removing reads derived from rRNA, metatranscriptomic approaches involve exactly the same analytical steps as used in metagenomic approaches after high-throughput sequencing. However, the starting material extracted from environmental samples is RNA instead of DNA. As a consequence, when obtaining information concerning the taxonomic affiliation of members of the microbial communities, the results reveal a snapshot of the active members of the community given RNA molecules are a product of gene transcription and these molecules have short half-lives compared to DNA. Half-lives of RNA samples vary from minutes to seconds (Deutscher, 2006). During sampling, it is essential to prevent RNA from being degraded. Procedures that are recommended for this purpose include snap-freezing samples at –80°C immediately after collection or using a commercial RNA preservation solution, such as the LifeGuard™ Soil Preservation Solution (MO BIO Laboratories, Inc., Carlsbad, CA). A common constraint of metatranscriptomic approaches is the isolation of sufficient high-quality RNA for the downstream analysis. Challenges arise mainly from RNA degradation by RNases, poor cell lysis and, in the case of soil samples, adsorption of RNA to soil particles. High salt concentration in extraction buffers are often used to inactivate RNases. Lowering the pH of extraction buffers prevents adsorption to soil particles (Chomczynski and Sacchi, 1987). Numerous extraction kits for RNA isolation of environmental samples are commercially available. These include E.Z.N.A.® Soil RNA Kit (Omega Bio-tek, Norcross, GA, USA), PowerSoil™ Total RNA Isolation Kit (MoBio Laboratories, Carlsbad, CA USA), FastRNA® Pro Soil-Indirect Kit (MP Biomedicals, Solon, OH, USA), FastRNA® Pro Soil-Direct Kit, ZR Soil/FEcal RNA MicroPrep™, ZR Plant RNA MiniPrep™ and Direct-zol™ RNA Kits (Zymo Research, Orange, CA, USA).

Another common attribute of soil-derived RNA is the presence of compounds that have similar chemical properties to nucleic acids and inhibit enzymes that act in subsequent steps to the RNA isolation, including reverse transcriptases. Examples of such compounds include humic and fulvic acids. Commercial kits are available for cleaning-up, removing inhibitors and concentrating RNA samples: RNA Clean & Concentrator™, OneStep™ PCR Inhibitor Removal Kit (Zymo Research, Orange, CA, USA) and PureLink® RNA Mini Kit (Ambion Inc., Austin, TX, USA). Other methods for removal of inhibitors are reviewed in Carvalhais *et al.* (2012).

Methods for isolating RNA from environmental samples usually extract total RNA, which is mostly composed of ribosomal RNA (95–99% of the total RNA; He *et al.*, 2010). For functional studies, it is paramount to enrich for messenger RNA (mRNA). Various methods have been developed for this purpose: (i) duplex specific nuclease (DSN; Yi *et al.*, 2011); (ii) exonuclease treatment (mRNA-ONLY Prokaryotic mRNA Isolation Kit, EPICENTRE Biotechnologies, Madison; USA); (iii) subtractive hybridisation (MICROBExpress Bacterial mRNA Enrichment Kit, Ambion, USA; Ribo-Zero rRNA Removal Kit (Bacteria), Illumina; Stewart *et al.*, 2010); (iv) size separation by gel electrophoresis (McGrath *et al.*, 2008). Messenger RNA amplification may be required as obtaining sufficient amounts of this molecule for downstream applications is challenging. An example of an efficient commercial kit available for mRNA amplification is the MessageAmp II-Bacteria aRNA amplification kit (Ambion, USA). This kit includes a polyadenylation step using a *Escherichia coli* poly(A) polymerase. The poly(A)-tailed RNA is then reverse-transcribed using an oligo(dT) primer containing a T7 promoter. Other commercial kits are also available, such as SMART®mRNA Amplification Kit

(Clontech, Palo Alto, CA, USA), Arcturus™ RiboAmp® PLUS Kit (Applied Biosystems, Foster City, CA, USA) and SeqPlex RNA Amplification Kit (Sigma-Aldrich). Even if mRNA enrichment methods are used, a substantial portion of the reads derive from rRNA after high-throughput sequencing (51–60%; Stewart *et al.*, 2010). Several bioinformatic tools are available for identifying and the removal of rRNA-derived sequences. These include riboPicker (Schmieder *et al.*, 2012), SortMeRNA (Kopylova *et al.*, 2012) and Infernal (Nawrocki and Eddy, 2013).

Biodiscovery of Compounds from Soil Microbiomes

Soil microorganisms are untapped resources for new biotechnological compounds which could be used in many industries. For example, plant growth-promoting rhizobacteria and other beneficial microbes are able to control plant diseases through mechanisms like competition, antagonism through the production of antibiotics, antimicrobial peptides and siderophores; induction of disease resistance and improvement of nutrient uptake (Figure 8.1; Anith *et al.*, 2004; Babalola, 2010; Maksimov *et al.*, 2011). While microorganisms are very abundant in soil samples, most are not culturable in standard laboratory culture media (Schmeisser *et al.*, 2007; Walsh and Duffy, 2013) and may only produce certain compounds in the presence of other microbes and their products. For this reason, it is possible that a significant amount of new bioactive compounds from unculturable microbes remain unknown. Recently, using a novel approach for culturing and screening of new antibiotics called "iChip" (see above), 25 antibiotics have been discovered. They have been shown to be effective against multiple drug-resistant *Staphylococcus aureus* or *Mycobacterium tuberculosis* (Ling *et al.*, 2015; Nichols *et al.*, 2010).

Bacteriocin-targeted PCR using degenerate primers

Proteins, enzymes and short peptides are encoded in the DNA. AMPs, which can have inhibitory effects against human and plant pathogens, are synthesised in a wide range of organisms from all kingdoms (Keymanesh *et al.*, 2009; Maróti *et al.*, 2011; Nakatsuji and Gallo, 2012). To find new AMPs synthesised by unculturable microbes, conserved regions in the coding DNA sequences can be used in PCR-based approaches using conserved degenerate primers. A phylogram can be generated using AMP-coding sequences from publically available databases by bioinformatics software and degenerate primers can be designed to target conserved regions in aligned sequences from the different clusters of the phylogram.

An example of a template derived from an environmental sample for PCR using degenerate primers is the DNA isolated from soil samples. PCR products can then be ligated to suitable vectors for sequencing, protein expression and/or further functional characterisation.

Cloning and screening of expression libraries

Targeting the soil microbial transcriptome for biodiscovery of bioactive proteins or peptides is recommended when treatments or environmental conditions favour the detection of coding sequences of molecules of interest via the induction of transcription. However, several challenges associated with this approach need to be taken into account. For instance, as previously discussed, RNA has a short half-life. Since transcription and translation of archaea and bacteria occur simultaneously, mRNA degradation starts quickly after translation (Deutscher, 2006); therefore, the chances of obtaining complete sequences of a target gene are reasonably low. To perform experiments targeting the environmental microbial transcriptomes, firstly total RNA needs to be extracted from soil or any other environmental samples, followed by cDNA synthesis. The latter includes a range of reverse degenerated primers which would target the gene of interest, for example, AMPs such as bacteriocins. Similar to the requirement for metatranscriptomics (see above), rRNA should be subtracted before cDNA synthesis. Finally, a suitable vector is essential to express the genes encoded in the cDNA in all reading frames and which should include an inducible promoter (e.g. IPTG-inducible T7 polymerase), and start and stop codons. We have recently developed a new vector on the basis of RNA polymerase slippage (Wagner *et al.*, 1990; Baranov *et al.*, 2005; Penno *et al.*, 2006). It contains a slippery region consisting of 12 thymine nucleotides and multiple stopping codons which allow for directional cloning and simultaneous expression of cDNA in all reading frames.

Functional assays will be required to screen the clone libraries. For example, to screen for AMPs, after obtaining colonies on selective medium, single colonies can be transferred to IPTG-containing a medium for inhibition tests. On this medium, AMP genes, even with incomplete sequences cloned in the vector may be expressed and the growth of the colonies could be visibly inhibited by the presence of large quantities of AMP in the cells. Colonies present in the IPTG-free medium but absent in IPTG-containing medium can then be sequenced and possible new antimicrobial genes and related compounds can be found. Further pathogen growth inhibition assays can then be performed so that functional AMPs are identified.

Conclusion

Soil microbiomes are arguably the most complex ecosystems on Earth. A combination of culture-dependent and -independent techniques can be used to elucidate the function of soil microbiomes. Many of the emerging DNA-, RNA-, protein- and metabolite-based tools have the potential to reveal a comprehensive picture of microbial community composition and its activity. These techniques, when coupled with high-throughput screening assays, provide excellent platforms for biodiscovery approaches. It can be expected that, as more and more knowledge on soil microbiomes is accumulated, well-developed bioinformatics tools will provide the path for future soil microbial ecology studies. However, many challenges remain in data mining and the elucidation of microbial functions. While correlations can be established relatively easily, to determine actual causality relationships requires well-defined experimental designs where specific questions can be tested experimentally.

References

Adl, S. M., Habura, A. and Eglit, Y. (2014). Amplification primers of SSU rDNA for soil protists. *Soil Biol. Biochem.* 69, 328–342.

Amann, R., Fuchs, B. M. and Behrens, S. (2001). The identification of microorganisms by fluorescence *in situ* hybridisation. *Curr. Opin. Biotechnol.* 12, 231–236.

Anith, K. N., Momol, M. T., Kloepper, J. W., Marois, J. J., Olson, S. M. and Jones, J. B. (2004). Efficacy of plant growth-promoting rhizobacteria, acibenzolar-S-methyl, and soil amendment for integrated management of bacterial wilt on tomato. *Plant Dis.* 88(6), 669–673.

Babalola, O. (2010). Beneficial bacteria of agricultural importance. *Biotechnol. Lett.* 32(11), 1559–1570.

Bai, Y., Müller, D. B., Srinivas, G., Garrido-Oter, R., Potthoff, E., Rott, M., *et al.* (2015). Functional overlap of the *Arabidopsis* leaf and root microbiota. *Nature* 528, 364–369.

Baranov, P., Hammer, A., Zhou, J., Gesteland, R. and Atkins, J. (2005). Transcriptional slippage in bacteria: distribution in sequenced genomes and utilization in IS element gene expression, *Genome Biol.* 6(3), R25.

Bartram, A. K., Lynch, M. D. J., Stearns, J. C., Moreno-Hagelsieb, G. and Neufeld, J. D. (2011). Generation of multimillion-sequence 16S rRNA gene libraries from complex microbial communities by assembling paired-end Illumina reads. *Appl. Environ. Microbiol.* 77, 3846–3852.

Bates, S. T., Clemente, J. C., Flores, G. E., Walters, W. A., Parfrey, L. W., Knight, R. and Fierer, N. (2013). Global biogeography of highly diverse protistan communities in soil. *ISME J.* 7(3), 652–659.

Bragg, L. and Tyson, G. W. (2014). Metagenomics using next-generation sequencing. *Meth. Mol. Biol.* 1096, 183–201.

Cao, Y., Zhang, Z., Ling, N., Yuan, Y., Zheng, X., Shen, B. and Shen, Q. (2011). *Bacillus subtilis* SQR 9 can control Fusarium wilt in cucumber by colonizing plant roots. *Biol. Fert. Soils* 47(5), 495–506.

Carvalhais, L. C., Dennis, P. G., Badri, D. V., Tyson, G. W., Vivanco, J. M. and Schenk, P. M. (2013b). Activation of the jasmonic acid plant defence pathway alters the composition of rhizosphere bacterial communities. *PLoS One* 8(2), e56457.

Carvalhais, L. C., Dennis, P. G., Tyson, G. W. and Schenk, P. M. (2012). Application of metatranscriptomics to soil environments. *J. Microbiol. Meth.* 91, 246–251.

Carvalhais, L. C., Muzzi, F., Tan, C. H., Hsien-Choo, J. and Schenk, P. M. (2013a). Plant growth in *Arabidopsis* is assisted by compost soil-derived microbial communities. *Front. Plant Sci.* 4, 235.

Chomczynski, P. and Sacchi, N. (1987). Single-step method of RNA isolation by acid guanidinium thiocyanate phenol chloroform extraction. *Anal. Biochem.* 162, 156–159.

Clardy, J., Fischbach, M. A. and Walsh, C. T. (2006). New antibiotics from bacterial natural products. *Nat. Biotechnol.* 24(12), 1541–1550.

Cooper, J. E. (2007). Early interactions between legumes and rhizobia: disclosing complexity in a molecular dialogue. *J. Appl. Microbiol.* 103(5), 1355–1365.

De Gregoris, T. B., Aldred, N., Clare, A. S. and Burgess, J. G. (2011). Improvement of phylum- and class-specific primers for real-time PCR quantification of bacterial taxa. *J. Microbiol. Meth.* 86, 351–356.

Delmont, T. O., Malandain, C., Prestat, E., Larose, C., Monier, J. M., Simonet, P. and Vogel T. M. (2011). Metagenomic mining for microbiologists. *ISME J.* 5, 1837–1843.

Deutscher, M. P. (2006). Degradation of RNA in bacteria: comparison of mRNA and stable RNA, *Nucl. Acids Res.* 34(2), 659–666.

Dinsdale, E. A., Edwards, R. A., Hall, D., Angly, F., Breitbart, M., Brulc, J. M., *et al.* (2008). Functional metagenomic profiling of nine biomes. *Nature* 452, 629–632.

Dirix, G., Monsieurs, P., Marchal, K., Vanderleyden, J. and Michiels, J. (2004). Screening genomes of Gram-positive bacteria for double-glycine-motif-containing peptides. *Microbiology* 150(5), 1121–1126.

Falkowski, P. G., Fenchel, T. and Delong, E. F. (2008). The microbial engines that drive Earth's biogeochemical cycles. *Science* 320, 1034–1039.

Fierer, N., Bradford, M. A. and Jackson, R. B. (2007). Toward an ecological classification of soil bacteria. *Ecology* 88(6), 1354–1364.

Fierer, N., Jackson, J. A., Vilgalys, R. and Jackson, R. B. (2005). Assessment of soil microbial community structure by use of taxon-specific quantitative PCR assays. *Appl. Environ. Microbiol.* 71, 4117–4120.

Fleischmann, R. D., Adams, M. D., White, O., Clayton, R. A., Kirkness, E. F., Kerlavage, A. R., Bult, C. J., Tomb, J. F., Dougherty, B. A. and Merrick, J. M. (1995). Whole-genome random sequencing and assembly of *Haemophilus influenzae* Rd. *Science* 269(5223), 496–512.

Harter, J., Krause, H. M., Schuettler, S., Ruser, R., Fromme, M., Scholten, T., *et al.* (2014). Linking N_2O emissions from biochar-amended soil to the structure and function of the N-cycling microbial community. *ISME J.* 8(3), 660–674.

He, S. M., Wurtzel, O., Singh, K., Froula, J. L., Yilmaz, S., Tringe, S. G., *et al.* (2010). Validation of two ribosomal RNA removal methods for microbial metatranscriptomics. *Nat. Meth.* 7, 807–858.

Heid, C. A., Stevens, J., Livak, K. J. and Williams, P. M. (1996). Real time quantitative PCR. *Genome Res.* 6, 986–994.

Hibbing, M. E., Fuqua, C., Parsek, M. R. and Peterson, S. B. (2010). Bacterial competition: surviving and thriving in the microbial jungle. *Nat. Rev. Microbiol.* 8(1), 15–25.

Hofmann, K. Y., Heuck, C. and Spohn, M. (2016). Phosphorus resorption by young beech trees and soil phosphatase activity as dependent on P availability. *Oecologia.* doi:10.1007/s00442-016-3581-x.

Holtsmark, I., Eijsink, V. G. and Brurberg, M. B. (2008). Bacteriocins from plant pathogenic bacteria. *FEMS Microbiol. Lett.* 280(1), 1–7.

Hugenholtz, P. and Tyson, G. W. (2008). Microbiology: metagenomics. *Nature* 455, 481–483.

Hugerth, L. W., Muller, E. E., Hu, Y. O., Lebrun, L. A., Roume, H., Lundin, D., Wilmes, P. and Andersson, A. F. (2014). Systematic design of 18S rRNA gene primers for determining eukaryotic diversity in microbial consortia. *PLoS ONE* 9(4), e95567.

Jones, R. T., Robeson, M. S., Lauber, C. L., Hamady, M., Knight, R. and Fierer, N. (2009). A comprehensive survey of soil acidobacterial diversity using pyrosequencing and clone library analyses. *ISME J. 3*(4), 442–453.

Kaeberlein, T., Lewis, K. and Epstein, S. S. (2002). Isolating "uncultivable" microorganisms in pure culture in a simulated natural environment. *Science* 296(5570), 1127–1129.

Kavoo-Mwangi, A. M., Kahangi, E. M., Ateka, E., Onguso, J., Mukhongo, R. W., Mwangi, E. K. and Jefwa J. M. (2013). Growth effects of microorganisms based commercial products inoculated to tissue cultured banana cultivated in three different soils in Kenya. *Appl. Soil Ecol.* 64, 152–162.

Keymanesh, K., Soltani, S. and Sardari, S. (2009). Application of antimicrobial peptides in agriculture and food industry. *World J. Microbiol. Biotechnol.* 25(6), 933–944.

Kim, M., Lee, K. H., Yoon, S. W., Kim, B. S., Chun, J. and Yi, H. (2013). Analytical tools and databases for metagenomics in the next-generation sequencing era. *Genom. Inform.* 11, 102–113.

Knight, R., Jansson, J., Field, D., Fierer, N., Desai, N., Fuhrman, J. A., *et al.* (2012). Unlocking the potential of metagenomics through replicated experimental design. *Nat. Biotechnol.* 30, 513–520.

Kopylova, E., Noe, L. and Touzet, H. (2012). SortMeRNA: fast and accurate filtering of ribosomal RNAs in metatranscriptomic data. *Bioinformatics* 28, 3211–3217.

Kunin, V., Copeland, A., Lapidus, A., Mavromatis, K. and Hugenholtz, P. (2008). A bioinformatician's guide to metagenomics. *Microbiol. Mol. Biol. Rev.* 72, 557–578.

Lauber, C. L., Hamady, M., Knight, R. and Fierer, N. (2009). Pyrosequencing-based assessment of soil pH as a predictor of soil bacterial community structure at the continental scale. *Appl. Environ. Microbiol.* 75(15), 5111–5120.

Levsky, J. M. and Singer, R. H. (2003). Fluorescence *in situ* hybridization: past, present and future. *J. Cell Sci.* 116, 2833–2838.

Lindgreen, S., Adair, K. L. and Gardner, P. P. (2016). An evaluation of the accuracy and speed of metagenome analysis tools. *Sci. Rep.* 6, 19233.

Ling, L. L., Schneider, T., Peoples, A. J., Spoering, A. L., Engels, I., Conlon, B. P., *et al.* (2015). A new antibiotic kills pathogens without detectable resistance. *Nature* 517(7535), 455–459.

Liu, H., Carvalhais, L., Crawford, M., Dang, Y., Dennis, P. and Schenk, P. M. (2016a). Strategic tillage on a Grey Vertisol after fifteen years of no-till management had no short-term impact on soil properties and agronomic productivity. *Geoderma* 267, 146–155.

Liu, H., Carvalhais, L. C., Rincon-Florez, V., Crawford, M., Dang, Y. P., Dennis, P. G., Schenk, P. M. (2016b). One-time strategic tillage does not cause major impacts on soil microbial properties in a no-till Calcisol. *Soil Tillage Res.* 158, 91–99.

Lundberg, D. S., Lebeis, S. L., Paredes, S. H., Yourstone, S., Gehring, J., Malfatti, S., *et al.* (2012). Defining the core *Arabidopsis thaliana* root microbiome. *Nature* 488(7409), 86–90.

Maksimov, I. V., Abizgil'dina, R. R. and Pusenkova, L. I. (2011). Plant growth promoting rhizobacteria as alternative to chemical crop protectors from pathogens (review). *Appl. Biochem. Microbiol.* 47(4), 333–345.

Maróti, G., Kereszt, A., Kondorosi, É. and Mergaert, P. (2011). Natural roles of antimicrobial peptides in microbes, plants and animals. *Res. Microbiol.* 162(4), 363–374.

McGrath, K. C., Thomas-Hall, S. R., Cheng, C. T., Leo, L., Alexa, A., Schmidt, S. and Schenk, P. M. (2008). Isolation and analysis of mRNA from environmental microbial communities. *J. Microbiol. Meth.* 75, 172–176.

Montesinos, E. (2007). Antimicrobial peptides and plant disease control. *FEMS Microbiol. Lett.* 270(1), 1–11.

Moter, A. and Göbel, U. B. (2000). Fluorescence *in situ* hybridization (FISH) for direct visualization of microorganisms. *J. Microbiol. Meth.* 41, 85–112.

Nakatsuji, T. and Gallo R. L. (2012). Antimicrobial peptides: old molecules with new ideas. *J. Invest. Dermatol.* 132(3), 887–895.

Nawrocki, E. P. and Eddy, S. R. (2013). Infernal 1.1: 100-fold faster RNA homology searches. *Bioinformatics* 29, 2933–2935.

Nes, I. F. and Johnsborg, O. (2004). Exploration of antimicrobial potential in LAB by genomics. *Curr. Opin. Biotechnol.* 15(2), 100–104.

Nichols, D., Cahoon, N., Trakhtenberg, E. M., Pham, L., Mehta, A., Belanger, A., *et al.* (2010). Use of Ichip for high-throughput *in situ* cultivation of "uncultivable" microbial species. *Appl. Environ. Microbiol.* 76(8), 2445–2450.

Paul, E. A. (2014). *Soil Microbiology, Ecology and Biochemistry*. Academic Press, London, UK.

Penno, C., Hachani, A., Biskri, L., Sansonetti, P., Allaoui, A. and Parsot, C. (2006). Transcriptional slippage controls production of type III secretion apparatus components in *Shigella flexneri*. *Mol. Microbiol.* 62(5), 1460–1468.

Pernthaler, A., Pernthaler, J. and Amann, R. (2002). Fluorescence *in situ* hybridization and catalyzed reporter deposition for the identification of marine bacteria. *Appl. Environ. Microbiol.* 68, 3094–3101.

Remus-Emsermann, M. N., Lücker, S., Müller, D. B., Potthoff, E., Daims, H. and Vorholt, J. A. (2014). Spatial distribution analyses of natural phyllosphere-colonizing bacteria on *Arabidopsis thaliana* revealed by fluorescence *in situ* hybridization. *Environ. Microbiol.* 16, 2329–2340.

Ritz, K. (2007). The plate debate: cultivable communities have no utility in contemporary environmental microbial ecology. *FEMS Microbiol. Ecol.* 60, 358–362.

Schmeisser, C., Steele, H. and Streit, W. (2007). Metagenomics, biotechnology with non-culturable microbes. *Appl. Microbiol. Biotechnol.* 75(5), 955–962.

Schmieder, R., Lim, Y. W. and Edwards, R. (2012). Identification and removal of ribosomal RNA sequences from metatranscriptomes. *Bioinformatics* 28, 433–435.

Schoch, C. L., Seifert, K. A., Huhndorf, S., Robert, V., Spouge, J. L., Levesque, C. A., *et al.* (2012). Nuclear ribosomal internal transcribed spacer (ITS) region as a universal DNA barcode marker for fungi. *Proc. Natl. Acad. Sci. USA* 109(16), 6241–6246.

Scholz, M. B., Lo, C. C. and Chain, P. S. (2012). Next generation sequencing and bioinformatic bottlenecks: the current state of metagenomic data analysis. *Curr. Opin. Biotechnol.* 23(1), 9–15.

Simon, C. and Daniel, R. (2011). Metagenomic analyses: past and future trends. *Appl. Environ. Microbiol.* 77, 1153–1161.

Smith, C. J. and Osborn, A. M. (2009). Advantages and limitations of quantitative PCR (Q-PCR)-based approaches in microbial ecology. *FEMS Microbiol. Ecol.* 67, 6–20.

Spohn, M., Carminati, A. and Kuzyakov, Y. (2013). Soil zymography — a novel *in situ* method for mapping distribution of enzyme activity in soil. *Soil Biol. Biochem.* 58, 275–280.

Spohn, M. and Kuzyakov, Y. (2013). Distribution of microbial-and root-derived phosphatase activities in the rhizosphere depending on P availability and C allocation — coupling soil zymography with ^{14}C imaging. *Soil Biol. Biochem.* 67, 106–113.

Spohn, M. and Kuzyakov, Y. (2014). Spatial and temporal dynamics of hotspots of enzyme activity in soil as affected by living and dead roots — a soil zymography analysis. *Plant Soil* 379, 67–77.

Spohn, M., Treichel, N. S., Cormann, M., Schloter, M. and Fischer, D. (2015). Distribution of phosphatase activity and various bacterial phyla in the rhizosphere of *Hordeum vulgare* L. depending on P availability. *Soil Biol. Biochem.* 89, 44–51.

Stewart, F. J., Ottesen, E. A. and DeLong, E. F. (2010). Development and quantitative analyses of a universal rRNA-subtraction protocol for microbial metatranscriptomics. *ISME J.* 4, 896–907.

Thauer, R. K., Kaster, A. K., Seedorf, H., Buckel, W. and Hedderich, R. (2008). Methanogenic archaea: ecologically relevant differences in energy conservation. *Nat. Rev. Microbiol.* 6, 579–591.

Thomas, T., Gilbert, J. and Meyer, F. (2012). Metagenomics — a guide from sampling to data analysis. *Microb. Informatics Exp.* 2, 1.

Torsvik, V. and Øvreås, L. (2002). Microbial diversity and function in soil: from genes to ecosystems. *Curr. Opin. Microbiol.* 5, 240–245.

Tringe, S. G., von Mering, C., Kobayashi, A., Salamov, A. A., Chen, K., Chang, H. W., *et al.* (2005). Comparative metagenomics of microbial communities. *Science* 308, 554–557.

Tyson, G. W., Chapman, J., Hugenholtz, P., Allen, E. E., Ram, R. J., Richardson, P. M., *et al.* (2004). Community structure and metabolism through reconstruction of microbial genomes from the environment. *Nature* 428, 37–43.

Van Guilder, H. D., Vrana, K. E. and Freeman, W. M. (2008). Twenty-five years of quantitative PCR for gene expression analysis. *Biotechniques* 44, 619.

Wagner, L. A., Weiss, R. B., Driscoll, R., Dunn, D. S. and Gesteland, R. F. (1990). Transcriptional slippage occurs during elongation at runs of adenine or thymine in *Escherichia coli*. *Nucl. Acid. Res.* 18(12), 3529–3535.

Walsh, F. and Duffy B. (2013). The culturable soil antibiotic resistome: a community of multi-drug resistant bacteria. *PLoS ONE* 8(6), e65567.

Watve, M. G., Tickoo, R., Jog, M. M. and Bhole, B. D. (2001). How many antibiotics are produced by the genus *Streptomyces*? *Arch. Microbiol.* 176(5), 386–390.

Wilkinson, D. G. (1998). *In situ Hybridization: A Practical Approach*. Oxford University Press, Oxford, UK.

Yi, H., Cho, Y. J., Won, S., Lee, J. E., Yu, H. J., Kim, S., Schroth, G. P., Luo, S. and Chun, J. (2011). Duplex-specific nuclease efficiently removes rRNA for prokaryotic RNA-seq. *Nucl. Acid. Res.* 39, E140.

9

Microbial Ecosystem Functions in Wetlands under Disturbance

Anne K. Steenbergh†, Annelies J. Veraart†, Adrian Ho and Paul L. E. Bodelier*

Department of Microbial Ecology,
Netherlands Institute of Ecology (NIOO-KNAW),
Droevendaalsesteeg 10, 6708 PB Wageningen, The Netherlands
**p.bodelier@nioo.knaw.nl*

Introduction

Wetlands are transition zones between land and water; their heterogeneous nature makes them hotspots for biodiversity and ecological productivity. They play an important role in the environment as transformers in the global cycling of various chemical elements, including nutrients and pollutants. Wetlands are formally described as ecosystems of which the soils are frequently inundated or water-saturated, and where the vegetation is predominantly hydrophytic-tolerating or adapted to life in water-saturated soil conditions.

We can distinguish several different types of wetlands, including coastal wetlands, swamps, bogs, boreal wetlands and mangroves. Due to their high productivity, wetlands have historically been important for human survival and development. They perform several key functions that are of importance to society: they store water and protect against both flooding and drought; improve water quality through retention and conversion of nutrients and contaminants;

† Both authors contributed equally.

sequester carbon through primary production; produce fuel, fibre and food — domestic wetlands like rice paddies are estimated to feed half of the global population (Mitsch and Gosselink, 2000); facilitate tourism; and lastly they harbour a huge biodiversity and a gene bank "with known and unknown functions" (De Groot *et al.*, 2002). The need for wetland ecosystem services is increasing due to anthropogenic activity (Millennium *et al.*, 2005); for example, an increase in extreme weather events can be mitigated by wetland water storage capacity and increased nutrient loading by denitrification that releases nitrogen to the atmosphere. Notwithstanding the importance of wetland functioning, the loss of wetlands and their species is even more rapid than for other ecosystems. The main driver of wetland loss is land-use change. For example, approximately half of European wetlands have been lost, mainly by drainage for agricultural use (Verhoeven, 2014). Due to this loss of wetland area, the remaining wetlands are under even stronger environmental pressure (Millennium *et al.*, 2005).

Wetland Biogeochemistry

Not only are wetlands hotspots for the diversity of animals and plants, they are also hotspots of biogeochemical functioning. A delicate balance exists in the complex network of biogeochemical processes performed in wetlands. These processes lead to an array of chemical conversions that are at the basis of all food webs. They link elemental cycles and are crucial for many of the above-mentioned ecosystem services, such as carbon storage and purification.

Compared to aerated soils, oxygen diffuses more slowly through waterlogged soils and sediments, causing steeper gradients of oxygen, redox conditions and nutrients. The oxic layer of wetland soils is commonly only a few millimetres thick; anoxic conditions prevail in most of the soil profile. Anoxic processes such as sulphate and iron reduction and methanogenesis therefore play an important role. Due to the lower mineralisation rates that can occur under anoxic conditions, the organic matter content in wetland soils can be up to 80%. The steep wetland gradients do not only occur at the water–sediment interface, but also around plant roots and due to bioturbation, leading to a highly heterogeneous environment, with ample oxic–anoxic interfaces.

The high microbial activity in wetlands makes them of major importance for greenhouse gas budgets. Many wetlands are sinks for carbon dioxide (CO_2) due to their high primary productivity (Mitsch *et al.*, 2013). However, the function of wetlands as a carbon sink is offset as they are globally the main source of the greenhouse gas methane (CH_4), produced by the final step of organic matter degradation under anoxic conditions (Petrescu *et al.*, 2015). Similarly, wetlands can be either sources or sinks of the potent greenhouse gas nitrous oxide (N_2O), which has a 300 times higher global warming potential than CO_2. Nitrous oxide is produced as a by-product of nitrification and denitrification and can be reduced to molecular nitrogen by denitrifying microbes when conditions are favourable. Small changes in environmental conditions can shift wetlands — and their biogeochemically active microbiomes — from performing ecosystem services to contributing to atmospheric pollution. Therefore, understanding wetland functioning under global change is essential for the understanding of global change feedbacks.

Main microbial-mediated carbon and nitrogen conversions in wetlands

Carbon

Wetland microbes play a crucial role in the global carbon cycle. Wetland plants and their associated rhizosphere create ample oxic–anoxic interfaces which are hotspots of microbial activity. In addition, wetland plants can produce large amounts of biomass and reduce flow rates which leads to the settling of large amounts of organic matter on wetland sediments. This organic matter can either be degraded aerobically in the surface layer of the sediment, through anaerobic respiration using nitrate, iron, manganese or sulphate as electron acceptors, or through fermentation in deeper sediment layers. Both aerobic respiration and anaerobic fermentation produce CO_2, whereas the products of fermentation are small hydrocarbons such as acetic acid. Under anoxic conditions, these compounds can be further reduced to CH_4 in a process called methanogenesis. This is the final step of anaerobic organic matter degradation. Methanogenesis is carried out by strictly anaerobic Archaea belonging to the *Euryarchaeota*, which are nearly exclusively dependent on methanogenesis for energy and growth. In wetlands, the most important substrates for methanogenesis are acetate and CO_2 and hydrogen, but other small hydrocarbons (methanol, methylamines) can be degraded to CH_4 as well (Bridgham *et al.*, 2013). Acetoclastic methanogens form methane from acetate, whereas hydrogenotrophic methanogens use hydrogen to reduce CO_2.

Methane can in turn be aerobically or anaerobically oxidised to CO_2. In wetlands, aerobic CH_4 oxidation is considered to be the dominant pathway of CH_4 oxidation, although recently it has been demonstrated that anaerobic oxidation of CH_4 can be substantial in wetland soils (Segarra *et al.*, 2015) and a recently discovered archeon oxidising CH_4 with nitrate as electron acceptor may be involved (Haroon *et al.*, 2013). Obligatory aerobic methanotrophic bacteria are most active at oxic–anoxic boundary layers in sediment and rhizosphere, in which both CH_4 and oxygen are available. Aerobic CH_4 oxidisers belong to different phylogenetic groups, including Alpha- and Gammaproteobacteria and Verrucomicrobia and display a strong biogeography hinting at traits or trait combinations facilitating the occupation of various environmental niches (Knief, 2015). The recently discovered candidate phylum NC10 is a special case of a CH_4 oxidiser which uses nitrite to form the oxygen required to oxidise CH_4 (Ettwig *et al.*, 2010).

Nitrogen

Nitrogen occurs in wetlands in many different forms. Organic nitrogen, ammonium, nitrate and nitrite can all be used and transformed by wetland microbiota. Organic nitrogen from biomass can be mineralised to ammonium in various degradation processes. Ammonium can be nitrified to nitrite, and subsequently to nitrate under oxic conditions. These two nitrification steps, ammonium oxidation and nitrite oxidation, were until recently thought to be carried out by distinct microbial guilds: ammonia-oxidising bacteria (AOB) and ammonia-oxidizing archaea (AOA), on the one hand, vs. nitrite-oxidising bacteria, on the other hand. However, in 2015, organisms carrying out complete nitrification (comammox) have been discovered within the Nitrospira genus of the Nitrospirae phylum, which deepens our understanding of the nitrogen cycle (van Kessel *et al.*, 2015; Daims *et al.*, 2015). Nitrate, derived from nitrification or hydrological influx, can be reduced in a stepwise fashion to dinitrogen gas in the denitrification process, which uses organic carbon (or sulphate) as terminal electron acceptor. The ability to denitrify is widespread in the microbial realm (Zumft, 1997). Most denitrifiers are facultative anaerobes and therefore denitrification is usually inhibited by the presence of oxygen. Different steps of denitrification (reduction of nitrate, nitrite, nitric oxide and N_2O) can be carried out by different members of the microbial community, but many denitrifiers are also capable of complete denitrification. Incomplete denitrification results in the emission of N_2O (Zumft, 1997), which is a strong greenhouse gas with a 298 times higher global warming potential than CO_2 (Solomon *et al.*, 2007). Therefore, there is a strong interest in maintaining low N_2O:N_2 ratios when optimising the denitrification efficiency of constructed wetlands. In wetland sediments, denitrifying organisms compete for nitrate and organic carbon with nitrate ammonifiers; these are organisms capable of dissimilatory nitrate reduction to ammonium (DNRA). They are active under the same conditions as denitrifiers, albeit favoured by different $C_{organic}$:N ratios (Tiedje, 1994; van de Leemput *et al.*, 2011). Furthermore, under anoxic conditions, anammox organisms can use nitrite to oxidise ammonium, which forms dinitrogen gas, and unlike denitrification and nitrification does not produce the greenhouse gas N_2O as a by-product (Mulder *et al.*, 1995). However, due to its fairly recent discovery and current *in situ* methodology, the contribution of anammox to wetland N-cycling remains unclear. Anammox could be of importance in anoxic and nitrite-rich sites like constructed wetlands or wetlands strongly impacted by nitrogenous pollution. Lastly, various microorganisms (diazotrophs) can fix dinitrogen gas into their biomass, although, as this process is energetically costly, nitrogen fixation usually only takes place under nitrogen-limited conditions. N-fixing microorganisms in wetlands can be either free-living or symbiotic, and both autotrophic (cyanobacteria, i.e. belonging to the nostocales, which can occur in dense mats on the sediment surface; Rejmánková and Komárková, 2000), methanotrophic (methanotrophs belonging to the alpha- and gammaproteobacteria; Vile *et al.*, 2014) and heterotrophic (i.e. sulphate reducers *Desulfovibrio* sp. associated with roots; Brown *et al.*, 2003). Because of the important role of oxygen and carbon and nitrogen substrate availability for these processes, the main drivers of biogeochemical processes in wetlands are hydrology and nutrient availability, with interacting effects of temperature, pH and vegetation.

Disturbances of Wetlands

According to Environmental Protection Agency (EPA) (http://www.epa.gov/wetlands) and World Wildlife Fund (WWF) (http://wwf.panda.org/about_our_earth/about_freshwater/intro/threats/), the major threats to wetlands are changes in hydrology, pollution and damage/change of vegetation, all of which are either the direct consequence of human action or indirect effects of climate change. Below we give an introduction to these wetland disturbances, outlining their causes and consequences for wetland ecosystem functioning.

Climate change: temperature

Climate models predict a global temperature rise of more than 2°C over this century (IPCC, 2014). Arctic areas, where natural wetlands are still abundant, will likely face twice this temperature increase (Serreze *et al.*, 2009; Winton, 2006). All processes occurring in wetlands are, to some extent, affected by altered temperatures, with most notable changes in wetland biogeochemistry. Temperature change can have both abiotic (such as substrate solubility and diffusion) and biotic effects, for example, by changing enzyme activity and altering competitive outcomes due to differential temperature optima and tolerances of competing populations. All biological processes are temperature-dependent, due to chemical kinetics and associated temperature effects on the enzymes involved. Rates of most biological processes double with a 10°C temperature increase, which is referred to as the Q_{10} value, albeit with process- and species-dependent variation, leading to Q_{10} values of 1.31–5.13 for a 0–40°C temperature range (Dell *et al.*, 2011). However, effects on individual processes are variable and depend on availability of substrates, which is in turn again affected by temperature, for example, by enhanced mineralisation (Kirschbaum, 1995), and changed solubility. In nature, many processes are linked and overall outcomes of temperature change are therefore difficult to predict. An example of amplified temperature dependence due to coupled processes is denitrification, the conversion of nitrate to gaseous nitrogen. Here, the sole temperature dependence of the denitrification enzymes is amplified by differential effects on two co-occurring processes: respiration and photosynthesis. Because respiration has a higher temperature dependence than photosynthesis, oxygen concentrations decrease, which stimulates denitrification. In addition, oxygen solubility is temperature-dependent: warmer water can contain less oxygen, which further increases the temperature effect on denitrification (Veraart *et al.*, 2011). Rising temperature can thus lead to increased denitrification rates and potentially increased release of its by-product N_2O.

Like denitrification, temperature effects on CH_4 fluxes can be complex, as different biological processes are involved. Methane production and CH_4 oxidation in peat slurries showed similar temperature optima at around 25°C, but CH_4 production showed a higher temperature dependence (Q_{10} = 5.3–16) than CH_4 oxidation (Q_{10} = 1.4–2.1), indicating a potential feedback of warming on CH_4 emissions (Dunfield *et al.*, 1993). In line with this observation, a recent meta-analysis shows that the temperature sensitivity of methanogenesis is considerably higher than that of respiration and photosynthesis (a 57-fold increase between 0 and 30°C), which can result in considerable increases in CH_4 fluxes from wetlands with global warming (Yvon-Durocher *et al.*, 2014).

A specific area of concern is the arctic permafrost, where small temperature increases lead to thawing, creating and expanding arctic wetland areas and changing their hydrology (Van Huissteden *et al.*, 2011; Walter *et al.*, 2006; see also Chapter 10). As permafrost soils thaw, moisture content increases, the organic-rich anaerobic zone increases and microbial activity rises. This alters local vegetation and increases sites of CH_4 production, while at the same time water tables rise, which decreases the size of the oxic layer in which CH_4 oxidation takes place. These processes combined could lead to significantly enhanced CH_4 fluxes from warmed arctic wetlands and lakes, although the extent of this temperature effect is still debated. For example, Wik *et al.* (2016) report that extension of ice-free seasons by 20 days can result in doubling of CH_4 emissions from lakes and ponds in northern latitudes by the end of this century. However, enhanced lake drainage and increased CH_4 oxidation could (partly) offset CH_4 emissions caused by temperature-induced increases in methanogenesis (Van Huissteden *et al.*, 2011; Lofton *et al.*, 2014). Hydrologic models show that lake drainage will also be enhanced as permafrost thaws, controlling the expansion of arctic lakes, thereby reducing the impact of thaw on the increase in CH_4 emissions (Van Huissteden *et al.*, 2011). Thus, warming of wetlands can have large effects on budgets of the potent greenhouse gases N_2O and CH_4 and can therefore become a considerable climatic feedback, but uncertainty around this feedback is fairly large. This contributes to the uncertainty of climate prediction models, which is even further complicated due to the co-occurrence of global warming with other wetland disturbances.

As described in the examples above, warming-induced changes in solubility and oxygen demand can have a considerable impact on the release of greenhouse gases, resulting in climatic feedbacks. Besides these effects, warming can also have numerous other effects on the functioning of wetlands. Here we highlight how warming-induced changes in oxygen concentration affects wetland biogeochemistry. As described previously, warming leads to enhanced biological demand and reduced oxygen solubility. The combination of these effects leads to anoxia, which will switch microbial communities from aerobic to anaerobic metabolism. This in turn results in a release of different products and intermediates (N_2, NH_4^+, N_2O, Mn^{2+}, Fe^{2+}, H_2S, CH_4), affecting a plethora of (micro)biota. Importantly, anoxia can result in the release of iron-bound P, increasing P-eutrophication, while at the same time denitrification can be enhanced, increasing N-limitation. Elemental imbalances (in this case, the reduced N:P ratio) can have cascading effects on the foodweb (Sterner and Elser, 2002). This type of interaction makes it difficult to predict overall competitive outcomes and wetland functioning from prolonged warming.

Climate change: drought and salinity

Earth system climate prediction models anticipate changes in precipitation patterns, leading to increased duration of drought and subsequent rewetting events (IPCC, 2014). Increased frequency in drought-rewetting events imposes stress on the wetland microorganisms, which may have adverse consequences on key microbially mediated biogeochemical processes by altering the soil quality (e.g. soil organic matter content, soil physical parameters) and nutrient concentrations (e.g. carbon, nitrogen) (Denef *et al.*, 2001; Mikha *et al.*, 2005). Although studies targeting broad microbial groups (phyla level) suggest the remarkable resilience of microorganisms to (episodic) drought (e.g. Fierer *et al.*, 2003; Evans and Wallenstein, 2012), recurring desiccation-rewetting events can have cumulative effects on specific soil processes like CH_4 oxidation over time (Ho *et al.*, 2016). Similarly, a prolonged drought event (>15 years) significantly reduces CH_4 oxidation upon rewetting (Collet *et al.*, 2015). These studies exemplify the vulnerability of specific microbial processes over time with the projected changes in precipitation patterns.

The salinity of wetlands is increased by rapidly evaporating surface water (drought) by concentrating dissolved ions (e.g. Na^+, Ca^{2+}, Cl^-), a process known as salinisation. In coastal wetlands, salinisation can also be caused by salt water intrusion from the marine environment, a phenomenon that can be exacerbated by increasing sea water levels. In addition, withdrawal of the ground water level may also result in salt accumulation on the surface soil layer. Although microorganisms possess different strategies to overcome osmotic stress induced by salinisation, the accumulation of salts has a pronounced effect on microbial processes. In soils, for instance, salinisation inhibits cell growth and has (in)direct effects on enzyme activity, for example, reduces dehydrogenase, while stimulating oxidoreductase activity (Pankhurst *et al.*, 2001). In addition, salinisation changes the soil physico-chemical characteristics, indirectly influencing the solubility of organic matter, and alters the availability of soil organic matter for decomposition. Considering more specific processes, high salinity poses an indirect effect on CH_4 production, lowering total CH_4 emissions in a wetland (see section below; Baldwin *et al.*, 2006). The consequences of salinisation are confounded by salinity-induced pH shifts and changes in vegetation coverage which changes the ecosystem water balances (Rath and Rousk, 2015).

Although the sensitivity of microorganisms to high salt concentrations is well known, currently little attention is given to the effects of less extreme salinisation levels on the response of microorganisms that perform wetland ecosystem functions (Laanbroek *et al.*, 2012).

Changes in vegetation

Wetlands have been called the "the kidneys of the landscape" and "biological supermarkets" referring to their crucial role in chemical and hydrological cycles and to the essential role they play in provision of food and supporting biodiversity (Daryadel and Talaei, 2014). The capacity of wetlands to retain water and to transform and retain nutrients is intimately linked with wetland vegetation and key to their productivity and their crucial role in modulating flows of carbon, nutrients and pollutants ending up in coastal and marine ecosystems. Wetland plants, whether emerging above the water surface (i.e. helophytes) or having their shoots and leaves under water (i.e. submerse macrophytes), play a crucial role in the functioning of wetlands and the ecosystem services they provide. They provide breeding and nursery sites, resting areas for migratory species, refuge from predators and, by being the most important source of organic matter, are an important food source for many invertebrates and fish both in the wetland itself and in associated aquatic systems. In addition, wetland plants physically reduce the amount and flow-rate of terrestrial runoff which in combination with their rooting activity reduces erosion of streambanks, shorelines and coastlines. Plants also play a crucial role in the global wetland budgets of greenhouse gases. Due to the total balance between sequestered carbon and emitted greenhouse gases (CO_2, CH_4, N_2O), wetlands are sinks of greenhouse gases, thereby contributing to cooling our planet (Mitsch *et al.*, 2013) when calculated over a 100-year time period. Fundamental to any of the important functions wetland plants have are the adaptations they have developed to equip them for colonisation and growth in anoxic flooded soils and sediments (Blom, 1999). The most important adaptation is the development of the so-called aerenchymatous tissue, which creates a gas-space continuum that stretches from the stomata in the leaves to the tips of the roots, facilitating the diffusion of gases from the atmosphere to the roots and vice versa. In this way, the roots are supplied with oxygen for respiration, but gases from the soil (i.e. CH_4, N_2O) can also escape to the atmosphere (Bodelier *et al.*, 2006). Roots do not only consume all oxygen, but also lose substantial amounts of oxygen to the surrounding anoxic soil, thereby creating additional oxic–anoxic interfaces in wetland soils. Combined with the input of carbon (exudation of photosynthetates or decomposing roots), this facilitates the simultaneous activity of aerobic as well as anaerobic microorganisms that catalyse redox reactions (see "Wetland biogeochemistry"; Figure 9.1) resulting in recycling of carbon, nitrogen, sulphur and iron, which is the basis of the high productivity in wetlands. Physical and chemical characteristics such as climate, topography, geology, hydrology and inputs of nutrients and sediments determine the rate of plant growth and plant community composition in wetlands (Mitsch and Gosselink, 2000; Lamers *et al.*, 2012). Almost all of these factors are influenced by human activity directly or indirectly through climate change. Human-made changes in hydrology (drainage, dredging, diking, damming, diversion of flow channels) and hydrology changes due to climate change (drying, flooding with fresh- or salt water) can eradicate total populations or modulate species composition and/or biomass (Janse *et al.*, 2015). Increases in nutrient load and/or pollutants as well as grazing by domestic animals and the establishment of invasive species are

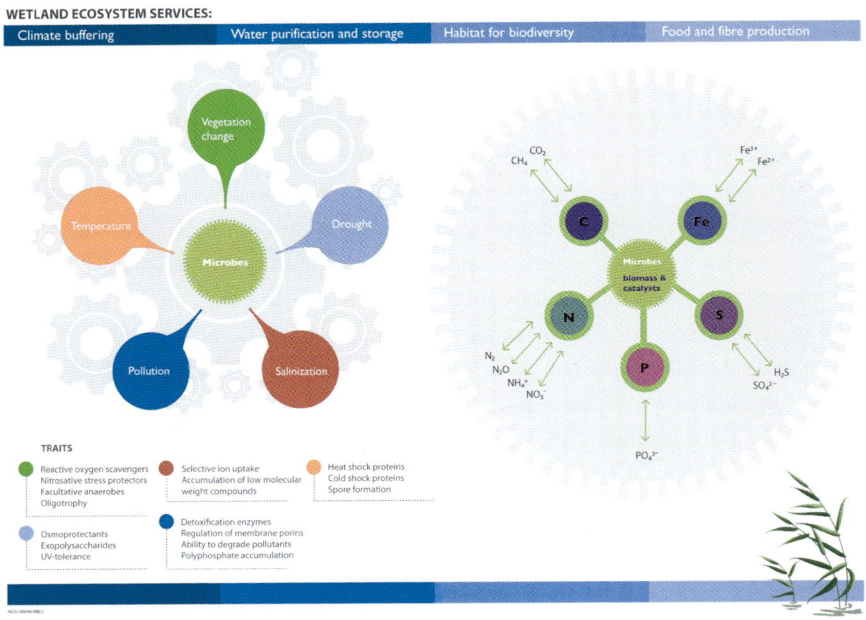

Figure 9.1 Threats to wetland ecosystem services and microbial traits involved in coping with the resulting stress. (a): Examples of microbial traits relevant for wetland disturbances and related ecosystem functions. (b): The main microbial carbon, nitrogen, phosphate, sulphur and iron conversions.

Source: NIOO-KNAW.

eminent threats to wetland vegetation (Daryadel and Talaei, 2014). It is expected that these stressors to freshwater wetlands will continue to dominate as human demand for water resources grows, accompanied by increased urbanisation and expansion of irrigated agriculture (Settele *et al.*, 2014). On top of that climate change will have significant impacts, from altered thermal regimes, altered precipitation and flow regimes, and, in the case of coastal wetlands, sea level rise (Settele *et al.*, 2014), all of which can either lead to local decrease or increase in vegetated wetland area. The combined effect of elevated CO_2 and temperature can have stimulating effects on wetland plant productivity and can lead to a range of shifts in species turning, for example, arctic wetlands into vascular plant-dominated wetlands with consequences for carbon sequestration and greenhouse gas emissions (Van Groenigen *et al.*, 2011; Settele *et al.*, 2014). Modelling the effects of climate change on wetland area and radiative forcing is prone to large uncertainties (Kirschke *et al.*, 2013; Melton *et al.*, 2013) which become even more uncertain when taking confounding effects into account, such as the unknown feedbacks of increased nitrogen deposition to wetland functioning and radiative forcing (Melton *et al.*, 2013).

In conclusion, wetland functioning will be significantly influenced by any factor affecting the presence, abundance and composition of wetland vegetation. Predicting implications for wetland biogeochemistry (nutrient cycling, carbon sequestration, greenhouse gas emissions) and forming appropriate management strategies will require assessment of the responses of the microbes catalysing these reactions upon changes in wetland vegetation.

Pollution

The effects of wetland pollution lead to loss of biodiversity, can hamper drinking water production, cause the loss of fishing grounds and impact the recreational use of wetlands (Millennium *et al.*, 2005). These effects can be exacerbated by the interaction with other wetland disturbances: increasing temperatures due to climate change can enhance the adverse effects of nutrient loading, leading to larger wetland areas with oxygen-depleted waters. The disruption of wetland ecosystem functioning due to pollution facilitates the success of invasive species, whether this be plants or animals, or antibiotic-resistant bacteria and faecal bacteria (e.g. Crooks *et al.*, 2011; Staley *et al.*, 2014). In addition, due to conversion of wetlands to agricultural land and associated water demand for irrigation, the effects of pollution on remaining wetlands will intensify (Millennium *et al.*, 2005).

Pollution can be from point sources (e.g. an oil spill, sewage overflow, chemical dump) or non-point sources (e.g. nutrients in agricultural runoff, atmospheric deposition) and occurs from a wide range of substances and even energy (light and heat), which have different effects on wetland ecosystem functioning. For many types of pollution, both natural and constructed wetlands act as a filter, thereby reducing the effect of the pollutant for downstream systems. Due to the high microbial activity and the functional breadth of microbial processes occurring in wetlands, many organic (carbon-containing) pollutants such as organic pesticides (e.g. glyphosate; Carlisle and Trevors, 1988), pharmaceuticals (hormones, antibiotics; Kümmerer, 2009a, b; Brausch *et al.*, 2012), household and personal care products (Matamoros *et al.*, 2009), and oil, can be degraded in wetlands. Other pollutants, such as heavy metals, can be retained in wetlands through bioaccumulation. Although bioaccumulation does not remove pollutants from the environment, the concentration of pollutants in organisms generally makes them less harmful to the environment and can facilitate their actual removal. However, this comes at the expense of organisms that are higher in the food chain (see Bertrand *et al.*, 2015). Another form of wetland pollution occurs through the introduction of viruses, faecal bacteria and antibiotic-resistant bacteria from, for example, wastewater treatment plants (Davies *et al.*, 1995; Fong and Lipp, 2005). Whether these bacteria can survive and multiply in wetlands depends on their ability to compete for resources with resident bacteria (Staley *et al.*, 2014).

Notwithstanding the large and diverse filtering capacity of wetlands, this capacity is finite. When pollution pressure increases, wetlands can reach a tipping point at which the normal functioning of the wetland ecosystem is largely lost. An increase in wetland eutrophication can lead to oxygen depletion, acidification and the production of toxic compounds during algal blooms. The transition from a seemingly healthy wetland to a degraded, eutrophic state can occur abruptly, and once this conversion has taken place, it is often difficult to return the system back to its original state (Scheffer *et al.*, 2001; Bayley and Prather, 2003).

In many environments, the most important source of wetland pollution is diffuse pollution from agriculture. Not only can agrochemicals such as pesticides and nitrification inhibitors (used in conjunction with fertilisers) place a burden on wetlands, but also nutrient loading leading to eutrophication is a major threat to wetland functioning. Wetlands can mitigate nutrient loading, mostly in the form of nitrogen and phosphorus, for downstream aquatic systems in several ways. When nitrogen or phosphorus is limiting primary productivity, an increase in these nutrients can result in higher biomass production in wetlands. When a part of this increased biomass is retained in the wetlands, the nutrient flux out of the wetland will be lower than its influx. In addition, denitrifying microorganisms in wetlands can convert reactive nitrogen (e.g. ammonium and nitrate) into molecular nitrogen (N_2), which is gaseous and can diffuse out of the water into the atmosphere. In this way, wetlands can reduce the nitrate concentration by more than 80% (Millennium *et al.*, 2005).

In contrast to nitrogen, phosphorus cannot easily be removed from wetland ecosystems through microbial activity. Wetlands mainly mitigate increased phosphorus inputs by storage in sediments through sorption, precipitation and accumulation of organic matter (see Dunne and Reddy, 2005 for a review on phosphorus biogeochemistry in wetlands). However, as stored phosphorus remains in the system, it can be released from the sediment under changing environmental conditions, leading to a eutrophic wetland state even if external phosphorus inputs are low. Although many mechanisms responsible for this internal loading are known, it is not clear which mechanisms are dominant under which environmental conditions (Søndergaard *et al.*, 2003). Microbial activity plays an important role in phosphorus cycling in sediments (Steenbergh, 2012); differences in the active microbial community between sediments may explain the idiosyncratic behaviour of internal phosphorus loading.

Because of the prominent role of wetlands in greenhouse gas cycling (CH_4 and N_2O, but also storage of carbon in waterlogged soils), wetland pollution is likely to impact global greenhouse gas budgets. However, the effects of increased nitrogen and phosphorus availability on greenhouse gas fluxes are difficult to predict. As for all life, nitrogen and phosphorus are needed by microorganisms for growth, but the nutrient requirements for optimal growth are specific to each microbial species. Changing nutrient availability can therefore lead to shifts in microbial community composition, which in turn can be accompanied by changes in microbial functioning. The effects of nitrogen are more difficult to predict than for phosphorus, as nitrogen is not only a nutrient, but also a substrate, inhibitor and toxicant for microorganisms involved in N_2O and CH_4 cycling (Bodelier and Steenbergh, 2014b). As the input of reactive nitrogen is projected to increase by approximately two-thirds by 2050 (on top of the doubling since pre-industrial times), a thorough understanding of the effect of nitrogen on wetland biogeochemical cycling is needed to predict, and mitigate, future greenhouse gas budgets.

Predicting Wetland Microbial Functioning under Disturbance

As described in the previous section, wetlands face various types of disturbances, which impact microbiota and their biogeochemical functioning. To predict and mitigate wetland functioning, we need a thorough understanding of wetland responses to disturbances. In this section, we will outline possible ways to achieve this.

Traditional (geochemical) wetland research has focused on changes in physicochemistry and hydrology to predict wetland functioning. In this top-down approach, all microbial processes are considered as a black box in which the input and output are of interest. The microbes are considered to be catalysts that will perform reactions when the conditions are thermodynamically favourable. This approach is based on the notion that due to the huge microbial diversity, the right microbe (thus the right catalyst) is always present (*sensu* Baas Becking; "everything is everywhere, but the environment selects" (O'Malley, 2008). Microbial communities are thought of as highly functional redundant and resilient, rapidly restoring functioning after disturbance. In this approach, changes in wetland functioning after disturbance are predicted only from changes in physicochemistry (such as substrate availability). Although this black-box approach is, to a large extent, effective for predicting the short-term effects of disturbance, for certain scenarios the contents of the black box cannot be ignored. For example, microbes perform more than one function; it is impossible to measure them all. Also, the extent of functional redundancy depends on the environmental conditions, as species roles and interactions differ between environments (Fetzer *et al.*, 2015). Lastly, using a black-box approach, it is difficult to predict feedback from disturbances and interactions between elemental cycles.

The field of microbial ecology is concerned with elucidating the microbial communities and processes in the black box. Wetland microbial functioning is studied using a bottom-up approach, traditionally by mainly focusing on small-scale predictions of single processes in culturing efforts. In the last decade, the toolbox of the microbial ecologist has rapidly expanded. For example, the advances in molecular techniques (i.e. next-generation sequencing) have led to an unprecedented availability of data on environmental microbial diversity and identity. These techniques facilitate upscaling from laboratory to environmental contexts. However, when predicting community responses, insights into microbial functions rather than identity are preferred. Currently, links between identity and function are often unknown; microbes may have unknown capabilities, and potential functions predicted from the genome may only be realised under the right environmental conditions. Such insights can be gained using techniques such as stable isotope probing (Dumont and Murrell, 2005) and novel visualisation techniques (i.e. nanoSIMS, RAMAN microscopy; Wagner, 2009). In addition, culturing of microorganisms remains an important way to link identity to function, study combinations of functions within each microorganism and ultimately observe their response to environmental change. Recently, interest in functional approaches has been rekindled, for example, using trait-based approaches or functional biodiversity concepts (Green *et al.*, 2008; Litchman and Klausmeier, 2008; Crowther *et al.*, 2014; Krause *et al.*, 2014). Below, we will briefly introduce trait concepts and their use to predict ecosystem functioning under disturbance.

Microbial traits

Many different trait definitions exist, but they all boil down to traits being any characteristic of an organism that in one way or the other influence their fitness (Violle *et al.*, 2007; Krause *et al.*, 2014). For example, the ability to fix dinitrogen is a useful trait for organisms that face nitrogen limitation, and drought resistance is a useful trait under dry conditions. Note that traits can be defined at different levels of organisation (gene, gene cluster, function) and that different cell mechanisms can result in similar functional traits. For macroorganisms, trait values are often directly measurable, for example, the leaf area of plants and tail length are straight forward to measure. Moreover, many traits show a correlation across species, for example, Crassulacean Acid Metabolism (CAM) metabolism in plants often co-occurs with swollen, fleshy, juicy leaves such as in succulent plants. As a result, the occurrence of one trait can predict the presence of a whole suite of related traits. Also, for macroorganisms, taxonomy is well defined, enabling predictions of traits and thereby functioning based on taxonomic knowledge. In contrast, for microorganisms, the species concept does not hold and therefore taxonomy is heavily debated, and traits of individual organisms are difficult to measure. Also the relation between function and phylogeny is weaker than for macroorganisms: some functions are taxonomically conserved, whereas others are scattered across the phylogenetic tree (Martiny *et al.*, 2013, 2015). Traits that do show a strong phylogenetic signal can be referred to as deep traits, for example, optimal pH, CH_4 oxidation and methanogenesis. On the other hand, shallow traits show a weak phylogenetic signal, for example, antibiotic resistance which can be obtained by lateral gene transfer, and the ability to grow on a particular carbon source which occurs in many different taxonomic groups (Martiny *et al.*, 2013). Due to the difficulty of measuring microbial traits, genetic information of microbes is currently often used as a proxy for traits (e.g. Litchman *et al.*, 2007; Martiny *et al.*, 2013, 2015). However, microbial gene annotation is still in its infancy, even of the best annotated microbial genomes, 30% of the predicted genes are of unknown function (Bork, 2000). In addition, the mere presence of a gene in a genome does not warrant its functionality, and trait functioning in itself is dependent on the environment. Hence, culturing approaches are needed to measure actual functioning in order to address these issues and if possible to validate the link between genetic information and traits.

Trait combinations

Microbes are often considered to be functionally redundant. However, while this can hold when only single functions are considered (e.g. respiration), the concept of functional redundancy does not hold for the multitude of functions that is harboured within each individual organism. Even if microbial communities are functionally redundant for functions related to their energy metabolism (such as denitrification, respiration, degradation of specific compounds), disturbances can cause major shifts for other functions (such as drought resistance, temperature tolerance).

As already mentioned above, many traits correlate positively with each other, but trade-offs between traits are also commonly observed. A classic example of such a trade-off in microbial cells is the SPANC balance (Self-Preservations And Nutritional Competence), which regulates the trade-off between self-preservation (pollution resistance) and nutrient compatibility (Ferenci, 2005). In addition to correlating traits, some traits co-occur in species without an apparent relation, making their occurrence within communities difficult to predict. Therefore, to predict community shifts after environmental change, knowledge of trait combinations within organisms is crucial. Such data on population-specific trait combinations can be obtained from culturing studies, in some situations combined with omics data.

Using trait-based approaches to improve modelling efforts

When assessing the impact of environmental change (climate change, land use, pollution, etc.) on ecosystems and ecosystem functioning, models of various scales and complexity are an indispensable tool to aid predictions used for possible policy and management measures. One of the most well-known institutions relying heavily on these models is the Intergovernmental Panel on Climate Change (http://www.ipcc.ch/). This panel synthesises the scientific knowledge available and makes predictions of climate change on earth systems and advises on possible mitigation strategies. The so-called Earth System Models (ESM) and Global Circulation Models (GCM) connect the Earth's biotic and abiotic environments with the atmosphere together for predicting the climate on Earth. Very important in these models is the amount of carbon sequestered or released and the amount of greenhouse gases emitted into the atmosphere, which are processes in which microbes play crucial roles. However, the properties of soil microbes are not explicitly incorporated in soil organic matter biogeochemistry and these models assume that biological responses to changes in the physical conditions are invariable across ecosystems and through time (Wieder *et al.*, 2015a), leading to estimates of carbon stored in soils that vary by sixfold (Wieder *et al.*, 2015b). The emission of CH_4 from global wetlands is an important factor in global warming and therefore is intensively measured and modelled. However, current wetland-CH_4 models vary by 40% in predicting the global annual emissions, and predictions for the year 2100 are highly uncertain (Kirschke *et al.*, 2013). One of the crucial factors in these uncertainties is the future wetland area. In addition, abiotic–biotic interactions causing nonlinear feedback on the climate system can be responsible for a large amount of uncertainty; temperature adaptation of microbes has consequences for decomposition; changes in plant performance due to elevated CO_2 changes below-ground processes with repercussions on biogeochemical cycles; enhanced nitrogen deposition in combination with elevated CO_2 and temperature causes changes in microbial carbon degradation and carbon-use efficiency in ways not incorporated in these models. Explicitly modelled microbial traits in carbon decomposition models increased their predictive power or at least performed as well as traditional ones (Treseder *et al.*, 2012; Wieder *et al.*, 2013, 2015a). For example, by incorporating copiotrophic (i.e. fast growing, adapted to high nutrient availability) and oligotrophic (i.e. slow growing adapted to low nutrient availability) functional groups (Wieder *et al.*, 2015b) or by including dormancy into carbon decomposition models (Wang *et al.*, 2015), an increased predictive value was obtained. Model predictions of the effects of water table manipulations on CH_4 emissions from peatlands were significantly improved by including a variable recovery time of methanogenic species after exposure to drought and oxygen (Kettunen, 2003). Incorporation of microbial traits in these examples perform particularly well in predicting processes or functions after environmental change or disturbance such as lowering or raising water table or nutrient flushes or deprivation, situations where the classical models perform poorly. However, when incorporating microbial traits into ESM and GCM models, detailed knowledge is necessary on the breadth of responses and trade-offs in various microbial groups and processes. A community-based model of a well-characterised microbial functional group, combining phylogeny, physiology and ecology of nitrifiers demonstrates that nitrification and associated N_2O production can be explained very well along a range of geochemical gradients (Bouskill *et al.*, 2012). In this study, community composition could predict process and vice versa. However, this example is extraordinary in the way that the diversity of cultivated and characterised nitrifying organisms were similar to that observed in the environment. For the vast majority of microbes, we do not yet have the information to link community structure to functioning. In a recent meta-analysis, community composition data (16S rRNA, diversity and abundance of functional genes) was used to predict carbon and nitrogen cycling processes. Most variation in processes was explained by environmental variables, however community composition data improved predictions significantly, especially for the so-called "narrow" phylogenetic guilds (e.g. nitrifiers) and processes carried out by facultative microbes (e.g. denitrifiers) (Graham *et al.*, 2016). This meta-analysis showed that combining environmental and community data is highly effective when predicting isolated processes in a one-to-one relationship with microbial community composition and abundance data. However, in order to predict the effect of disturbances for processes that are integrated in ecosystems and therefore interact with each other, a good understanding of community composition in relation to traits is necessary. This is, for example, the case in aquatic foodweb-based models used to predict clear vs turbid water states (Janssen *et al.*, 2015; Kuiper *et al.*, 2015), or in global dynamic vegetation models where inclusion of plant-trait variation reduces the projected global soil carbon sink (Verheijen *et al.*, 2015).

On top of improving model predictions, the strength of trait-based approaches lies in informing process-based (black box) models of unknown processes and their modulators. In this way, by including the relevant environmental variables in the models, their predictive power can be increased.

The knowledge that most dominant cyanobacteria in oceans need more iron than other phytoplankton to facilitate nitrogen fixation and subsequent CO_2 fixation (Chappell *et al.*, 2012) has led to iron fertilisation experiments of oceans to increase carbon fixation from the atmosphere (Blain *et al.*, 2007). Conversely, process-based observations have led to trait-based studies of the microbes involved. For example, methanogenesis in lakes has been shown to occur at oxic conditions (Grossart *et al.*, 2011) and CH_4 has been shown to be oxidised under anaerobic conditions (Ettwig *et al.*, 2010; Haroon *et al.*, 2013), while nitrate production from ammonia has recently been shown to be carried out by one microbe (Daims *et al.*, 2015; van Kessel *et al.*, 2015). The organisms involved have been enriched or isolated and a range of all traits assessed which are not yet included in models. As challenging as it may be, inclusion of these novel but also known traits and trait variation into models will generate novel insights into the dynamics and feedbacks of processes facilitating hypotheses generation and design of experiments and sampling campaigns, from which wetland management and conservation will also profit.

Traits of Importance to Wetland Functioning under Disturbance

Microbial traits and trait sets involved in temperature responses

How temperature affects wetlands and their microbiomes

As described in this section about temperature effects on wetland processes, all biological reactions are in some way affected by temperature. Most reaction rates can be described by the Boltzmann–Arrhenius model, which describes an exponential increase in reaction rates with rising temperature. Cold- and warm-adapted organisms form exceptions, having high process rates outside of the linear Arrhenius range. Microorganisms adapted to cold conditions are especially important in wetlands, as many wetland areas occur in temperate and boreal regions. We can distinguish cold-loving "Psychrophyles", which grow best at temperatures below 20°C, and many of these tolerate sub-zero temperatures, while cold-tolerating "Psychrotolerants" survive in cold conditions but grow best above 20°C. Both psychrophyle and psychrotolerants microbes have several ways to deal with low temperatures, including altered membrane lipids, proteins, nutrient uptake systems, synthesis of cryoprotectants and adapted energy metabolism (reviewed in Margesin *et al.*, 2006). These adaptations enable them to function at low temperatures. Psychrotolerants species dominate in areas with large diurnal or seasonal temperature fluctuations such as continental areas and will be abundant in temperate and arctic wetlands.

General traits dealing with temperature stress

Temperature tolerance of microorganisms is regulated in several different ways. Best described for microbial processes is their possible protection or tolerance against heat, which may be relevant for microbes living in rice paddies (both due to their geographic location and solarisation practises) and in areas with strong diurnal fluctuations, which occur in semi-arid continental regions and at higher altitudes, where wetland areas may be found along river beds or shallow lakes. Heat can denature proteins and disrupt cell structures. Heat-shock proteins help the cells cope with sudden temperature increases (Neidhardt *et al.*, 1984; Gross, 1996; Tobor-Kapłon *et al.*, 2006). Induction of these proteins is not only triggered by temperature, but also by other external cues (osmotic shock, heavy metal presence, presence of aromatic compounds; Ramos *et al.*, 2001; Griffiths and Philippot, 2013), which hints at cross-protection. Similarly, microbes can cope with lower temperatures by the production of cold-shock proteins (involved in supercoiling DNA and initiation of translation, which is likely to be the limiting step when temperature suddenly decreases), and the repression of heat-shock proteins (Jones and Inouye, 1994). For example, when temperature abruptly decreases (from 37 to 10°C), *Escherichia coli* enters a lag-phase after which it resumes growth, albeit at lower growth rates (Jones *et al.*, 1987). Other ways to cope with temperature stress are spore formation, structural adaptations such as the possession of thick cell walls (Griffiths and Philippot, 2013) and the ability to alter membrane proteins to maintain fluidity at low temperatures (Methé *et al.*, 2005).

Community effects hinting at traits of importance in wetlands

Temperature can affect microbial communities in wetlands in numerous ways. Warming will affect microbial functioning, as described above, which alters biogeochemistry and will likely shift community composition in the long-term. Cold-adapted species may be outcompeted by mesophyles. Note that different traits and functions within one microorganism can have different temperature dependence, but also the same trait can have a different temperature effect in different organisms (Dell *et al.*, 2011), which complicates predicting overall temperature effects. In estimating effects, it is also important at which level traits are described, that is, if the trait is "methanogenesis" or "acetoclastic-methanogenesis" (which is a specific pathway for CH_4 formation). Schulz *et al.* (1997) observed that temperature dependence of CH_4 production depends on the source material. In sediments of Lake Constance, which has an average temperature of 4°C, acetoclastic methanogenesis occurred at all temperatures, whereas hydrogenotrophic methanogenesis only occurred above 22°C due to substrate (H_2) limitation at lower temperatures. Thus, different classes of methanogens are affected by temperature in different ways. In another study performed in arctic peat, warming increased overall methanogen diversity, abundance and activity, with some specific populations of methanogens reacting to temperature and others controlled by other factors (Høj *et al.*, 2008). Thus, temperature shifts could lead to loss of specific species, which may lead to loss of other functional traits harboured by these species.

Heat disturbance, in the form of solarisation, can be used as a management strategy to decrease disease incidence in rice-cropping systems. For example, Culman *et al.* (2006) assessed effects of such management in Nepal's rice–wheat cropping system, in which solarisation results in an effective temperature increase of 10°C. They found clear effects on fungal community composition. However, over time communities became more and more similar, regardless of solarisation, indicating selection for species resilient to the prevailing conditions.

It is unclear to which extent microbes adapt their temperature optimum to changed conditions, or in other words it is unclear if temperature tolerance is a plastic or static trait, and if it differs amongst microbial guilds. Abdollahi and Nedwell (1979) described temperature effects on bacterial sulphate reduction in a saltmarsh sediment. They found no evidence for temperature adaptation throughout the year. Over a seasonal range, sulphate reduction had a constant Q_{10} of 3.5.

Trait combinations (temperature)

Little is known about specific trait combinations involved in temperature responses. However, there are a few traits that seem to give cross-protection, as revealed by protein expression. For example, heat-shock proteins are expressed not only at elevated temperatures, but also upon chemical stress, which broadens the stress tolerance of the involved organisms (Ramos *et al.*, 2001; Griffiths and Philippot, 2013).

Cold-adapted microbes exhibit trait combinations that help to protect them against freezing and other adverse effects of living in cold environments, such as desiccation and salination (Margesin *et al.*, 2006) and protection against reactive oxygen species (ROS; Methé *et al.*, 2005). For example, cold-adapted *Pseudoalteromonas haloplanktis* lacks pathways that produce ROS, which forms protection against oxidative stress but at the same time limits metabolic versatility. Organisms with such trait combinations may therefore have a competitive advantage during times of salinity or drought, but can experience a competitive penalty at elevated temperatures.

Microbial traits involved in drought and salinity

Moisture content is a key driver of microbial processes. Drought is characterised by decreased moisture content which imposes physiological stress on wetland microbes, selecting for communities best adapted to cope with the stress (i.e. shift in the community composition) or causing a shift in activity, while maintaining the community composition (Schimel and Gulledge, 1998). Accordingly, different wetlands harbour inherently different microbial communities performing a similar function (e.g. communities in bogs and fens are more similar to each other than to communities in a riparian wetland), causing different responses to drought (Kim *et al.*, 2008). Moreover, how microorganisms respond to (recurring) drought events may depend on the site history; a diverse reservoir of the seed bank community will enable some communities to emerge and thrive under specific conditions.

As such, the response of the microbial community provides clues to important traits pertinent to survival and recovery from drought. Among these traits is the production of exopolysaccharides to trap moisture and confer protection against desiccation (Roberson and Firestone, 1992). In a grassland soil, rapid responders (e.g. Actinobacteria) to rewetting after drought were found to possess more ribosomes than other taxa; it was hypothesised that a higher ribosomal copy number was required to sustain a high protein turnover rate during revival from drought (Placella *et al.*, 2012). Similarly, recurring desiccation-rewetting events favours resistant fast-growing microbial populations capitalising on the labile substrates released upon rewetting (Ho *et al.*, 2016). Drought also exposes microorganisms to other forms of stress as an indirect effect. For instance, a drought-induced change in the above-ground vegetation community may cause shifts in below-ground microbial community structure by altering the litter quality and the physiochemical soil status (e.g. nutrient availability; Tilman and Downing, 1994). Although many studies explicitly focused on specific responses and microbial traits to drought, a single trait may not be sufficient for the survival and persistence of microorganisms. Frequently, it is likely that a combination of traits affords a more effective means of survival.

Combinations of traits pertinent under a specific condition may also become important under other conditions. For instance, microorganisms produce osmoprotectant or compatible solutes to maintain cellular activity during desiccation. This trait, expressed during drought, is also important to overcome osmotic stress in high-salinity environments. Salinisation inhibits microbial growth/biomass (Pankhurst *et al.*, 2001) and alters the enzymatic activity in soils and sediments, affecting key biogeochemical processes. As a specialised strategy among microorganisms inhabiting high-salinity environment, halophiles actively take up ions (e.g. K^+) to maintain cell turgor and to prevent dehydration. Selective uptake of ions is achievable via an antiport system where K^+ is preferentially taken up more than Na^+ through a pump (Ventosa *et al.*, 1998). A more generalised and widespread strategy is for cells to accumulate low-molecular-weight organic compounds like amino acids to overcome the concentration gradient and prevent outflow of water. The microbial community in a saline soil showed higher resistance to desiccation-rewetting, suggesting that the ability to tolerate desiccation was induced by prior exposure to high salinity (Baumann and Marschner, 2013). Indeed, decomposer communities in soil employed similar strategies to overcome both desiccation and salinity stress (Rath and Rousk, 2015), which reflect on essential shared traits relevant under multiple global change scenarios. Very recently, it was demonstrated that the microbial response to a combined disturbance of drought and nitrogen fertilisation in soil was phylogenetically conserved among fungi and bacteria indicating that combinations of traits indeed facilitate resilience or resistance against disturbances (Amend *et al.*, 2016).

Traits of importance for change in vegetation

Environmental changes/stresses relevant for wetland microbiomes

Natural as well as anthropogenic disturbances described above will provoke changes in wetland vegetation that will manifest itself in either changes in abundance or composition of the vegetation. From a microbial perspective, this will either change redox conditions (more or less release of oxygen into the anoxic soil/sediment), amount and type of carbon exuded by the plants and/or the amount of nutrients taken up by the plants themselves. Wetland plants, in common with terrestrial vegetation, have a natural effect on soil/sediment microbes. This results in microbial communities which in terms of abundance and composition are different when associated with roots (rhizosphere, rhizoplane, inside roots = endosphere), as compared to the root-free surrounding soil/sediment (Edwards *et al.*, 2015; He *et al.*, 2015). This "selective effect" of the plants differs between plant species (He *et al.*, 2015) and indicates the differences in physicochemical conditions in, on or surrounding the plant roots where microbes are, for example, exposed to gradients in oxygen, pH, carbon, nutrients (Blossfeld *et al.*, 2011). The high relative abundance of anaerobic microbes is typical for wetland microbial communities that utilise electron acceptors other than oxygen (e.g. sulphate-reducer, methanogens) or aerobes thriving on products of anaerobic metabolism (e.g. CH_4; methane-oxidisers, Fe^{2+}; iron oxidisers) utilising oxygen released by roots. Methanogens can make up 17% of the total microbial community (Prasse *et al.*, 2015) in the rhizosphere of wetland plants species, while sulphate reducers and denitrifiers have been shown to reach abundances ranging from 7 to 17% (He *et al.*, 2015). Aerobic CH_4 oxidisers can reach up to 4% and iron oxidisers colonising iron oxides on rice roots made up 59% of the total bacterial community (Schmidt and Eickhorst, 2014). This modulating effect of wetland plants is also reflected in the processes catalysed by microbes which can be highly stimulated (Bodelier *et al.*, 1996, 2000; Conrad *et al.*, 2008) in the root environment. This can subsequently lead, for example, to higher CH_4 emissions from wetlands (Bouchard *et al.*, 2007; Kao-Kniffin *et al.*, 2010). Well-studied modulators of wetlands vegetation with a high likelihood of becoming more important in future are grazing and invasive species. Grazing has been demonstrated to lead to changes in soil carbon, nitrogen and phosphorus in wetlands (Wang *et al.*, 2014a, b) as well as to higher emissions of greenhouse gases (Dingemans *et al.*, 2011; Chen *et al.*, 2013). In arctic wetlands, the absence of muskoxen grazing led to a shift in vegetation to more vascular plants followed by a shift from CH_4 formation from acetate (acetotrophic) to hydrogen (hydrogenotrophic) due to changes in substrate concentrations (acetate), with lower subsequent CH_4 emissions (Falk *et al.*, 2015; Strom *et al.*, 2015). Invasive wetland plant species (e.g. *Phragmites australis*, *Typha latifolia*, *Phalaris arundinacea*, *Spartina alterniflora*) have invaded and replaced native wetlands vegetation globally with a range of biogeochemical consequences ranging from increases in soil organic carbon (Yang *et al.*, 2013), higher ammonium, nitrate and phosphorus (Angeloni *et al.*, 2006) to enhanced decomposition (Duke *et al.*, 2015) and the increased presence of pathogens (Nelson and Karp, 2013). In a number of cases, it has been shown that biogeochemical changes caused by invasion led to selection of specific microbes. In a *S. alterniflora* marsh, the diversity of archaeal AOB was substantially increased

(Zhang *et al.*, 2011) which hints at their capability for utilising other substrates, except for ammonium which may be provided by the invading plants. Invasion of a freshwater wetland with *Typha latofolia* led to a shift towards denitrifiers containing the nirS gene (Angeloni *et al.*, 2006), coding for nitrite reductase also indicating that these nirS denitrifiers have traits which give them advantage in the rhizosphere of *Typha* as compared to the denitrifiers which contain the nirK gene.

Relevant microbial traits for functioning under vegetation change

The most obvious stress that anaerobic microbes in wetland environments will face is redox stress due to exposure to too much oxygen. In the case of aerobic microbes, they can be subjected to limiting amounts of oxygen. Other stressors can be intermediates or end products of anaerobic respiration (ammonia, NO, nitrite, S^{2-}, Fe^{2+}) that can also be toxic to other microbes. Anaerobes like methanogens and sulphate reducers have enzymes (flavour and metalloproteins) which reduce oxygen partially, resulting in the formation of ROS like O_2^- and H_2O_2 which can damage proteins and DNA in the cells. Protective mechanisms are at

Figure 9.2 Co-occurrence of traits within microorganisms. When only one trait of a community is measured (such as denitrification rate; see "measured trait" in legend), the microbial functioning before and after the disturbance appears to be the same (suggesting functional redundancy of the microbial community). However, when all traits are taken into consideration, the functioning of the community before and after the disturbance is different.

the level of oxygen sensing and coupled regulation of transcription of redox sensitive enzymes producing ROS, for example, MsvR transcriptional regulator in methanogens (Sheehan *et al.*, 2015). Most anaerobes are able to remove ROS by means of ROS-scavenging molecules like superoxide dismutase and reductase, ruberythrin and rubredoxin without production of oxygen (Zhou *et al.*, 2011; Horne and Lessner, 2013). Methanogens have been demonstrated to be surprisingly resistant against oxygenated conditions and even able to produce CH_4 in oxic desert soils (Angel *et al.*, 2011). Here, hydrogenotrophic methanogens belonging to the genera *Methanosarcina* and *Methanocella* produced CH_4 under oxic conditions. The genome sequence of *Methanocella palidicola*, isolated from rice soils, reveals many adaptations to deal with ROS which may explain why these methanogens dominate on rice roots (Conrad *et al.*, 2008) and produce most of the CH_4 from rice worldwide (Lu and Conrad, 2005). Besides ROS oxygen species, the methanogens in rice soils also have a number of traits protecting them from nitrosative stress (NO_2^-, NO, N_2O) (Erkel *et al.*, 2006), which can occur due to the oxic–anoxic interfaces in wetlands that promote nitrate production and denitrification in close proximity.

Just as anaerobes have to be able to tolerate oxygen in wetland rhizospheres, aerobic bacteria need to be active at low oxygen levels. Hence, having traits to operate at sub-oxic conditions may enable survival and give competitive benefits. Neutrophilic iron-oxidising bacteria (catalysing Fe^{2+} to Fe^{3+}) related to the genus *Gallionella* not only have to compete with other microbes for oxygen, but also with chemical oxidation. This is possible at sub-oxic conditions where chemical oxidation is slow. It has been shown that iron oxidisers can actively move towards these sub-oxic zones by chemotaxis (see Figure 9.3; Wang *et al.*, 2009) and up to 10% of their genomes contain genes involved in oxygen sensing and binding involving hemerythrins and globins (Emerson *et al.*, 2013). Among the 10 most abundant gene categories are genes involved in environmental sensing, signal transduction and chemotaxis (Emerson *et al.*, 2013) which is a trait typical for bacteria living in fluctuating redox environments or low nutrient habitats. Although the biochemistry and physiology of iron oxidisers is largely unresolved, they seem to be very competitive in wetland habitats even outcompeting CH_4-oxidising bacteria (Wang *et al.*, 2012a), which have also been shown to be adapted to sub-oxic conditions (Reim *et al.*, 2012). In the case of CH_4 oxidisers, representatives of the genus *Methylobacter* perform better than other CH_4 oxidisers in wetland habitats at low oxygen concentrations (Reim *et al.*, 2012; Hernandez *et al.*, 2015), which is thought to be connected to their ability to utilise nitrate as an electron acceptor.

The facultative metabolism (using either oxygen or other electron acceptors) that has been suggested for methanotrophs is a widespread trait within microbes

Figure 9.3 (a) Microbial iron oxidation at the soil surface and around the roots of wetland plants grown in microcosms. The orange deposits are iron oxides formed microbially by oxidation of Fe^{2+} with oxygen coming either from the atmosphere (soil surface) or from roots. (b) Gradient tubes to enrich and isolate iron-oxidising bacteria. The tubes have saturated FeS solution at the bottom which releasing Fe^{2+} into the overlying solution. Oxygen diffuses into the tubes from the atmosphere. The iron will either be oxidised chemically (control) or by microbes inoculated from rhizosphere soil or attached to pieces of root. Iron oxidisers in the rhizosphere and root tubes have positioned themselves in the tubes at suboxic conditions (orange bands), outcompeting chemical oxidation. The combination of redox sensing and chemotaxis are typical traits of these bacteria to occupy this niche in wetland systems. (Adapted from Wang *et al.*, 2009.)

and of major importance in wetland environments with highly variable (both temporally and spatially) availability of electron acceptors. The ability to use nitrate or a fermentative metabolism is widespread in microbes also occurring in wetland soils and sediments, while the facultative use of metals (e.g. Fe^{3+}, Mn) is much more restricted and executed mainly by representatives from the Genus *Shewanella* (Todorova and Costello, 2006).

It has quite recently been discovered that special traits are needed to make a living in anoxic environments where the anaerobic oxidation of CH_4 and ammonium occurs. *Methylomirabilis oxyfera* generates its own oxygen, from NO produced by denitrification of nitrite (Ettwig *et al.*, 2010), while members belonging to archaeal candidate lineage *Methanoperedens* perform reverse methanogenesis with nitrate as an electron acceptor (Haroon *et al.*, 2013). Also fairly recently, it has been shown that these processes contribute significantly to CH_4 cycling in wetlands (Segarra *et al.*, 2015). This is also the case for the anaerobic oxidation of ammonium in combination with denitrification of nitrite performed by the so-called Anammox microbes (Zhu *et al.*, 2015).

Next to fluctuating redox conditions, variation in the abundance and composition of wetland plants will lead to highly variable availability of energy source as well as macro nutrients like nitrogen, phosphorus and potassium which are all taken up by the plant. There is a huge body of literature on microbial adaptations to energy and nutrient starvation and as it is not within the scope of this chapter to discuss this further, the reader is referred to excellent papers on this topic (Fierer *et al.*, 2007; Lever *et al.*, 2015). Overall, adaptations of cell size, cell membrane, environmental sensing, substrate transport systems, internal substrate regulation, respiratory kinetics and electron transport chains and adaptation of motility or adhesive properties may all be modified to cope with energy or nutrient stress. When focusing on wetland microbes, there are examples of methanogens adapted to function at low substrate concentrations. For example, *Methanosaeta* adapted to grow at very low acetate concentrations which explains its observed occurrence outside of the rhizosphere (He *et al.*, 2015), while representatives of the genus *Methanoregula* are adapted to the low nutrient concentrations in peat (Brauer *et al.*, 2015). Many wetland microbes can shift to nitrogen fixation when mineral nitrogen is limited, as exemplified by the high nitrogen-fixing activity on wetland plant roots (Moseman, 2007), and the high diversity of nitrogen-fixation genes (i.e. nifH) being expressed there (Bao *et al.*, 2014). A potential trait of relevance in this respect may be the oxygen sensitivity of the nitrogenase enzyme, which has been reported to vary between different CH_4 -oxidising species (Bodelier and Steenbergh, 2014a) possibly explaining the dominance of alpha-proteobacterial nifH transcripts associated with rice roots (Bao *et al.*, 2014).

Also, acquiring phosphorus in the vicinity of plant roots may require specific traits (e.g. exophosphatases, high-affinity P uptake systems, P sensing systems). Recently, it has been demonstrated that in wetlands, especially arctic permafrost areas, CH_4 cycling microbes can be P-limited. They differ substantially in their trait potential for low P acquisition (Veraart *et al.*, 2015), with consequences for CH_4 emissions from these areas.

The typical strong fluctuations in redox and nutrient conditions in wetlands may also be tolerated by accumulating reserve materials (e.g. polyphosphates, polyhydroxybutyrate) which can be used as endogenous energy/nutrient sources to maintain basic cellular activity or to regain activity after a period of starvation. In methanotrophic bacteria, production of PHB is induced by N or P limitation and is mainly a trait of alpha proteobacterial methane oxidisers (Pfluger *et al.*, 2011).

Trait combinations and evidence for trade-offs with vegetation change

As described above, the multitude of changes taking place in wetlands with their temporally or spatially variable patterns of partial/complete presence or absence of plants requires a large set of traits to be able to perform under these conditions. In contrast, when there are stable oxic/anoxic zones in wetland soils, the communities of microbes inhabit these zones with very different sets of traits and with surprisingly little redundancy in the active members (Kim and Liesack, 2015). In the case of transitions or environmental perturbation, the SPANC balance within the available trait packages will be an important determinant for which species will be active. The very different community composition of methanogens in various wetland systems can be explained by differences in trait packages. The more redox-stable, nutrient poor and often acidic wetlands are dominated by methanogens belonging to the specialised members of the families *Methanosaetaceae* and *Methanoregulaceae* (Bridgham *et al.*, 2013). These are active at very low acetate concentrations in the case of *Methanosaeta* and at low pH and low ionic concentrations in the case of *Methanoregula* (Brauer *et al.*, 2015). The more dynamic and nutrient-rich environments with periodic exposure to oxygen are dominated by representatives of the orders *Methanosarcinales* and *Methanocellales* (Bridgham *et al.*, 2013). Apparently, the adaptations required to withstand redox and nitrosative stress do not go hand in hand with growth at low pH and at low substrate concentrations. These hypothetical "trade-offs" can have serious consequences for CH_4 production, as evidenced by orders of magnitude higher CH_4 production from soils planted with rice dominated by *Methanosarcinales* and *Methanocellales*, as compared to soils dominated by *Methanoregula*-related species (Conrad *et al.*, 2008). Representatives of the *Acidobacteria*, which are one of the major phyla also present in wetlands (Hartman *et al.*, 2008; Wang *et al.*, 2012b) and involved in degradation of complex carbon molecules, seem to combine resistance against low pH, low nutrients and fluctuating redox conditions

(Ward *et al.*, 2009). This is mirrored by their very low growth rates. Unexpected traits can cross-cut trait packages defining niches of microbes, with large consequences for also functioning after disturbance. In the case of methanotrophs, there is a strong biogeography and link between diversity, phylogeny and habitat type (Knief, 2015) which can be partially explained by differences in traits. Furthermore, there is a clear difference in preference of protozoa for grazing on certain types of methanotrophs. Gamma proteobacterial methanotrophs are grazed upon much more (Murase and Frenzel, 2008; Murase *et al.*, 2011) with substantial consequences for CH_4 consumption since these are predominantly active Methane Oxidizing Bacteria (MOB) in many habitats (Bodelier *et al.*, 2013; Daebeler *et al.*, 2014). A rich source of potential trait packages with relevance for global carbon degradation models in wetlands will be sulphate-reducing microbes. This group of anaerobes displays a strong biogeography even in freshwater wetlands with fluctuating redox conditions and maintain a high diversity and responsiveness to substrate availability after long periods of exposure to oxygen and deprivation form sulphate (Miletto *et al.*, 2008, 2010; Pester *et al.*, 2012). Also, in peatlands with low pH and nutrients, a very specific sulphate-reducing species (*Desulfosporosinus*) that was a very minor component of the community was responsible for most of the sulphate reduction in this habitat (Pester *et al.*, 2012). Despite the enormous resilience and versatility within this group of microbes, pH in combination with low nutrient availability is a strong selecting trait set.

Microbial traits of importance for wetland pollution

The strategies of microorganisms to deal with pollutants are as diverse as the pollutants themselves. From a microbe's perspective, the occurrence of what we consider to be a pollutant does not need to be negative. For example, most organic pollutants can be used as a substrate by one or another microbe. Microbes do not only degrade naturally formed pollutants such as oil, but also xenobiotic compounds. The ability of microorganisms to degrade specific compounds can often be transferred from microbe to microbe by horizontal gene transfer, resulting in a weak link between this trait and microbial phylogeny (i.e. shallow trait; Martiny *et al.*, 2013). Horizontal gene transfer is facilitated by the fact that the responsible genes often occur on mobile genetic elements such as plasmids, enabling the rapid adaptation of microbial communities to novel compounds (Top and Springael, 2003), including antibiotics.

Another way in which microorganisms can lower levels of pollution is through co-metabolism of the pollutant. In this case, degradation of the pollutant does not yield energy or carbon for the microbe involved; the pollutant passes through the microbe's enzymatic machinery due to a lack of specificity of the enzymes for their intended substrate. Pollutants can be completely degraded in this way, or only partially degraded. In the best case, the microbe experiences only a slight loss of efficiency, but co-metabolism can also lead to enzyme blockage or the production of toxic intermediates. This is the case, for example, in the co-metabolism of ammonia by CH_4-oxidising bacteria due to the small specificity of the enzyme methane monooxygenase (Stein and Klotz, 2011). Ammonia oxidation leads to a range of toxic intermediates, of which hydroxylamine is the most toxic. Methane oxidising bacteria differ in their ability to handle these toxic intermediates, mainly through further oxidation of the N_2O to nitrate, which is only toxic at high concentrations. The effect of ammonia on CH_4-oxidising bacteria also depends on the ratio between available CH_4 and ammonia. Methane-oxidising bacteria living in environments with high ammonia concentrations relative to the CH_4 concentration will benefit from a higher specificity of the methane monooxygenase and enzymes that catalyse the oxidation of the intermediates of ammonia oxidation.

Although the microbial community as a whole shows a remarkable capability to degrade pollutants, specific members of the community can be negatively affected by pollutants. For example, the active sites of microbial enzymes catalysing specific reactions can be blocked when a pollutant shows morphological similarity with a substrate, or pollutants only affect microbes with a specific structure of their membranes. Ecosystem functions that are performed by a narrow selection of microbes, in particular, can be affected in this way, for example, nitrification, whereas functions that are performed by a wide range of microbes (respiration) can remain unaffected.

In addition to the direct effects pollutants can have on microbial communities and functioning, pollutants can affect microorganisms indirectly in many ways. For example, microbial populations can grow when the availability of readily degradable organic matter increases from organisms killed by a pollutant, such as when fish kills occur. Fast-growing microorganisms (ruderals or r-strategists; Grime, 1977; Fierer *et al.*, 2007) will benefit most from these sudden bountiful conditions. In addition, a reduced competition for resources between groups of microorganisms can occur when microorganisms are differentially affected by a pollutant. Pollutants can also have a major effect on wetland biogeochemical functioning by changing redox conditions, as redox conditions in part determine which metabolic routes can be used by microorganisms. Lowering of redox conditions can, for example, be caused by loss of bioturbation due to poisoning of macrofauna, or reduced input of oxygen from plant roots due to herbicide use. However, a more common way in which pollution affects redox conditions in wetlands is through eutrophication.

Eutrophication is caused by excessive loading of nutrients, most importantly nitrogen and phosphorus. In this section, microbial traits related to changes in the availability of these nutrients in relation to wetland functioning are discussed. Eutrophication does not only affect redox conditions, but can also influence microbial activity in several other ways. As for all life, microorganisms require nitrogen and phosphorus for growth, thereby linking the nitrogen and phosphorus cycles to all microbially mediated processes. For phosphorus cycling, a vicious cycle exists between anoxia and a higher release of phosphorus from aquatic sediments. In addition to abiotic processes, such as reduced binding of phosphates to iron in the sediment under anoxic conditions (Dunne and Reddy, 2005), microbial activity is involved in this feedback. To be able to predict the response of wetlands to changed phosphorus input, more knowledge of microbially mediated phosphorus cycling is needed. Currently, internal phosphorus loading is a big obstacle to the remediation of wetlands from a eutrophied state, and it is often not clear which mechanisms of internal loading are most important under given environmental conditions (Søndergaard *et al.*, 2003). Microorganisms catalyse the mineralisation of organic phosphorus to phosphate and incorporate phosphorus in their biomass during growth. They can also produce refractory phosphorus-containing organic molecules that can build-up in wetland soils and sediments and by their combined activity greatly influence the redox conditions and concentrations and oxidation state of compounds that can interact with phosphate (iron, sulphur, etc.) (Gächter and Meyer, 1993). In addition, when phosphate and energy are freely available but microbial growth is limited by nitrogen availability, for example, many bacteria, archaea and fungi can store phosphorus in their cells, mostly as polyphosphates. Although many microorganisms have the potential to synthesise polyphosphates, the amount of polyphosphate that is stored in individual cells differs across species (Orell *et al.*, 2012). With polyphosphates, microorganisms not only store phosphorus, but also energy, as the phosphorus bonds in polyphosphates are high in energy. The stored phosphorus can be used to produce new biomass when conditions for growth turn more favourable, or to overcome periods of limited availability of energy, for example, when redox conditions change from oxic to anoxic. Interestingly, polyphosphate metabolism is not only involved in the storage of energy and phosphorus, but also renders microbes more resistant to other stressors, such as heavy metals, salinity, acidity, heat, oxidants, antibiotics and UV radiation (Orell *et al.*, 2012), although the exact mechanisms behind this stress protection remain to be elucidated. The resistance against heavy metals is likely to be dependent on the intracellular adsorption of the metal to phosphate and subsequent transport out of the cell. The intracellular phosphate concentration is kept high by hydrolysis of polyphosphates, which can cause a correlation between a microorganism's ability to store polyphosphates and their resistance against heavy metals. In this way, an environment in which polyphosphate-accumulating organisms have an advantage due to fluctuating redox conditions (which occur along the sediment–water interface of healthy wetlands) could also select for organisms that are more resistant to heavy-metal pollution. If the redox conditions become permanently anoxic due to eutrophication, this reduces the fitness advantage of polyphosphate accumulation, and thereby potentially the heavy-metal resistance of the microbial community.

Because organisms need both phosphorus and nitrogen for the production of new biomass, microorganisms capable of fixing dinitrogen are at an advantage when phosphorus is available in surplus, which can lead to shifts in community composition (Ward *et al.*, 2013). Nitrogen fixation is a strictly microbial trait. Although the genetic basis for this trait is more complex than for antibiotic resistance or the ability to degrade specific compounds (Martiny *et al.*, 2015), nitrogen fixation is distributed widely across the microbial phylogenetic tree. Lateral gene transfer of the whole cluster of genes necessary for nitrogen fixation is possible as they are commonly located in a tightly conserved operon (Raymond *et al.*, 2004).

The competitive ability of microorganisms to acquire fixed nitrogen is linked to their ability to acquire other nutrients, such as phosphorus. This correlation between the microbial traits related to nutrient acquisition can be either positive or negative. An example of a three-way trade-off between the competitive capabilities for phosphorus and nitrogen and cell size is described by Edwards *et al.* (2011) and others, where phosphorus and nitrogen acquisition capabilities correlate positively when organisms of different cell sizes are considered. This is because small organisms are at an advantage over large organisms when nutrient concentrations are low and limited by diffusion. Large organisms, by contrast, can be at an advantage in environments with fluctuating nutrient concentrations, as they have more storage capacity. For a given cell size, however, an organism's capability to acquire nitrogen and phosphorus is negatively correlated. The potential mechanism behind this negative correlation is a trade-off between investing in energy, matter and even cell surface area for uptake proteins for either nutrient. For freshwater plankton, the trade-off between nitrogen and phosphorus uptake capabilities is more prominent than the trade-off between cell size and overall nutrient uptake (Edwards *et al.*, 2011). For wetlands, this can result in a shift in community composition when the balance between available phosphorus and nitrogen changes.

The effect of nitrogen pollution on microbial processes in wetlands is even more complex: in addition to nitrogen being a nutrient and able to indirectly influence redox conditions, nitrogen can also act as an inhibitor or toxicant to microorganisms. Nitrogen toxicity especially occurs from the intermediate products of nitrification, denitrification and dissimilatory nitrate reduction to ammonia, for example, as described above for ammonium co-metabolism by CH_4-oxidising bacteria. Nitrite is also toxic to many microorganisms and macroorganisms, as it prevents functioning of many metal-dependent enzymes (including human haemoglobin).

Some pathways in the nitrogen cycle, such as denitrification, can be performed by a broad range of microorganisms, whereas other pathways are only performed by specialised microorganisms, as for nitrification. However, also in the case of ammonia oxidation, several groups of microorganisms are involved. The two main groups of nitrifying microorganisms are AOB and AOA. Since the discovery of AOA in 2005 (Könneke *et al.*, 2005), many studies have been undertaken to unravel the environmental conditions that select for either AOB or AOA (Prosser and Nicol, 2012). Although from a process view it may be trivial which of these two groups of microorganisms catalyses ammonia oxidation, from an ecosystem perspective the dominance of either bacteria or archaea can have major implications. For example, archaeal ammonia oxidisers are much smaller than bacterial ammonia oxidisers, possibly rendering them more prone to being grazed, thereby providing a substrate for bacterivorous organisms.

In addition to the fact described above that some nitrogen transformations can be performed by physiologically highly distinct groups of microorganisms for which the environmental drivers are not fully understood, the environmental drivers that determine which nitrogen transformations take place are also not fully understood. This is, for example, the case for the balance between denitrification by respiratory organisms and DNRA by fermentative organisms. As the end product of these pathways is either dinitrogen or ammonium, the ratio between denitrification and DNRA has a major impact on nitrogen loss or retention in wetlands. Although the balance between these two pathways, in general, is determined by the availability of carbon substrates and nitrate, in some situations the dominant pathway cannot be predicted based on thermodynamics, which indicates that the traits of the organisms involved can be confounding factors (van de Leemput *et al.*, 2011).

These two examples of AOB vs. AOA and of denitrification vs. DNRA clearly show that even when at process level (i.e. nitrification rate and rate of nitrate loss) no change is observed after a disturbance, the shifts in microbial community structure can have profound effects on wetland ecosystem functioning. This clearly demonstrates the need to determine the co-occurrence of microbial traits within microbial species in order to predict wetland ecosystem functioning under disturbance (Figure 9.2).

Synthesis and Outlook

Global environmental change and increasing anthropogenic pressure on wetland ecosystems make these systems among the most threatened ecosystems worldwide. Disturbances due to changes in climate (hydrology, temperature), changes in vegetation by grazing or invasive species, pollution and land-use change and overexploitation contribute to this alarming status. Considering the crucial ecosystem services they provide, it is of the utmost importance to obtain comprehensive understanding of the functioning of these ecosystems in order to inform politicians, policy-makers and wetland managers to take the proper actions to protect, conserve and restore these ecosystems. Predictive tools and models have been developed to predict future wetland area and functioning, which are important modules for predicting global climate scenarios (IPCC, 2014). However, there are large uncertainties in these predictions and they are not designed to predict short-term changes after acute disturbances. The traditional process-based top-down approach in these models and the process-based approaches in wetland research do not include microbial ecology, while microbes are the catalysts of all biogeochemical processes fundamental to wetland functioning. A stronger integration of geochemistry and microbial ecology is necessary in order to understand wetland functioning at the level of model predictions. The need to understand feedback and interactions between elemental cycling and climate in multiple disturbance scenarios is especially important, so knowledge on microbial responses and behaviour is essential. There are already very good examples in which geochemistry informs microbiology to look for specific microbes and microbial properties, but there are also cases where novel microbes are found or microbial properties point towards important geochemical implications, which were not considered before. The future lies in developing ways to integrate geochemistry and microbiology (see Chapter 6). Trait-based approaches, as have already been developed for aquatic foodwebs and plant communities, are a very promising avenue to unite wetland biogeochemical processes and trait-based, more mechanistic models.

However, to understand and deal with nonlinear and often idiosyncratic processes and responses, it will be crucial to single out the active species and take their traits into account. This touches upon the central problem in microbial ecology, which is the large number of unknown species in complex environments like wetland soils and sediments, and the inability to pinpoint the active species and connect them to a specific process. Methodological developments in past decades have resulted in an avalanche of techniques, which have resulted in deep insights into microbial functional diversity. However, techniques linking diversity to function cannot yet be applied via high-throughput screening for applications in an environmental context. To solve these issues, microbial trait-based approaches have been used to assess links between phylogeny and functioning, in order to use a taxonomy or gene-based framework to predict functioning from the community composition. This opens up high-throughput screening and assessment of trait distribution given the modern sequencing methods. However, we should only see this as one of the tools in the box we need. Analogous to trait-based approaches in plant ecology, it is never a single trait that will determine the functioning of a particular species after disturbance. It is a "package" of traits in interaction with other organisms and the abiotic environment which will determine functioning. This aspect of trait-based approaches is as yet not considered in microbial ecology, where often only one function is considered (e.g. denitrification), and trait packages are dealt with by looking for correlating traits (i.e. genes). However, the non-correlating traits may be of equal importance in determining the functioning of an organism. Nevertheless, modelling frameworks have been developed that incorporate microbial traits and trade-offs in ecosystem models, improving and informing predictions. Models combining phylogeny, ecology and physiology of well-characterised functional groups (i.e. nitrifiers) have proved to be very effective in predicting processes, pointing to where the culturing of microbes and measuring traits and trait responses will be of crucial importance in order to realise integrated biogeochemistry. For parameterising models, knowledge of the breadth of trait responses and especially the way they interact with other traits is essential, which can only be obtained from culturing efforts. Now that genomics, transcriptomics, proteomics and metabolomics are available, a comprehensive understanding can be obtained regarding trait packages (interactions and responses) under various disturbances, which will generate knowledge that can inform process-based studies on what to look for or measure. These attributes can be fed into phylogeny-based, trait-based approaches by looking for specific gene combinations. Of course, the organisms in culture do not cover the full environmental diversity, but it is not expected that "trait packages" will be so exotic that extrapolations are not possible. These integrated culture studies yielding trait package information have still to be generated and should focus, in the first instance, on species of which environmental relevance has been demonstrated, and where links between traits and function can be expected (e.g. methanogens, nitrifiers, methanotrophs) to obtain an environmental framework that can subsequently be extended. Next to culturing, more factorial disturbance studies have to be executed in combination with omics assessment, giving mechanistic information on regulating factors and responses. Combined with the genetic information, this will, in time, be complemented by the culturing trait package information.

Integrated biogeochemistry, necessary to predict wetland functioning after disturbance, is feasible but will require a stronger focus on culturing and lab-based trait response studies, integrated in an environmental modelling framework. To some, this may seem like taking a step back … but it is a necessary one to move forward.

Acknowledgements

The authors of this paper were financially supported by research grants (823.001.008) of the Netherlands Organisation for Scientific Research (NWO) (823.001.008) and the BE-Basic R&D Program, which was granted an FES subsidy from the Dutch Ministry of Economic Affairs, Agriculture and Innovation (EL&I). This publication is publication number 6183 of the Netherlands Institute of Ecology (NIOO-KNAW).

References

Abdollahi, H. and Nedwell, D. B. (1979). Seasonal temperature as a factor influencing bacterial sulfate reduction in a saltmarsh sediment. *Microb. Ecol.* 5, 73–79.

Amend, A. S., Martiny, A. C., Allison, S. D., Berlemont, R., Goulden, M. L., Lu, Y., *et al.* (2016). Microbial response to simulated global change is phylogenetically conserved and linked with functional potential. *ISME J.* 10, 109–118.

Angel, R., Matthies, D. and Conrad, R. (2011). Activation of methanogenesis in arid biological soil crusts despite the presence of oxygen. *PLoS ONE* 6(5), e20453. doi: 10.1371/journal.pone.0020453.

Angeloni, N. L., Jankowski, K. J., Tuchman, N. C. and Kelly, J. J. (2006). Effects of an invasive cattail species (*Typha* × *glauca*) on sediment nitrogen and microbial community composition in a freshwater wetland. *FEMS Microbiol. Lett.* 263, 86–92.

Baldwin, D. S., Rees, G. N., Mitchell, A. M., Watson, G. and Williams, J. (2006). The short-term effects of salinization on anaerobic nutrient cycling and microbial community structure in sediment from a freshwater wetland. *Wetlands* 26, 455–464.

Bao, Z. H., Okubo, T., Kubota, K., Kasahara, Y., Tsurumaru, H., Anda, M., *et al.* (2014). Metaproteomic identification of diazotrophic methanotrophs and their localization in root tissues of field-grown rice plants. *Appl. Environ. Microb.* 80, 5043–5052.

Baumann, K. and Marschner, P. (2013). Effects of salinity on microbial tolerance to drying and rewetting. *Biogeochemistry* 112, 71–80.

Bayley, S. E. and Prather, C. M. (2003). Do wetland lakes exhibit alternative stable states? Submersed aquatic vegetation and chlorophyll in western boreal shallow lakes. *Limnol. Oceanogr.* 48, 2335–2345.

Bertrand, J.-C., Doumenq, P., Guyoneaud, R., Marrot, B., Martin-Laurent, F., Matheron, R., *et al.* (2015). Applied microbial ecology and bioremediation. *Environmental Microbiology: Fundamentals and Applications.* Springer, Dordrecht, The Netherlands. pp. 659–753.

Blain, S., Queguiner, B., Armand, L., Belviso, S., Bombled, B., Bopp, L., *et al.* (2007). Effect of natural iron fertilization on carbon sequestration in the Southern Ocean. *Nature* 446, 1070-U1.

Blom, C. (1999). Adaptations to flooding stress: from plant community to molecule. *Plant Biol.* 1, 261–273.

Blossfeld, S., Gansert, D., Thiele, B., Kuhn, A. J. and Losch, R. (2011). The dynamics of oxygen concentration, pH value, and organic acids in the rhizosphere of *Juncus* spp. *Soil Biol. Biochem.* 43, 1186–1197.

Bodelier, P. L. E., Frenzel, P., Drake, H. L., Hurek, T., Kusel, K., Lovell, C., *et al.* (2006). Wetlands and natural resource management. In Verhoeven, J. T. A., Beltman, B., Bobbink, R. and Whigham, D. F. (eds.), *Ecological Aspects of Microbes and Microbial Communities Inhabiting the Rhizosphere of Wetland Plants.* Springer, Berlin. pp. 205–238.

Bodelier, P. L. E., Hahn, A. P., Arth, I. R. and Frenzel, P. (2000). Effects of ammonium-based fertilisation on microbial processes involved in methane emission from soils planted with rice. *Biogeochemistry* 51, 225–257.

Bodelier, P. L. E., Libochant, J. A., Blom, C. and Laanbroek, H. J. (1996). Dynamics of nitrification and denitrification in root-oxygenated sediments and adaptation of ammonia-oxidizing bacteria to low-oxygen or anoxic habitats. *Appl. Environ. Microb.* 62, 4100–4107.

Bodelier, P. L. E., Meima-Franke, M., Hordijk, C. A., Steenbergh, A. K., Hefting, M. M., Bodrossy, L., *et al.* (2013). Microbial minorities modulate methane consumption through niche partitioning. *ISME J.* 7, 2214–2228.

Bodelier, P. L. E. and Steenbergh, A. K. (2014a). Metagenomics of the microbial nitrogen cycle: theory, methods and applications. In Marco, D. (ed.), *Interactions Between Methane and Nitrogen Cycling: Current Metagenomic Studies and Future Trends.* Caister Academic Press, Poole, UK. pp. 33–63.

Bodelier, P. L. E. and Steenbergh, A. K. (2014b). Interactions between methane and the nitrogen cycle in light of climate change. *Curr. Opin. Environ. Syst.* 9–10, 26–36.

Bork, P. (2000). Powers and pitfalls in sequence analysis: the 70% hurdle. *Genome Res.* 10, 398–400.

Bouchard, V., Frey, S. D., Gilbert, J. M. and Reed, S. E. (2007). Effects of macrophyte functional group richness on emergent freshwater wetland functions. *Ecology* 88, 2903–2914.

Bouskill, N. J., Tang, J., Riley, W. J. and Brodie, E. L. (2012). Trait-based representation of biological nitrification: model development testing, and predicted community composition. *Front. Microbiol.* 3(364), 1–17.

Brauer, S., Cadillo-Quiroz, H., Kyrpides, N., Woyke, T., Goodwin, L., Detter, C., *et al.* (2015). Genome of *Methanoregula boonei* 6A8 reveals adaptations to oligotrophic peatland environments. *Microbiol. Sgm.* 161, 1572–1581.

Brausch, J. M., Connors, K. A., Brooks, B. W. and Rand, G. M. (2012). Human pharmaceuticals in the aquatic environment: a review of recent toxicological studies and considerations for toxicity testing. *Rev. Environ. Contam. T.* 218, 1–99.

Bridgham, S. D., Cadillo-Quiroz, H., Keller, J. K. and Zhuang, Q. L. (2013). Methane emissions from wetlands: biogeochemical, microbial, and modeling perspectives from local to global scales. *Glob. Change Biol.* 19, 1325–1346.

Brown, M. M., Friez, M. J. and Lovell, C. R. (2003). Expression of nifH genes by diazotrophic bacteria in the rhizosphere of short form *Spartina alterniflora. FEMS Microbiol. Ecol.* 43, 411–417.

Carlisle, S. and Trevors, J. (1988). Glyphosate in the environment. *Water Air Soil Poll.* 39, 409–420.

Chappell, P. D., Moffett, J. W., Hynes, A. M. and Webb, E. A. (2012). Molecular evidence of iron limitation and availability in the global diazotroph Trichodesmium. *ISME J.* 6, 1728–1739.

Chen, H., Zhu, Q. A., Peng, C. H., Wu, N., Wang, Y. F., Fang, X. Q., *et al.* (2013). The impacts of climate change and human activities on biogeochemical cycles on the Qinghai-Tibetan Plateau. *Glob. Change Biol.* 19, 2940–2955.

Collet, S., Reim, A., Ho, A. and Frenzel, P. (2015). Recovery of paddy soil methanotrophs from long term drought. *Soil Biol. Biochem.* 88, 69–72.

Conrad, R., Klose, M., Noll, M., Kemnitz, D. and Bodelier, P. L. E. (2008). Soil type links microbial colonization of rice roots to methane emission. *Glob. Change Biol.* 14, 657–669.

Crooks, J. A., Chang, A. L. and Ruiz, G. M. (2011). Aquatic pollution increases the relative success of invasive species. *Biol. Invasions* 13, 165–176.

Crowther, T. W., Maynard, D. S., Crowther, T. R., Peccia, J., Smith, J. R. and Bradford, M. A. (2014). Untangling the fungal niche: the trait-based approach. *Front. Microbiol.* 5(579), 1–12.

Culman, S., Duxbury, J., Lauren, J. and Thies, J. (2006). Microbial community response to soil solarization in Nepal's rice–wheat cropping system. *Soil Biol. Biochem.* 38, 3359–3371.

Daebeler, A., Bodelier, P. L. E., Yan, Z., Hefting, M. M., Jia, Z. J. and Laanbroek, H. J. (2014). Interactions between Thaumarchaea, Nitrospira and methanotrophs modulate autotrophic nitrification in volcanic grassland soil. *ISME J.* 8, 2397–2410.

Daims, H., Lebedeva, E. V., Pjevac, P., Han, P., Herbold, C., Albertsen, M., *et al.* (2015). Complete nitrification by Nitrospira bacteria. *Nature* 528, 504–509.

Daryadel, E. and Talaei, F. (2014). Analytical study on threats to wetland ecosystems and their solutions in the framework of the ramsar convention. *Int. J. Soc. Behav. Educ. Econ. Business Ind. Eng.* 8, 2100–2110.

Davies, C. M., Long, J. A., Donald, M. and Ashbolt, N. J. (1995). Survival of fecal microorganisms in marine and freshwater sediments. *Appl. Environ. Microb.* 61, 1888–1896.

De Groot, R. S., Wilson, M. A. and Boumans, R. M. (2002). A typology for the classification, description and valuation of ecosystem functions, goods and services. *Ecol. Econ.* 41, 393–408.

Dell, A. I., Pawar, S. and Savage, V. M. (2011). Systematic variation in the temperature dependence of physiological and ecological traits. *Proc. Natl. Acad. Sci. USA* 108, 10591–10596.

Denef, K., Six, J., Paustian, K. and Merckx, R. (2001). Importance of macroaggregate dynamics in controlling soil carbon stabilization: short-term effects of physical disturbance induced by dry-wet cycles. *Soil Biol. Biochem.* 33, 2145–2153.

Dingemans, B. J. J., Bakker, E. S. and Bodelier, P. L. E. (2011). Aquatic herbivores facilitate the emission of methane from wetlands. *Ecology* 92, 1166–1173.

Duke, S. T., Francoeur, S. N. and Judd, K. E. (2015). Effects of *Phragmites australis* invasion on carbon dynamics in a freshwater marsh. *Wetlands* 35, 311–321.

Dumont, M. G. and Murrell, J. C. (2005). Stable isotope probing — linking microbial identity to function. *Nat. Rev. Microbiol.* 3, 499–504.

Dunfield, P., Dumont, R. and Moore, T. R. (1993). Methane production and consumption in temperate and subarctic peat soils: response to temperature and pH. *Soil Biol. Biochem.* 25, 321–326.

Dunne, E. J. and Reddy, K. R. (2005). Phosphorus biogeochemistry of wetlands in agricultural watersheds. In Dunne, E. J., Reddy, K. R. and Carton, O. T. (eds.), *Nutrient Management in Agricultural Watersheds: A Wetlands Solution.* Wageningen Academic Publishers, Wageningen. pp. 105–119.

Edwards, J., Johnson, C., Santos-Medellin, C., Lurie, E., Podishetty, N. K., Bhatnagar, S., *et al.* (2015). Structure, variation, and assembly of the root-associated microbiomes of rice. *Proc. Natl. Acad. Sci. USA* 112, E911–E920.

Edwards, K. F., Klausmeier, C. A. and Litchman, E. (2011). Evidence for a three-way trade-off between nitrogen and phosphorus competitive abilities and cell size in phytoplankton. *Ecology* 92, 2085–2095.

Emerson, D., Bellows, W., Keller, J. K., Moyer, C. L., Sutton-Grier, A. and Megonigal, J. P. (2013). Anaerobic metabolism in tidal freshwater wetlands: ii. Effects of plant removal on archaeal microbial communities. *Estuar. Coast.* 36, 471–481.

Erkel, C., Kube, M., Reinhardt, R. and Liesack, W. (2006). Genome of rice cluster I archaea — the key methane producers in the rice rhizosphere. *Science* 313, 370–372.

Ettwig, K. F., Butler, M. K., Le Paslier, D., Pelletier, E., Mangenot, S., Kuypers, M. M. M., *et al.* (2010). Nitrite-driven anaerobic methane oxidation by oxygenic bacteria. *Nature* 464, 543–548.

Evans, S. E. and Wallenstein, M. D. (2012). Soil microbial community response to drying and rewetting stress: does historical precipitation regime matter? *Biogeochemistry* 109, 101–116.

Falk, J. M., Schmidt, N. M., Christensen, T. R. and Strom, L. (2015). Large herbivore grazing affects the vegetation structure and greenhouse gas balance in a high arctic mire. *Environ. Res. Lett.* 10(045001), 1–11.

Ferenci, T. (2005). Maintaining a healthy SPANC balance through regulatory and mutational adaptation. *Mol. Microbiol.* 57, 1–8.

Fetzer, I., Johst, K., Schäwe, R., Banitz, T., Harms, H. and Chatzinotas, A. (2015). The extent of functional redundancy changes as species' roles shift in different environments. *Proc. Natl. Acad. Sci. USA* 112, 14888–14893.

Fierer, N., Bradford, M. A. and Jackson, R. B. (2007). Toward an ecological classification of soil bacteria. *Ecology* 88, 1354–1364.

Fierer, N., Schimel, J. P. and Holden, P. A. (2003). Influence of drying-rewetting frequency on soil bacterial community structure. *Microb. Ecol.* 45, 63–71.

Fong, T.-T. and Lipp, E. K. (2005). Enteric viruses of humans and animals in aquatic environments: health risks, detection, and potential water quality assessment tools. *Microbiol. Mol. Biol. Rev.* 69, 357–371.

Gächter, R. and Meyer, J. S. (1993). The role of microorganisms in mobilization and fixation of phosphorus in sediments. *Proceedings of the Third International Workshop on Phosphorus in Sediments*. Springer, Dordrecht, the Netherlands. pp. 103–121.

Graham, E. B., Knelman, J. E., Schindlbacher, S., Siciliano, S., Breulmann, M., Yannarell, A., *et al.* (2016). Microbes as engines of ecosystem function: when does community structure enhance predictions of ecosystem processes? *Front. Microbiol.* 7(241), 1–10.

Green, J. L., Bohannan, B. J. M. and Whitaker, R. J. (2008). Microbial biogeography: from taxonomy to traits. *Science* 320, 1039–1043.

Griffiths, B. S. and Philippot, L. (2013). Insights into the resistance and resilience of the soil microbial community. *FEMS Microbiol. Rev.* 37, 112–129.

Grime, J. P. (1977). Evidence for existence of three primary strategies in plants and its relevance to ecological and evolutionary theory. *Am. Nat.* 111, 1169–1194.

Gross, C. (1996). Function and regulation of the heat shock proteins. *Escherichia coli and Salmonella: Cellular and Molecular Biology*, 2nd edn. ASM Press, Washington, DC, pp. 1382–1399.

Grossart, H. P., Frindte, K., Dziallas, C., Eckert, W. and Tang, K. W. (2011). Microbial methane production in oxygenated water column of an oligotrophic lake. *Proc. Natl. Acad. Sci. USA* 108, 19657–19661.

Haroon, M. F., Hu, S. H., Shi, Y., Imelfort, M., Keller, J., Hugenholtz, P., *et al.* (2013). Anaerobic oxidation of methane coupled to nitrate reduction in a novel archaeal lineage. *Nature* 500, 567–570.

Hartman, W. H., Richardson, C. J., Vilgalys, R. and Bruland, G. L. (2008). Environmental and anthropogenic controls over bacterial communities in wetland soils. *Proc. Natl. Acad. Sci. USA* 105, 17842–17847.

He, S. M., Malfatti, S. A., Mcfarland, J. W., Anderson, F. E., Pati, A., Huntemann, M., *et al.* (2015). Patterns in wetland microbial community composition and functional gene repertoire associated with methane emissions. *mBio* 6(3), e00066–15. doi:10.1128/mBio.00066-15.

Hernandez, M. E., Beck, D. A. C., Lidstrom, M. E. and Chistoserdova, L. (2015). Oxygen availability is a major factor in determining the composition of microbial communities involved in methane oxidation. *Peerj* 3, e801, 1–13. doi: 10.7717/peerj.801.

Ho, A., Van Den Brink, E., Reim, A., Krause, S. M. B. and Bodelier, P. L. E. (2016). Recurrence and frequency of disturbance have cumulative effect on methanotrophic activity, abundance, and community structure. *Front. Microbiol.* 6, 1493, 1–10.

Høj, L., Olsen, R. A. and Torsvik, V. L. (2008). Effects of temperature on the diversity and community structure of known methanogenic groups and other archaea in high Arctic peat. *ISME J.* 2, 37–48.

Horne, A. J. and Lessner, D. J. (2013). Assessment of the oxidant tolerance of *Methanosarcina acetivorans*. *FEMS Microbiol. Lett.* 343, 13–19.

IPCC. (2014). Climate change 2013: the physical science basis. *Working Group I Contribution to the Fifth Assessment Report of the Intergovernmental Panel on Climate Change*. Cambridge University Press, UK, p. 1535.

Janse, J. H., Kuiper, J. J., Weijters, M. J., Westerbeek, E. P., Jeuken, M., Bakkenes, M., *et al.* (2015). GLOBIO-Aquatic, a global model of human impact on the biodiversity of inland aquatic ecosystems. *Environ. Sci. Policy* 48, 99–114.

Janssen, A. B. G., Arhonditsis, G. B., Beusen, A., Bolding, K., Bruce, L., Bruggeman, J., *et al.* (2015). Exploring, exploiting and evolving diversity of aquatic ecosystem models: a community perspective. *Aquat. Ecol.* 49, 513–548.

Jones, P. G. and Inouye, M. (1994). The cold-shock response — a hot topic. *Mol. Microbiol.* 11, 811–818.

Jones, P. G., Vanbogelen, R. A. and Neidhardt, F. C. (1987). Induction of proteins in response to low temperature in *Escherichia coli*. *J. Bacteriol.* 169, 2092–2095.

Kao-Kniffin, J., Freyre, D. S. and Balser, T. C. (2010). Methane dynamics across wetland plant species. *Aquat. Bot.* 93, 107–113.

Kettunen, A. (2003). Connecting methane fluxes to vegetation cover and water table fluctuations at microsite level: a modeling study. *Glob. Biogeochem. Cycles* 17(2), 1051, 1–19. doi: 10.1029/2002GB001958.

Kim, S. Y., Lee, S. H., Freeman, C., Fenner, N. and Kang, H. (2008). Comparative analysis of soil microbial communities and their responses to the short-term drought in bog, fen, and riparian wetlands. *Soil Biol. Biochem.* 40, 2874–2880.

Kim, Y. and Liesack, W. (2015). Differential assemblage of functional units in paddy soil microbiomes. *PLoS ONE* 10(4), e0122221, 1–20. doi: 10.1371/journal.pone.0122221.

Kirschbaum, M. U. (1995). The temperature dependence of soil organic matter decomposition, and the effect of global warming on soil organic C storage. *Soil Biol. Biochem.* 27, 753–760.

Kirschke, S., Bousquet, P., Ciais, P., Saunois, M., Canadell, J. G., Dlugokencky, E. J., *et al.* (2013). Three decades of global methane sources and sinks. *Nat. Geosci.* 6, 813–823.

Knief, C. (2015). Diversity and habitat preferences of cultivated and uncultivated aerobic methanotrophic bacteria evaluated based on pmoA as molecular marker. *Front. Microbiol.* 6, 1346, 1–38.

Könneke, M., Bernhard, A. E., José, R., Walker, C. B., Waterbury, J. B. and Stahl, D. A. (2005). Isolation of an autotrophic ammonia-oxidizing marine archaeon. *Nature* 437, 543–546.

Krause, S., Le Roux, X., Niklaus, P. A., Van Bodegom, P. M., Lennon, J. T., Bertilsson, S., *et al.* (2014). Trait-based approaches for understanding microbial biodiversity and ecosystem functioning. *Front. Microbiol.* 5, 251, 1–10.

Kuiper, J. J., Van Altena, C., De Ruiter, P. C., Van Gerven, L. P. A., Janse, J. H. and Mooij, W. M. (2015). Food-web stability signals critical transitions in temperate shallow lakes. *Nat. Commun.* 6, 7727, 1–7. doi: 10.1038/ncomms8727.

Kümmerer, K. (2009a). Antibiotics in the aquatic environment — a review. Part I. *Chemosphere* 75, 417–434.

Kümmerer, K. (2009b). Antibiotics in the aquatic environment — a review. Part II. *Chemosphere* 75, 435–441.

Laanbroek, H. J., Keijzer, R. M., Verhoeven, J. T. A. and Whigham, D. F. (2012). The distribution of ammonia-oxidizing betaproteobacteria in stands of Black mangroves (*Avicennia germinans*). *Front. Microbiol.* 3, 153, 1–11.

Lamers, L. P. M., Van Diggelen, J. M. H., Op Den Camp, H. J. M., Visser, E. J. W., Lucassen, E., Vile, M. A., *et al.* (2012). Microbial transformations of nitrogen, sulfur, and iron dictate vegetation composition in wetlands: a review. *Front. Microbiol.* 3, 156, 1–12.

Lever, M. A., Rogers, K. L., Lloyd, K. G., Overmann, J., Schink, B., Thauer, R. K., *et al.* (2015). Life under extreme energy limitation: a synthesis of laboratory- and field-based investigations. *FEMS Microbiol. Rev.* 39, 688–728.

Litchman, E. and Klausmeier, C. A. (2008). Trait-based community ecology of phytoplankton. *Annu. Rev. Ecol. Evol. Syst.* 39, 615–639.

Litchman, E., Klausmeier, C. A., Schofield, O. M. and Falkowski, P. G. (2007). The role of functional traits and trade-offs in structuring phytoplankton communities: scaling from cellular to ecosystem level. *Ecol. Lett.* 10, 1170–1181.

Lofton, D. D., Whalen, S. C. and Hershey, A. E. (2014). Effect of temperature on methane dynamics and evaluation of methane oxidation kinetics in shallow Arctic Alaskan lakes. *Hydrobiology* 721, 209–222.

Lu, Y. H. and Conrad, R. (2005). *In situ* stable isotope probing of methanogenic archaea in the rice rhizosphere. *Science* 309, 1088–1090.

Margesin, R., Neuner, G. and Storey, K. B. (2006). Cold-loving microbes, plants, and animals — fundamental and applied aspects. *Naturwissenschaften* 94, 77–99.

Martiny, J. B. H., Jones, S. E., Lennon, J. T. and Martiny, A. C. (2015). Microbiomes in light of traits: a phylogenetic perspective. *Science* 350(6261), 1–8. doi: 10.1126/science.aac9323.

Martiny, A. C., Treseder, K. and Pusch, G. (2013). Phylogenetic conservatism of functional traits in microorganisms. *ISME J.* 7, 830–838.

Matamoros, V., Arias, C., Brix, H. and Bayona, J. M. (2009). Preliminary screening of small-scale domestic wastewater treatment systems for removal of pharmaceutical and personal care products. *Water Res.* 43, 55–62.

Melton, J. R., Wania, R., Hodson, E. L., Poulter, B., Ringeval, B., Spahni, R., *et al.* (2013). Present state of global wetland extent and wetland methane modelling: conclusions from a model inter-comparison project (WETCHIMP). *Biogeosciences* 10, 753–788.

Methé, B. A., Nelson, K. E., Deming, J. W., Momen, B., Melamud, E., Zhang, X., *et al.* (2005). The psychrophilic lifestyle as revealed by the genome sequence of *Colwellia psychrerythraea* 34H through genomic and proteomic analyses. *Proc. Natl. Acad. Sci. USA* 102, 10913–10918.

Mikha, M. M., Rice, C. W. and Milliken, G. A. (2005). Carbon and nitrogen mineralization as affected by drying and wetting cycles. *Soil Biol. Biochem.* 37, 339–347.

Miletto, M., Loeb, R., Antheunisse, A. M., Bodelier, P. L. E. and Laanbroek, H. J. (2010). Response of the sulfate-reducing community to the re-establishment of estuarine conditions in two contrasting soils: a mesocosm approach. *Microb. Ecol.* 59, 109–120.

Miletto, M., Loy, A., Antheunisse, A. M., Loeb, R., Bodelier, P. L. E. and Laanbroek, H. J. (2008). Biogeography of sulfate-reducing prokaryotes in river floodplains. *FEMS Microbiol. Ecol.* 64, 395–406.

Millennium, Ecosystem & Assessment. (2005). Ecosystems and human well-being: wetlands and water synthesis. *Millennium Ecosystem Assessment.* World Resources Institute, Washington, DC, pp. 1–80.

Mitsch, W. J., Bernal, B., Nahlik, A. M., Mander, U., Zhang, L., Anderson, C. J., *et al.* (2013). Wetlands, carbon, and climate change. *Landscape Ecol.* 28, 583–597.

Mitsch, W. and Gosselink, J. (2000). *Wetlands.* Wiley, New York, p. 571.

Moseman, S. M. (2007). Opposite diel patterns of nitrogen fixation associated with salt marsh plant species (*Spartina foliosa* and *Salicornia virginica*) in southern California. *Mar. Ecol. Evol. Perspect.* 28, 276–287.

Mulder, A., Graaf, A., Robertson, L. and Kuenen, J. (1995). Anaerobic ammonium oxidation discovered in a denitrifying fluidized bed reactor. *FEMS Microbiol. Ecol.* 16, 177–184.

Murase, J. and Frenzel, P. (2008). Selective grazing of methanotrophs by protozoa in a rice field soil. *FEMS Microbiol. Ecol.* 65, 408–414.

Murase, J., Hordijk, K., Tayasu, I. and Bodelier, P. L. E. (2011). Strain-specific incorporation of methanotrophic biomass into eukaryotic grazers in a rice field soil revealed by PLFA-SIP. *FEMS Microbiol. Ecol.* 75, 284–290.

Neidhardt, F., Vanbogelen, R. and Vaughn, V. (1984). The genetics and regulation of heat-shock proteins. *Annu. Rev. Genet.* 18, 295–329.

Nelson, E. B. and Karp, M. A. (2013). Soil pathogen communities associated with native and non-native *Phragmites australis* populations in freshwater wetlands. *Ecol. Evol.* 3, 5254–5267.

O'Malley, M. A. (2008). 'Everything is everywhere: but the environment selects': ubiquitous distribution and ecological determinism in microbial biogeography. *Stud. Hist. Phil. Biol. Biomed. Sci.* 39, 314–325.

Orell, A., Navarro, C. A., Rivero, M., Aguilar, J. S. and Jerez, C. A. (2012). Inorganic polyphosphates in extremophiles and their possible functions. *Extremophiles* 16, 573–583.

Pankhurst, C. E., Yu, S., Hawke, B. G. and Harch, B. D. (2001). Capacity of fatty acid profiles and substrate utilization patterns to describe differences in soil microbial communities associated with increased salinity or alkalinity at three locations in South Australia. *Biol. Fert. Soils* 33, 204–217.

Pester, M., Knorr, K. H., Friedrich, M. W., Wagner, M. and Loy, A. (2012). Sulfate-reducing microorganisms in wetlands — fameless actors in carbon cycling and climate change. *Front. Microbiol.* 3(71), 1–19.

Petrescu, A. M. R., Lohila, A., Tuovinen, J.-P., Baldocchi, D. D., Desai, A. R., Roulet, N. T., *et al.* (2015). The uncertain climate footprint of wetlands under human pressure. *Proc. Natl. Acad. Sci. USA* 112, 4594–4599.

Pfluger, A. R., Wu, W. M., Pieja, A. J., Wan, J., Rostkowski, K. H. and Criddle, C. S. (2011). Selection of Type I and Type II methanotrophic proteobacteria in a fluidized bed reactor under non-sterile conditions. *Bioresour. Technol.* 102, 9919–9926.

Placella, S. A., Brodie, E. L. and Firestone, M. K. (2012). Rainfall-induced carbon dioxide pulses result from sequential resuscitation of phylogenetically clustered microbial groups. *Proc. Natl. Acad. Sci. USA* 109, 10931–10936.

Prasse, C. E., Baldwin, A. H. and Yarwood, S. A. (2015). Site history and edaphic features override the influence of plant species on microbial communities in restored tidal freshwater wetlands. *Appl. Environ. Microb.* 81, 3482–3491.

Prosser, J. I. and Nicol, G. W. (2012). Archaeal and bacterial ammonia-oxidisers in soil: the quest for niche specialisation and differentiation. *Trends Microbiol.* 20, 523–531.

Ramos, J. L., Gallegos, M. A.-T., Marqués, S., Ramos-González, M.-I., Espinosa-Urgel, M. and Segura, A. 2001. Responses of Gram-negative bacteria to certain environmental stressors. *Curr. Opin. Microbiol.* 4, 166–171.

Rath, K. M. and Rousk, J. (2015). Salt effects on the soil microbial decomposer community and their role in organic carbon cycling: a review. *Soil Biol. Biochem.* 81, 108–123.

Raymond, J., Siefert, J. L., Staples, C. R. and Blankenship, R. E. (2004). The natural history of nitrogen fixation. *Mol. Biol. Evol.* 21, 541–554.

Reim, A., Luke, C., Krause, S., Pratscher, J. and Frenzel, P. (2012). One millimetre makes the difference: high-resolution analysis of methane-oxidizing bacteria and their specific activity at the oxic-anoxic interface in a flooded paddy soil. *ISME J.* 6, 2128–2139.

Rejmánková, E. and Komárková, J. (2000). A function of cyanobacterial mats in phosphorus-limited tropical wetlands. *Hydrobiology* 431, 135–153.

Roberson, E. B. and Firestone, M. K. (1992). Relationship between desiccation and exopolysaccharide production in a soil *Pseudomonas* sp. *Appl. Environ. Microb.* 58, 1284–1291.

Scheffer, M., Carpenter, S., Foley, J. A., Folke, C. and Walker, B. (2001). Catastrophic shifts in ecosystems. *Nature* 413, 591–596.

Schimel, J. P. and Gulledge, J. (1998). Microbial community structure and global trace gases. *Glob. Change Biol.* 4, 745–758.

Schmidt, H. and Eickhorst, T. (2014). Detection and quantification of native microbial populations on soil-grown rice roots by catalyzed reporter deposition-fluorescence *in situ* hybridization. *FEMS Microbiol. Ecol.* 87, 390–402.

Schulz, S., Matsuyama, H. and Conrad, R. (1997). Temperature dependence of methane production from different precursors in a profundal sediment (Lake Constance). *FEMS Microbiol. Ecol.* 22, 207–213.

Segarra, K. E. A., Schubotz, F., Samarkin, V., Yoshinaga, M. Y., Hinrichs, K. U. and Joye, S. B. (2015). High rates of anaerobic methane oxidation in freshwater wetlands reduce potential atmospheric methane emissions. *Nat. Commun.* 6, 7477, 1–8. doi: 10.1038/ncomms8477.

Serreze, M., Barrett, A., Stroeve, J., Kindig, D. and Holland, M. (2009). The emergence of surface-based Arctic amplification. *Cryosphere* 3, 11–19.

Settele, J. R., Scholes, R., Betts, R. A., Bunn, S., Leadley, P., Nepstad, D., *et al.* (2014). Terrestrial and inland water systems. In Field, C. B., Barros, V. R., Dokken, D. J., Mach, K. J., Mastrandea, M. D., Bilir, T. E., Chatterjee, M., Ebi, K. L., Estrada, Y. O., Genova, R. C., Girma, B., Kissel, E .S., Levy, A. N., MacCracken, S., Mastrandrea, P. R. and White, L. L. (eds.), *Climate Change 2014: Impacts, Adaptation, and Vulnerability. Part A: Global and Sectoral Aspects*. Contribution of Working Group II to the Fifth Assessment Report of the Intergovernmental Panel on Climate Change. Cambridge University Press, Cambridge, UK and New York, NY, USA.

Sheehan, R., Mccarver, A. C., Isom, C. E., Karr, E. A. and Lessner, D. J. (2015). The *Methanosarcina acetivorans* thioredoxin system activates DNA binding of the redox-sensitive transcriptional regulator MsvR. *J. Ind. Microbiol. Biot.* 42, 965–969.

Solomon, S., Qin, D., Manning, M., Chen, Z., Marquis, M., Averyt, K., *et al.* (2007). *IPCC, 2007: Climate Change 2007: The Physical Science Basis*. Contribution of Working Group I to the fourth assessment report of the Intergovernmental Panel on Climate Change. Cambridge University Press, Cambridge.

Søndergaard, M., Jensen, J. P. and Jeppesen, E. (2003). Role of sediment and internal loading of phosphorus in shallow lakes. *Hydrobiology* 506, 135–145.

Staley, Z. R., Rohr, J. R., Senkbeil, J. K. and Harwood, V. J. (2014). Agrochemicals indirectly increase survival of *E. coli* O157: H7 and indicator bacteria by reducing ecosystem services. *Ecol. Appl.* 24, 1945–1953.

Steenbergh, A. K. (2012). *The Microbial Control of Phosphorus Fluxes in Marine Sediments*. Netherlands Institute of Ecology, PhD Thesis, Utrecht University, Utrecht, the Netherlands.

Stein, L. Y. and Klotz, M. G. (2011). Nitrifying and denitrifying pathways of methanotrophic bacteria. *Biochem. Soc. Trans.* 39, 1826–1831.

Sterner, R. W. and Elser, J. J. (2002). *Ecological Stoichiometry: The Biology of Elements from Molecules to the Biosphere*. Princeton University Press, Princeton, NJ, p. 464.

Strom, L., Falk, J. M., Skov, K., Jackowicz-Korczynski, M., Mastepanov, M., Christensen, T. R., *et al.* (2015). Controls of spatial and temporal variability in CH4 flux in a high arctic fen over three years. *Biogeochemistry* 125, 21–35.

Tiedje, J. M. (1994). Denitrifiers. In R. W. Weaver, J. S. Angle, and P. J. Bottomley (eds.), *Methods of Soil Analysis: Part 2 — Microbiological and Biochemical Properties.* Soil Science Society of America, Madison, WI, pp. 245–267.

Tilman, D. and Downing, J. A. (1994). Biodiversity and stability in grasslands. *Nature* 367, 363–365.

Tobor-Kapłon, M. A., Bloem, J. and De Ruiter, P. C. (2006). Functional stability of microbial communities from long-term stressed soils to additional disturbance. *Environ. Toxicol. Chem.* 25, 1993–1999.

Todorova, S. G. and Costello, A. M. (2006). Design of Shewanella-specific 16S rRNA primers and application to analysis of Shewanella in a minerotrophic wetland. *Environ. Microbiol.* 8, 426–432.

Top, E. M. and Springael, D. (2003). The role of mobile genetic elements in bacterial adaptation to xenobiotic organic compounds. *Curr. Opin. Biotech.* 14, 262–269.

Treseder, K. K., Balser, T. C., Bradford, M. A., Brodie, E. L., Dubinsky, E. A., Eviner, V. T., *et al.* (2012). Integrating microbial ecology into ecosystem models: challenges and priorities. *Biogeochemistry* 109, 7–18.

Van De Leemput, I. A., Veraart, A. J., Dakos, V., De Klein, J. J., Strous, M. and Scheffer, M. (2011). Predicting microbial nitrogen pathways from basic principles. *Environ. Microbiol.* 13, 1477–1487.

Van Groenigen, K. J., Osenberg, C. W. and Hungate, B. A. (2011). Increased soil emissions of potent greenhouse gases under increased atmospheric CO_2. *Nature* 475, 214-U121.

Van Huissteden, J., Berrittella, C., Parmentier, F., Mi, Y., Maximov, T. and Dolman, A. (2011). Methane emissions from permafrost thaw lakes limited by lake drainage. *Nat. Clim. Change* 1, 119–123.

Van Kessel, M. A., Speth, D. R., Albertsen, M., Nielsen, P. H., Den Camp, H. J. O., Kartal, B., *et al.* (2015). Complete nitrification by a single microorganism. *Nature* 528, 555–559.

Ventosa, A., Nieto, J. J. and Oren, A. (1998). Biology of moderately halophilic aerobic bacteria. *Microbiol. Mol. Biol. Rev.* 62, 504–544.

Veraart, A. J., De Klein, J. J. and Scheffer, M. (2011). Warming can boost denitrification disproportionately due to altered oxygen dynamics. *PLoS ONE* 6, e18508.

Veraart, A. J., Steenbergh, A. K., Ho, A., Kim, S. Y. and Bodelier, P. L. E. (2015). Beyond nitrogen: the importance of phosphorus for CH_4 oxidation in soils and sediments. *Geoderma* 259, 337–346.

Verheijen, L. M., Aerts, R., Brovkin, V., Cavender-Bares, J., Cornelissen, J. H. C., Kattge, J., *et al.* (2015). Inclusion of ecologically based trait variation in plant functional types reduces the projected land carbon sink in an earth system model. *Glob. Change Biol.* 21, 3074–3086.

Verhoeven, J. T. A. (2014). Wetlands in Europe: perspectives for restoration of a lost paradise. *Ecol. Eng.* 66, 6–9.

Vile, M., Kelman Wieder, R., Živković, T., Scott, K., Vitt, D., Hartsock, J., *et al.* (2014). N2-fixation by methanotrophs sustains carbon and nitrogen accumulation in pristine peatlands. *Biogeochemistry* 121(2), 317–328.

Violle, C., Navas, M. L., Vile, D., Kazakou, E., Fortunel, C., Hummel, I., *et al.* (2007). Let the concept of trait be functional! *Oikos* 116, 882–892.

Wagner, M. (2009). Single-cell ecophysiology of microbes as revealed by Raman microspectroscopy or secondary ion mass spectrometry imaging. *Annu. Rev. Microbiol.* 63, 411–429.

Walter, K. M., Zimov, S., Chanton, J. P., Verbyla, D. and Chapin, F. S. (2006). Methane bubbling from Siberian thaw lakes as a positive feedback to climate warming. *Nature* 443, 71–75.

Wang, G. S., Jagadamma, S., Mayes, M. A., Schadt, C. W., Steinweg, J. M., Gu, L. H., *et al.* (2015). Microbial dormancy improves development and experimental validation of ecosystem model. *ISME J.* 9, 226–237.

Wang, J. J., Krause, S., Muyzer, G., Meima-Franke, M., Laanbroek, H. J. and Bodelier, P. L. E. (2012a). Spatial patterns of iron- and methane-oxidizing bacterial communities in an irregularly flooded, riparian wetland. *Front. Microbiol.* 3(64), 1–13.

Wang, L. X., Liu, H. M., Liu, Y. H., Li, J. W., Shao, H. B., Wang, W., *et al.* (2014a). Soil characteristic comparison of fenced and grazed riparian floodplain wetlands in the typical steppe region of the inner Mongolian Plateau, China. *Sci. World J.* 10(765907), 1–10. doi: org/10.1155/2014/765907.

Wang, J. J., Muyzer, G., Bodelier, P. L. E. and Laanbroek, H. J. (2009). Diversity of iron oxidizers in wetland soils revealed by novel 16S rRNA primers targeting Gallionella-related bacteria. *ISME J.* 3, 715–725.

Wang, W., Sardans, J., Zeng, C., Zhong, C., Li, Y. and Penuelas, J. (2014b). Responses of soil nutrient concentrations and stoichiometry to different human land uses in a subtropical tidal wetland. *Geoderma* 232, 459–470.

Wang, Y., Sheng, H. F., He, Y., Wu, J. Y., Jiang, Y. X., Tam, N. F. Y., *et al.* (2012b). Comparison of the levels of bacterial diversity in freshwater, intertidal wetland, and marine sediments by using millions of illumina tags. *Appl. Environ. Microb.* 78, 8264–8271.

Ward, N. L., Challacombe, J. F., Janssen, P. H., Henrissat, B., Coutinho, P. M., Wu, M., *et al.* (2009). Three genomes from the phylum acidobacteria provide insight into the lifestyles of these microorganisms in soils. *Appl. Environ. Microb.* 75, 2046–2056.

Ward, B. A., Dutkiewicz, S., Moore, C. M. and Follows, M. J. (2013). Iron, phosphorus, and nitrogen supply ratios define the biogeography of nitrogen fixation. *Limnol. Oceanogr.* 58, 2059–2075.

Wieder, W. R., Allison, S. D., Davidson, E. A., Georgiou, K., Hararuk, O., He, Y. J., *et al.* (2015a). Explicitly representing soil microbial processes in Earth system models. *Glob. Biogeochem. Cycles* 29, 1782–1800.

Wieder, W. R., Bonan, G. B. and Allison, S. D. (2013). Global soil carbon projections are improved by modelling microbial processes. *Nat. Clim. Change* 3, 909–912.

Wieder, W. R., Grandy, A. S., Kallenbach, C. M., Taylor, P. G. and Bonan, G. B. (2015b). Representing life in the Earth system with soil microbial functional traits in the MIMICS model. *Geosci. Model Dev.* 8, 1789–1808.

Wik, M., Varner, R. K., Anthony, K. W., Macintyre, S. and Bastviken, D. (2016). Climate-sensitive northern lakes and ponds are critical components of methane release. *Nat. Geosci.* 9, 99–105.

Winton, M. (2006). Amplified Arctic climate change: what does surface albedo feedback have to do with it? *Geophys. Res. Lett.* 33, L03701, 1–4. doi: 10.1029/2005GL025244.

Yang, W., Zhao, H., Chen, X. L., Yin, S. L., Cheng, X. L. and An, S. Q. (2013). Consequences of short-term C-4 plant *Spartina alterniflora* invasions for soil organic carbon dynamics in a coastal wetland of Eastern China. *Ecol. Eng.* 61, 50–57.

Yvon-Durocher, G., Allen, A. P., Bastviken, D., Conrad, R., Gudasz, C., St-Pierre, A., *et al.* (2014). Methane fluxes show consistent temperature dependence across microbial to ecosystem scales. *Nature* 507, 488–491.

Zhang, Q. F., Peng, J. J., Chen, Q., Li, X. F., Xu, C. Y., Yin, H. B., *et al.* (2011). Impacts of *Spartina alterniflora* invasion on abundance and composition of ammonia oxidizers in estuarine sediment. *J. Soils Sediments* 11, 1020–1031.

Zhou, J., He, Q., Hemme, C. L., Mukhopadhyay, A., Hillesland, K., Zhou, A., *et al.* (2011). How sulphate-reducing microorganisms cope with stress: lessons from systems biology. *Nat. Rev. Microbiol.* 9, 452–466.

Zhu, G. B., Wang, S. Y., Zhou, L. L., Wang, Y., Zhao, S. Y., Xia, C., *et al.* (2015). Ubiquitous anaerobic ammonium oxidation in inland waters of China: an overlooked nitrous oxide mitigation process. *Sci. Rep.* 5, 10.

Zumft, W. G. (1997). Cell biology and molecular basis of denitrification. *Microbiol. Mol. Biol. R.* 61, 533–616.

10

Arctic Soil Microbial Sensitivity to Seasonal Dynamics and Climate Change

Anders Priemé*, Daan Blok, Merian Skouw Haugwitz, Jana Voříšková and Bo Elberling

Center for Permafrost (CENPERM),
Department of Geosciences and Natural Resource Management,
University of Copenhagen, Øster Voldgade 10,
1350 Copenhagen K, Denmark
**aprieme@bio.ku.dk*

Introduction

Soil environmental factors, such as temperature and water availability, can exhibit large fluctuations in the Arctic and are known to affect microbial activity, biomass, community structure and thereby ecosystem functioning. But the sensitivity of soil microorganisms to these fluctuations is poorly constrained, and general patterns need to be assessed in order to improve modelling of greenhouse gas flux rates and other ecosystem responses to future climate conditions. Early work from the Arctic (Cheng and Virginia, 1993) confirmed that the fumigation–extraction and substrate-induced respiration methods, developed to study microbial biomass in mostly temperate agricultural soils, can also be successfully applied to study microbial biomass dynamics in arctic tundra soils. Subsequently, these methods have also been widely applied in arctic investigations.

Challenges Facing Arctic Soil Microorganisms

Cold-region soils that are exposed to mean annual air temperatures below freezing differ from soils at lower latitudes by having an active layer and a permafrost layer. Permafrost is defined as soil or sediment kept at temperatures below 0°C for at least two consecutive years. It has been recently reported that permafrost can remain frozen for as long as 3 million years (Jansson and Taş, 2014). Most permafrost is located at high latitudes but it also occurs at lower latitudes in alpine ecosystems. The active layer, that is, the top layer of soil that thaws during summer and freezes again during the autumn, varies in thickness from less than 30 cm to a few metres and experiences annual freeze–thaw cycles (Figure 10.1).

Despite Arctic soils being exposed to harsh and challenging environmental conditions, they are inhabited by diverse and numerous bacteria, archaea, fungi and other microorganisms. The sub-freezing temperatures prevailing in permafrost year-round and in the active layer during the cold season affect microbial life by

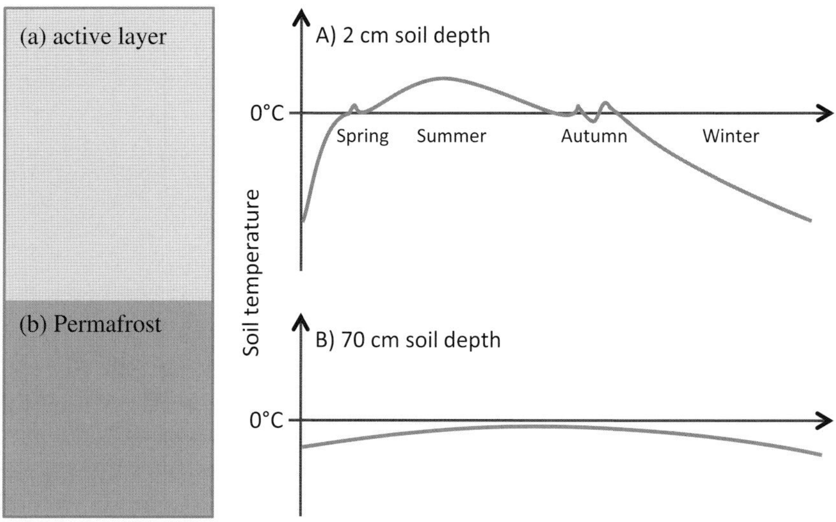

Figure 10.1 A simplified arctic soil profile with an active layer (including roots) and permafrost. The temperature changes during a full year in (a) the uppermost active layer and (b) the uppermost permafrost. The uppermost active layer is exposed to marked changes from typically well below –20 to +10°C, while uppermost permafrost layers remain below 0°C and fairly constant reaching annual maximum temperature later in the year compared to the active layer. The active layer depth ranges from less than 30 cm in the extreme high Arctic to more than 2 m at the southern limit of permafrost distribution or under very dry soil conditions.

reducing enzymatic reaction rates, cell membrane fluidity, transport of substrates and rates of DNA replication, gene expression and protein synthesis compared to microorganisms in soils exposed to higher temperatures (Chattopadhyay, 2006; D'Amico *et al.*, 2006; de Maayer *et al.*, 2014). Despite the challenges that arctic microorganisms are facing, they can maintain their metabolic processes even under freezing conditions by modifying the fluidity of their cell membranes, accumulating osmolytes and antifreeze proteins, producing cold-active enzymes and exploiting isozymes that are active over different temperature ranges (Walker *et al.*, 2006; Ayala-Del-Rio *et al.*, 2010; Hall *et al.*, 2010; Timling and Taylor, 2012; Mykytczuk *et al.*, 2013; Ewert and Deming, 2014; Feng *et al.*, 2014; Nunn *et al.*, 2015).

Besides low temperatures, arctic soil microorganisms are also faced with the indirect effects of low temperatures, which involve freezing of soil water, resulting in low water availability. All microorganisms require liquid water to maintain growth, although the water-stress threshold below which microorganisms cease their activity is highly variable and species-specific (Manzoni *et al.*, 2011). Moreover, the relationship between the liquid soil water fraction is not linearly related to temperature as not all soil water freezes below 0°C (Figure 10.2), with brine veins, pockets or films of water remaining liquid at temperatures well below 0°C (Gilichinsky, 2002; Schimel and Mikan, 2005). The freezing point of water varies with soil type, vegetation cover and soil biogeochemistry (Öquist *et al.*, 2009). During freezing of soil water, solutes are excluded from the ice. This leads to a decrease in soil water potential, estimated to correspond to 1.1 MPa per °C

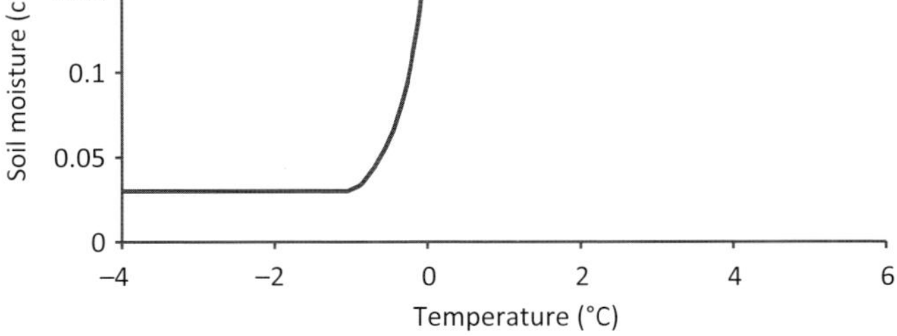

Figure 10.2 Unfrozen water may exist in otherwise frozen soil and allows for microbial activity at sub-zero temperatures. Adapted from Aanderud *et al.* (2013).

decrease in temperature (Jefferies *et al.*, 2010). As an extreme example, isolates of *Psychrobacter* and *Exiguobacterium* from Siberian permafrost display metabolic activity at −11 MPa (Ponder *et al.*, 2008). Hence, active permafrost microorganisms need to be halotolerant as well as psychrotolerant, and osmotic concentration rather than temperature extremes may represent the greater stress to the microorganisms (Ewert and Deming, 2014).

Unfrozen water in otherwise frozen soil enables mass transfer of microbial substrates, with the highest rates of mass exchange found in microzones with low ice content (Ostroumov and Siegert, 1996). Microorganisms use specific enzymes to facilitate active transport of substrates from the environment to the cell. At low temperatures, the microorganisms experience a reduction in the specific affinity for substrates taken up by active transport (Nedwell, 1999). In other words, substrates taken up by active transport may become less available at low temperatures, because the ability of an organism to sequester the substrate declines with temperature. Besides low substrate diffusion rates and valuable substrates being locked in the frozen soil, microorganisms in arctic soils also face limited abilities to take up substrates at low concentrations. However, permafrost microorganisms may circumvent this limitation by using substrates like acetate, which can diffuse into the cell without an energetically costly transport system. This has, for example, been reported for the permafrost bacterium *Psychrobacter arcticus*, which uses acetate as the basis for its biosynthesis and energy metabolism (Ayala-Del-Rio *et al.*, 2010).

Diversity and Biomass of Arctic Soil Microorganisms

Despite the harsh living conditions, the structure and diversity of microbial communities in arctic soils are similar to soils in less extreme environments. The bacterial community in active layer soils is dominated by *Proteobacteria*, *Actinobacteria*, *Acidobacteria*, *Verrucomicrobia* and *Gemmatimonadetes* (Yergeau *et al.*, 2010; Tveit *et al.*, 2013; Schostag *et al.*, 2015) and does not differ fundamentally in terms of diversity from surface soils in low-latitude ecosystems (Fierer and Jackson, 2006; Chu *et al.*, 2010). Also, the fungal communities in arctic soils include representatives of all the major fungal phyla (Wallenstein *et al.*, 2007; Timling and Taylor, 2012; Timling *et al.*, 2014).

Chloroform fumigation–extraction and substrate-induced respiration were originally used to study the microbial biomass in mostly temperate agricultural soils (Wardle, 1992) and were soon applied in studies of microbial biomass dynamics in arctic soils (Cheng and Virginia, 1993). In order to provide an overview of the average amount of C bound in the microbial biomass across the arctic

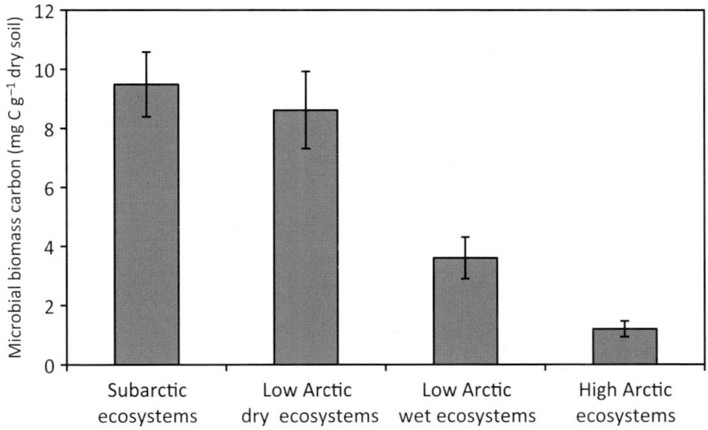

Figure 10.3 Soil microbial biomass carbon content in subarctic ecosystems, low arctic dry and wet ecosystems, and high arctic ecosystems. The dry ecosystems were mostly dwarf-shrub tundra and dry tussock tundra, whereas wet ecosystems were typically sedge meadows and riparian areas. Studies used are listed in Appendix 1. The bars represent one standard error.

region, values were derived from 18 studies (Figure 10.3). Microbial biomass values were divided into subarctic, low arctic and high arctic ecosystems, with low arctic ecosystems being further divided into dry and wet ecosystems depending on the site descriptions in the study. From this gradient from subarctic to high arctic ecosystems, it was evident that the soil microbial biomass decreased with increasing latitude (Figure 10.3), which may be due to climatic restrictions acting directly on soil microorganisms and/or low substrate availability caused by the small plant biomass restricting carbon input into high arctic soil.

Effects of Summer Warming on Arctic Soil Microorganisms

Short growing seasons and low temperatures characterise arctic summers. Since low temperatures are one of the primary determinants of arctic ecosystems, increases in summer temperature may have profound effects on ecosystem structure and functioning. Climate change scenarios project that before the end of the century, arctic land mass may warm between 4 and 8°C during summer (IPCC, 2013). These higher temperatures may affect the active layer soil microbial communities directly because soil microorganisms, and the processes they perform are temperature-sensitive (Bradford, 2013; Frey *et al.*, 2013; Hagerty *et al.*,

2014). Furthermore, the temperature dependency of microbial respiration rates in arctic soils has been shown to be higher compared to soils from other latitudes (Bekku *et al.*, 2003; Karhu *et al.*, 2014). This suggests that increasing temperatures in the Arctic may increase element turnover through higher soil organic matter mineralisation rates, more rapidly than in other biomes.

Additionally, summer warming may result in a longer growing season for plants, which may affect microorganisms, as plant communities are strong drivers of the distribution and functionality of soil microorganisms (Broeckling *et al.*, 2008; Deslippe and Simard, 2011). Based on satellite observations of tundra ecosystems, warming increased net primary production and led to an increased carbon dioxide (CO_2) uptake by tundra plants (Sitch *et al.*, 2007). In particular, warming has been shown to enhance the abundance of shrubs in the Arctic at the expense of non-vascular plants, primarily by increasing the abundance of deciduous shrubs (Aerts, 2006; Myers-Smith *et al.*, 2011; Elmendorf *et al.*, 2012), but also due to increasing distributions of evergreen ericaceous shrubs (Sorensen *et al.*, 2012). Shrub expansion in the arctic tundra may promote mycorrhizal colonisation by shrub roots (Clemmensen *et al.*, 2006; Deslippe and Simard, 2011) and lead to increased shrub nutrient uptake (Rustad *et al.*, 2001; Hobbie and Hobbie, 2006). Furthermore, enhanced plant activity during the summer may lead to an increased rhizodeposition of labile carbon, thereby stimulating the microbial biomass and possibly inducing a faster turnover of organic matter, driving a higher soil nutrient availability (Nadelhoffer *et al.*, 1992; Aerts *et al.*, 2006). Depending on the future turnover rate of organic matter relative to the carbon input to the soil, warming may either increase or decrease carbon losses from arctic tundra surface soils, with associated feedback effects on the climate system (Wookey *et al.*, 2009).

Current Knowledge from *In Situ* Warming Experiments

In order to investigate the effects of increasing summer temperatures in the Arctic, many experiments have been setup in the region to mimic predicted future warmer arctic summers. Although the harshness and impassability of the arctic landscape make fieldwork challenging and at times impossible, a set of standardised experiments (e.g. snow addition, fertilisation, watering) have been established to investigate the consequences of climate changes in the Arctic. A well-known example of a coordinated tundra manipulation experiment is the International Tundra Experiment (ITEX), a network of *in situ* warming experiments that have produced long-term datasets across the tundra biome (Molau and Mølgaard, 1996; Henry and Molau, 1997). Warming is achieved by transparent

Figure 10.4 An experimental setup at Disko Island (West Greenland), where open-top ITEX chambers increase the temperature during the summer in an arctic heath.

open-top chambers that passively warm the air inside the chambers during the plant-growing season (Figure 10.4).

Long-term experimental warming studies have shown that the microbial biomass in arctic ecosystems is relatively stable throughout the growing season (Jonasson *et al.*, 1999; Sistla and Schimel, 2013) and that microbial biomass varies less between years during summer compared to winter (Edwards and Jefferies, 2013). In mid-latitude ecosystems, warming effects on soil microbial activity are short-lived due to a depletion of labile soil carbon (Bradford *et al.*, 2008). However, at a subarctic heath, no effects from warming were observed on microbial biomass following 10 years of warming (Rinnan *et al.*, 2007a) with a significant decrease in microbial biomass only becoming apparent after 15 years. In contrast, the microbial biomass increased compared to ambient non-warmed plots after two decades of warming, possibly as a result of increased vascular plant cover (Rinnan *et al.*, 2013). Thus, changes in the microbial biomass in arctic soils due to warming can be very slow (Rinnan *et al.*, 2007a, 2013). Similarly, long-term warming may be required to alter the soil fungal and bacterial community structure in tundra soils (Deslippe *et al.*, 2011, 2012; Geml *et al.*, 2015; Morgado *et al.*, 2015; Semenova *et al.*, 2015). Such warming-induced shifts in microbial community composition may lead to changes in ecosystem functioning through changes in functional traits of the soil organisms and their role in rate-limiting or rate-controlling steps in the turnover of organic matter (Schimel and Schaeffer, 2012). For example, two decades of summer warming in an Alaskan tundra ecosystem changed the structure and activity of the soil microbial community in the deeper organic soil (below 5 cm depth) and shifted the vertical distribution and turnover of soil carbon to the mineral layer. These changes occurred without changing the ecosystem net carbon storage as increased plant photosynthesis was counterbalanced by increased soil respiration (Sistla *et al.*, 2013).

Changes in temperature are often coupled with changes in soil water availability. Since water availability is a key factor controlling soil microbial communities, this might explain the inconsistent microbial community responses between short- and long-term warming manipulations. Furthermore, litter decomposition in tundra ecosystems warmed in open top chambers has been shown to decrease, probably as a result of water limitation (Robinson *et al.*, 1995; Pries *et al.*, 2013; Blok *et al.*, 2015). Therefore, while soil microbial biomass increases with temperature in long-term studies in the Arctic, microbial responses to warming may be different in water-limited ecosystems or in ecosystems experiencing an altered future soil–water balance.

Tundra Soil Freeze-up and Soil Thaw: The Importance of "Shoulder Seasons" Dynamics on Microbial Biomass and Functioning

Autumn soil freeze-up

In arctic and alpine tundra ecosystems, the transition from summer into winter occurs more rapidly than in other ecosystems. During autumn, soil temperatures often drop below 0°C in a matter of days. During periods with little snow cover and/or early frost events, soil microorganisms need to adapt swiftly to survive the rapid physical and chemical changes that occur in their environment during soil freezing. Soil temperatures remain close to 0°C as long as enough liquid water is present and continues to release latent heat, effectively buffering the soil from freezing (Romanovsky and Osterkamp, 2000). When substantial snowfall occurs early during autumn before major frost events, soils may stay at this "zero degree curtain" for an even longer period of time. In freezing soils, the temperature sensitivity of respiration is markedly different from unfrozen soils (Mikan *et al.*, 2002), possibly due to effects of water limitation on microbial respiration

(Öquist *et al.*, 2009) and microbial growth rates (Manzoni *et al.*, 2011). While heterotrophic respiration rates decline exponentially with temperature, microbial substrate use shifts from detritus material towards dissolved and recycled microbial biomass (Schimel and Mikan, 2005). Surprisingly, this shift in substrate use may already have started above 0°C, suggesting that tundra soil microorganism physiology anticipates forthcoming freezing conditions (Schimel and Mikan, 2005), a phenomenon termed the "near-freezing response" (Kandror *et al.*, 2004). The idea that tundra microorganisms may be adapted to freezing conditions is further supported by the observed resistance of microbial community structure to frequent freeze–thaw cycles (Männistö *et al.*, 2009; Stres *et al.*, 2010), although fungal growth rates have been observed to increase with freeze–thaw events in boreal soil, while bacterial growth rates are diminished (Haei *et al.*, 2011). This suggests that fungi may be more resistant to freeze–thaw cycles than bacteria.

Spring soil thaw

Microbial biomass has been observed to decline during the winter–spring transition due to the severe stresses imposed by the gradual depletion of substrates by the end of the winter (Lipson *et al.*, 2000) and by the rapid changes in osmotic concentration and water availability during freeze–thaw cycles (Jefferies *et al.*, 2010). Even though soil microorganisms may be adapted and resilient to extreme arctic climate conditions (Männistö *et al.*, 2009; Stres *et al.*, 2010), the physiological costs involved in surviving multiple freeze–thaw cycles may limit their capacity for growth, as microorganisms need to invest energy in protective structures and fine-tune their osmolyte pool to match changing soil water availability (Schimel *et al.*, 2007). Microorganisms with a lower tolerance to freeze–thaw cycles are likely to be killed during a freeze–thaw cycle, in some cases this may be as high as 50% of the microbial biomass (Jefferies *et al.*, 2010). This can lead to a respiration pulse upon tundra soil thawing (Schimel and Clein, 1996) that may be related to surviving microorganisms feeding and growing on dead microorganisms. This respiration pulse declines with the number of subsequent freeze–thaw cycles (Schimel and Clein, 1996), possibly due to a declining and increasingly resistant microbial biomass pool (Grogan *et al.*, 2004). The frequently observed crash in microbial biomass during early spring is associated with a nutrient pulse being released from ruptured microbial cells supporting the subsequent plant productivity peak during the short tundra growing season (Jefferies *et al.*, 2010). However, such a decline in microbial biomass during spring may not be directly related to a decrease in microbial activity. Although microbial biomass and activity may be uncoupled due to the very small proportion of total microbial biomass being active under low substrate availability (<5%), the potentially active microbial pool may be much greater (10–60%) during a period of high labile substrate input (Blagodatskaya and Kuzyakov, 2013), for example, during a spring thaw event. A recent study conducted at an alpine meadow site suggested that a higher frequency of freeze–thaw cycles may increase leaching of dissolved nutrients during spring, but only when accompanied by fluxes of water from melting snow (Wipf *et al.*, 2015). However, some alpine plants have evolved a strategy to bypass these potential ecosystem nitrogen losses that occur during snow runoff by growing "snow roots" upwards into the snowpack during winter and thus directly taking up air-deposited nitrogen in the snowpack before snowmelt (Onipchenko *et al.*, 2009). These complex plant–soil–microbial interactions and the competition for nutrients between soil microorganisms and plants may be greatly affected by legacy effects from changes in winter snow conditions during and after snow melt. For example, a 5-year increase in winter snow depth was found to have carry-over effects on microbial activity during the snow-free season, leading to a shift from net nitrogen immobilisation to net nitrogen mineralisation during the growing season (Schimel *et al.*, 2004). Greater soil nitrogen mineralisation during winter may in turn increase nitrogen availability during summer, thereby increasing plant productivity and carbon uptake (Natali *et al.*, 2012). However, this observed increase in plant productivity was more than offset by increased winter soil respiration, resulting in a net carbon loss from soil to atmosphere in response to soil winter warming (Natali *et al.*, 2014).

Effects of Winter Warming on Arctic Soil Microorganisms

The arctic winter is characterised by snow-covered landscapes for almost two-thirds of the year (Zhang *et al.*, 1996) and is often considered a cold, dark, dormant and biologically inactive period of the year. However, a wide range of soil microbial processes occur below the snow in tundra ecosystems and soil microbial activity may be sustained throughout the winter (Elberling, 2003). Experimental snow addition experiments show that soil microbial respiration rates may be greatly increased during winter by the thermal insulation provided by the snow (Welker *et al.*, 2000; Schimel *et al.*, 2004; Nobrega and Grogan, 2007; Morgner *et al.*, 2010; Natali *et al.*, 2011). Moreover, due to the long duration of the arctic snow season, these processes may represent a considerable contribution to the annual ecosystem activity. Winter CO_2 emission rates indicate that significant microbial activity may occur under the snowpack (Zimov *et al.*, 1993; Fahnestock *et al.*, 1999). This was further supported by Schadt *et al.* (2003), who demonstrated that tundra soil microbial biomass reaches its annual peak during winter under snow-covered conditions, with fungi accounting for most of the microbial biomass. The dominance of fungi, which are better decomposers of recalcitrant components compared to bacteria (de Boer *et al.*, 2005), suggests that microbial growth under snow occurs through decomposition of organic polymers and phenolic compounds. Soil microbial biomass that has been accumulating under the snow cover during winter sharply declines at the end of snow-covered season (Larsen *et al.*, 2007) as a result of microbial cell lysis caused by soil freeze–thaw event (see previous paragraph).

Effects of increased snow cover

Seasonal snow cover has an impact on the terrestrial ecosystem beneath and alters winter processes in several ways. The depth and duration of snow cover seem to be the most important parameters affecting tundra soil microbial activity (Schimel *et al.*, 2004; Buckeridge and Grogan, 2008), both of which are highly influenced by climate change. Climate change projections suggest that the most pronounced effects will occur in high latitudes, with increased winter precipitation in the form of snow being forecasted for these regions (IPCC, 2013). The snow pack serves as an insulating layer for the soil and vegetation (Edwards *et al.*, 2007), protecting both from extreme low air temperatures during the arctic winter. A thick snow cover may enhance the isolation effect and boost soil microbial activity resulting in faster nutrient cycling, larger CO_2 fluxes and greater soil nitrogen availability (Schimel *et al.*, 2004; Rogers *et al.*, 2011). Increased winter precipitation may also influence the growing season length. A deeper snow pack will take longer to thaw, postponing the snowmelt date, which in turn affects the start of the growing season and thus delays plant growth and photosynthesis.

In order to simulate future snow cover scenarios, snow manipulation treatments are used in field experiments (Wipf and Rixen, 2010). The construction of snow fences where wind-drifted snow accumulates on the lee side of the fence (Figure 10.5) is a relatively easy and robust way of mimicking increasing winter precipitation. This approach has been recently used by Morgner *et al.* (2010) (Figure 10.5) to study respiration of seasonally snow-covered ecosystems in a heath and mesic meadow in Svalbard. The total growing season respiration rate

(a)

(b)

Figure 10.5 The experimental snow manipulation site at Adventdalen, Svalbard, during (a) winter and (b) spring (late April and mid-June, respectively). The prevailing wind direction during winter at the site is from right to left leading to accumulation of snow at the lee side of the fence.

was smaller behind fences because of the delayed snowmelt, but measurements in early winter—the period when the treatment effect (difference in snow depth) is most pronounced—showed larger CO_2 flux rates in fenced areas compared to the control. In total, a higher snow pack increased non-growing season CO_2 flux rates substantially in both habitats. Similar observations were made by Lupascu *et al.* (2014) in Northwest Greenland monitoring summer ecosystem respiration in vegetated and bare areas after ~10 years of a snow-manipulation treatment. In the latter experiment, medium and high snow fences were used to mimic intermediate and deep snow cover. Neither of the treatments had an effect on ecosystem respiration during summer but growing season respiration rates were positively correlated to the soil temperature and negatively to the soil water content. Thus, elevated winter precipitation has a carry-over effect during the ensuing summer with reduced summer respiration on snow-manipulated plots compared to control plots by up to 40% as a consequence of shortened growing season length caused by the increased snowpack.

A moderate increase in snow cover for one winter season was used by Larsen *et al.* (2007) to mimic future snow depth in a Scandinavian sub-arctic dwarf shrub heath and a birch forest. These authors found that increased snow depth enhanced ecosystem respiration in late winter but not during spring-thaw. Microbial carbon content was larger in both ecosystems studied in the snow-covered period compared to summer (Jonasson *et al.*, 1999). However, neither an effect of increased snow depth on microbial biomass nor a correlation between microbial biomass and ecosystem respiration was observed. These findings affirm results of previous studies (Wardle, 1992; Michelsen *et al.*, 2004) suggesting that soil microbial biomass is not a reliable indicator of microbial activity. A short-term snow manipulation experiment carried out in low arctic tundra in Canada by Buckeridge and Grogan (2008) affirmed the conclusions from Larsen *et al.* (2007) that there was no effect of deeper snow cover on soil microbial biomass. However, Buckeridge and Grogan (2008) demonstrated that soil microbial communities in the early stages of spring thaw are affected by increased snow cover, even after only 1 year of snow manipulation. Deep snow cover reduced bacterial abundance, but also increased cell volume resulting in no significant changes in bacterial biomass. Responses of microbial communities to snow manipulation treatments differ between short- and long-term experiments (Buckeridge and Grogan, 2010; Olofsson *et al.*, 2011). Thus, Buckeridge and Grogan (2010) investigated the effect of a 3-year snow manipulation treatment on microorganisms in Canadian tundra. Their observations revealed a significant increase in microbial biomass as well as in soil solutions of carbon, nitrogen and phosphorus during a late snow-thaw period as a response to the deeper snow pack during winter. These results suggest that soil biogeochemistry processes during thaw are controlled by winter conditions and may, therefore, be sensitive to future changes in winter precipitation.

These studies indicate that our knowledge about the effects of increased snow cover on soil microorganisms is limited to knowledge on ecosystem respiration and microbial biomass. Knowledge on the effects of winter snow regimes on soil microbial population structure is largely lacking. Future studies should focus on the effects of winter snow dynamics on soil microbial structure and functioning, for example, using metagenomic or metatranscriptomic approaches that may lead to a better understanding of both microbial structure and microbial functioning in response to changes in future winter climate.

Responses of Permafrost Microorganisms to Thaw

Permafrost covers approximately 25% of the land area in the northern hemisphere not covered by ice sheets (Zhang *et al.*, 2008) and is estimated to contain 1470 Pg of organic carbon, corresponding to nearly half of the estimated global soil organic carbon pool (Tarnocai *et al.*, 2009). The Arctic is greatly affected by global warming (IPCC, 2013), which may lead to an increase in permafrost thaw. In turn, this may lead to enhanced microbial activity (Elberling *et al.*, 2013; Knoblauch *et al.*, 2013) and elevated release to the atmosphere of the greenhouse gases CO_2 (Schuur *et al.*, 2009), methane (CH_4) (Olefeldt *et al.*, 2013) and nitrous oxide (N_2O) (Elberling *et al.*, 2010). The magnitude of CO_2 emissions from thawing permafrost is estimated to correspond to the contribution of greenhouse gas emissions from global land-use changes (Schuur *et al.*, 2015).

Microbial biomass and activity in permafrost

Being mindful of the sub-freezing temperatures in permafrost, it is not surprising that the microbial biomass (Gittel *et al.*, 2014a), number of bacteria and fungi (Steven *et al.*, 2008) and microbial CO_2 production (Waldrop *et al.*, 2010) are smaller in permafrost soils compared to the overlying active layer. Although data are sparse, fungi presumably occur in greatly reduced numbers in permafrost soils (Gittel *et al.*, 2014b; Hultman *et al.*, 2015). The low fungal biomass in permafrost soils compared with active layer soils may alter permafrost soil functioning upon thawing, for example, by limiting the quantity of enzymes involved in degradation of complex organic matter. In addition, compared to active layer soil, microbial communities in permafrost soils have a lower general functional potential (Hultman *et al.*, 2015) and a lower abundance and diversity of genes involved in key ecological functions (Yergeau *et al.*, 2010). This may also influence the degradation of organic matter in thawed permafrost soil.

Very little is known about the activity of microorganisms in intact permafrost soil. However, it is becoming increasingly clear that permafrost microorganisms are active at sub-zero temperatures (Mikan *et al.*, 2002; Steven *et al.*, 2008; Karhu *et al.*, 2010; Tilston *et al.*, 2010). So far, the lowest temperature where soil bacterial activity was reported is −39°C (Panikov *et al.*, 2006). Two-thirds of bacterial isolates from Siberian permafrost samples grew at −2.5°C, indicating that they are psychroactive (Vishnivetskaya *et al.*, 2006). Growth of fungi isolated from permafrost has been reported at −35°C (Panikov and Sizova, 2007) and of bacteria at -20°C (Tuorto *et al.*, 2014). Interestingly, the latter study reported several members of the common soil phyla Acidobacteria, Actinobacteria and Proteobacteria, which were only able to grow at temperatures below -6°C.

The activity of microorganisms in frozen soil increases with increasing temperature and an increasing amount of unfrozen water (Panikov *et al.*, 2006). Thus, it is likely that microorganisms in permafrost will respond to increases in temperature well below 0°C. However, it should be noted that CO_2 production at low temperatures may not represent ecologically relevant processes like turnover of soil organic matter because CO_2 may also be produced during cold shock responses and other microbial survival strategies. Accordingly, transcriptional profiling of microbial activity in permafrost soil under frozen conditions suggests a dominance of stress responses, survival strategies and maintenance processes (Coolen and Orsi, 2015).

Microbial responses to permafrost thaw

Permafrost microorganisms respond within days to a complete thaw of the soil by increasing their activity related to degradation of soil organic matter and production of greenhouse gases. Mackelprang *et al.* (2011) provided the first comparison of active layer and permafrost metagenomes before and after thaw and revealed a rapid (within 2 days) shift in the structure and function of the microbial community post-thaw. Also, 11 days after thawing, a permafrost microbial community expressed more mRNA transcripts for enzymes involved in the decomposition of soil organic matter (Coolen and Orsi, 2015).

Soon after thaw of permafrost soil samples, an initial burst of CO_2 and CH_4 release may be observed (e.g. Dutta *et al.*, 2006; Mackelprang *et al.*, 2011).

This may be due to a mineralisation flush based on easily degradable organic carbon (Moni *et al.*, 2015) or release of gases trapped in the frozen soil (Rivkina *et al.*, 1998; Albert and Perron, 2000; Gilichinsky, 2002; Lee *et al.*, 2010; Mackelprang *et al.*, 2011). It is currently unclear what fraction of this initial gas release actually represented gases produced by microorganisms upon thawing. The initial gas burst is short-lived, and most studies on gas release from thawed permafrost involve incubation of permafrost samples for weeks or months. However, rates from these studies may overestimate rates of gas release because carbon loss decreases exponentially over time (Elberling *et al.*, 2013; Schädel *et al.*, 2014).

CO_2 efflux from soil samples shows a strong dependence on temperature (Fang and Moncrieff, 2001; Davidson and Janssens, 2006). Decomposition of organic matter in mineral soils, for example, permafrost soils, may be controlled by the quality of organic matter and its availability to microorganisms (Dungait *et al.*, 2012; Schimel and Schaeffer, 2012) or by the size and structure of the microbial decomposer communities (Schmidt *et al.*, 2011; Wieder *et al.*, 2013). Schädel *et al.* (2014) showed that vulnerability of permafrost organic matter to loss increased in soils with a high carbon-to-nitrogen ratio. Otherwise, very little is known about the extent to which decomposition of organic matter in thawed permafrost soil is controlled by the quality and availability to microorganisms of the organic matter or by the size and structure of the microbial decomposer communities.

The release of CO_2 and CH_4 from permafrost soil depends not only on temperature, but also on redox conditions in the soil (Elberling *et al.*, 2013; Treat *et al.*, 2015). Using an unprecedented long incubation period of 12 years, Elberling *et al.* (2013) found that 58–77% of soil organic carbon was lost during aerobic conditions, while only 9% was lost during anaerobic incubation. Anaerobic conditions slow CO_2 production, but increase production of CH_4 by methanogenic archaea. The amount of CH_4 produced depends on soil type with organic permafrost soils releasing more CH_4 compared to mineral soils (Treat *et al.*, 2015). CH_4 has a larger atmospheric warming potential than CO_2 and may be responsible for a considerable fraction of the global warming potential of greenhouse gases released from thawing permafrost under anaerobic conditions. However, Schuur *et al.* (2015) concluded that the total amount of CO_2 and CH_4 released from newly thawed permafrost will have a greater impact on global warming if the permafrost thaws under dry, aerobic conditions compared to thawing under waterlogged, anaerobic conditions, due to the enhanced production of CO_2 under aerobic conditions.

The release of greenhouse gases from thawing permafrost soil may form an important positive feedback on global warming. On the other hand, an overlooked negative feedback mechanism involving permafrost thawing may be the release of nutrients during degradation of organic matter in the thawed permafrost. To our knowledge, no data exist that quantify the amount of nitrogen, phosphorus or other nutrients released during mineralisation of organic matter in thawing permafrost and later transported to the root zone in areas downhill from the thawing permafrost. This mechanism may provide a locally important input of nutrients to arctic plants (Keuper *et al.*, 2012), which currently show enhanced growth rates for some plant functional types such as shrubs (Elmendorf *et al.*, 2012), but future growth may at least locally be limited by lack of available nutrients.

Discussion

Effects of seasonal fluctuations on arctic soil microbial biomass

In general, arctic soils can be characterised as resource-limited. This is mainly due to a low decomposition rate of soil organic material. Since arctic soil microorganisms constitute the main decomposer community, they also immobilise large amounts of nutrients and create a strong microbial–plant competition for nitrogen and phosphorus (Jonasson *et al.*, 1996). Fluctuations of the microbial biomass, especially in connection with spring thaw of arctic soils (Edwards *et al.*, 2006; Jefferies *et al.*, 2010), can release large amounts of nutrients into the soil and facilitate growth of plants and microorganisms (Figure 10.6). In a future changing climate, freeze–thaw events may even occur during the winter season, when plants and most microorganisms are inactive. This may result in leaching of nutrients from arctic soils, thereby potentially further limiting nutrient availability and increasing plant–microbial competition (Wipf *et al.*, 2015).

Temperature response at near-freezing temperature

The Q_{10} temperature coefficient is a measure of the change in the rate of a biological process when temperature is increased by 10°C and is often used in climate models to parameterise microbial temperature sensitivity. However, the focus on this relationship masks other parameters that influence the temperature sensitivity of microbial processes in arctic soils. Thus, Q_{10} values for microbial

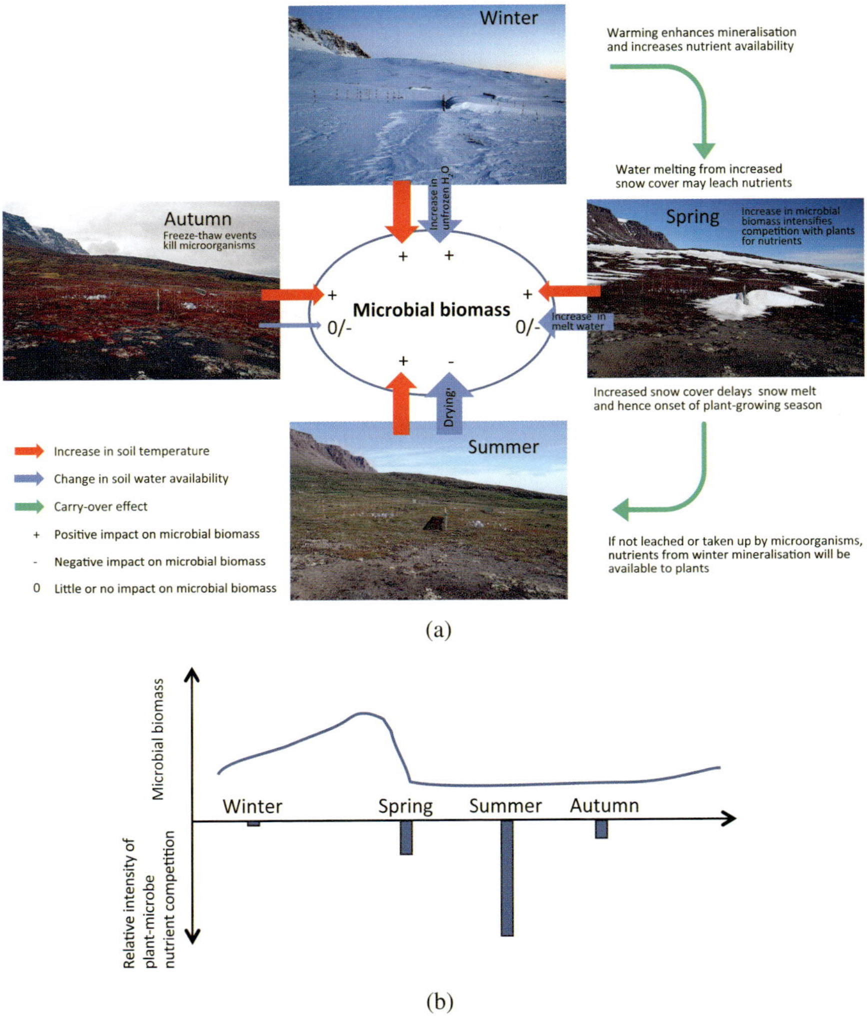

Figure 10.6 (a) Conceptual figure summarising seasonal effects on microbial biomass in arctic active layer soils under future climate change, which is expected to lead to drier summers and more snowy winters. Winter: increases in air temperature and snow cover lead to increased soil temperatures, more liquid soil water and higher microbial activity, which may include a shift from microbial nitrogen immobilisation to nitrogen mineralisation. Spring: freeze–thaw events kill microorganisms leading to a decrease in microbial biomass and increased release of nutrients. Increased snow cover during winter has a carry-over effect in spring as water from the melting snow may leach nutrients. Summer: the increased winter snow cover even has an effect in summer by delaying onset of

Figure 10.6 (*Continued*) the plant-growing season. Increases in future summer temperature will have a direct positive effect on soil microbial activity, but have a possible larger positive effect by prolonging the plant-growing season. However, the increase in microbial activity may be offset by decreased soil water availability. Autumn: freeze–thaw events kill microorganisms leading to a decrease in microbial biomass. Width of red and blue arrows indicates estimated relative impact by changes in soil temperature and water availability, respectively. (b) Seasonal variation in microbial biomass in arctic soil (adapted from Jefferies *et al.*, 2010) and estimates of relative competition for soil nutrients between microorganisms and plants. Photos are taken by an automated camera located at the experimental site at Disko Island, West Greenland (for details, see Blok *et al.*, 2015).

activity in arctic soils become more uncertain as soils approach freezing point (Mikan *et al.*, 2002; Bekku *et al.*, 2003; Elberling and Brandt, 2003; Elberling 2005; Moni *et al.*, 2015), and the shift in unfrozen water content upon freezing seems to be more important for microbial CO_2 production than the direct effect of temperature (Elberling and Brandt, 2003). A better understanding of microbial Q_{10} values in the Arctic is needed to decrease the uncertainty of predictions on future microbial activity in arctic soils. There is some evidence that soil microbial communities in the Arctic are more sensitive to warming compared to other biomes (Bekku *et al.*, 2003), but the interactions between temperature and water availability suggest that the responses of microbial communities and organic matter decay rates to warming also depend on other factors such as freeze–thaw cycles and drainage.

Effects of increasing winter and summer temperatures on soil microbial CO_2 production

In general, microbial communities are able to undergo rapid changes on short time scales and shift from active to dormant stages at any point in time, and these changes in microbial activity regulate the overall ecosystem response to environmental variation. Microbial communities may vary more across the seasons than in response to long-term climate manipulations (Yuste *et al.*, 2014). However, available data regarding seasonality of microbial abundance and activity in the Arctic are sporadic, so it is impossible to make any unequivocal conclusion about the contribution of individual seasons to, for example, soil organic matter degradation and annual greenhouse gas emissions. Several studies have reported relatively high CO_2 flux rates from winter soil indicating that microorganisms are not dormant at this time of the year (Zimov *et al.*, 1993;

Fahnestock *et al.*, 1998, 1999), however in comparison with those measured during the growing season, daily winter CO_2 emission rates are substantially lower (Björkman *et al.*, 2010; Morgner *et al.*, 2010). On the other hand, due to the long duration of the arctic winter, winter emissions may significantly contribute to the total annual CO_2 emissions (Morgner *et al.*, 2010). Contradictory observations have resulted from the effect of increased snow cover on winter CO_2 fluxes (Björkman *et al.*, 2010; Morgner *et al.*, 2010), indicating that snow cover thickness is not the only factor determining winter ecosystem CO_2 emission. If, as predicted, climate change alters both winter snow deposition and summer temperatures, warmer summer temperatures will increase CO_2 flux rates (Jones *et al.*, 1998; Welker *et al.*, 2000), but increased snow depth may decrease CO_2 summer emissions due to delayed snowmelt and thus shortening of the growing season (Björkman *et al.*, 2010).

Short-term vs. long-term effects of temperature increase and effects on microbial activity in different soil layers

Climate change impacts on soil microbial composition and functioning are likely to evolve over time with microorganisms adapting to new environmental conditions. As tundra plants generally respond slowly to shifts in climate, their impact on soil microorganisms may take many years to become apparent as they, for example, alter litter inputs into the soil organic matter pool. Even longer time scales may be needed to detect the impacts of climatic changes on microbial activity in deeper soil layers. However, an opposite response has been observed after 20 years of warming in Alaskan arctic tundra where, surprisingly, microbial decomposer activity in the surface soil layer was suppressed while deep mineral soil microbial activity was stimulated (Sistla *et al.*, 2013). A similar response to long-term warming was observed in a subarctic peatland in Northern Sweden, where most of the increase in soil respiration was found to be derived from the bottom of the active layer (Dorrepaal *et al.*, 2009), termed the "biotic awakening of soil microorganisms". This may be initiated by roots growing into the deeper mineral soil layers releasing carbon-rich exudates and by cryoturbation of litter material, that is, mixing of litter into deeper soil layers due to freezing and thawing (Sistla *et al.*, 2013). Also, responses of microbial communities to snow manipulation treatments differ between short- and long-term experiments (Buckeridge and Grogan, 2010; Olofsson *et al.*, 2011). These observations are helping to better understand the effect of warming on CO_2 emission from arctic soils. This includes the fate of permafrost organic pools upon thawing, especially if this warming-induced stimulation of deep soil microbial activity is not transitory but instead sustained on long time scales.

Effect of extreme weather events on arctic soil microorganisms

Extreme weather events are common, and climate change increases the frequency and magnitude of extreme weather events (Hansen *et al.*, 2012). In the Arctic, extreme winter warming events appear to be increasing in frequency (Shabbar and Bonsal, 2003; Bokhorst *et al.*, 2011). This could be a few days of warm weather during winter involving, for example, a change from –20 to +5°C in 24 h (Phoenix and Lee, 2004), resulting in snow melting and possible bare ground freezing when winter weather returns. It could also be dry spells during summer or summer drought due to lack of precipitation. Even short-lived extremes are known to impact soil microbial–plant interactions in the Arctic (Bokhorst *et al.*, 2009, 2011; Semenchuk *et al.*, 2013). Similarities are seen from studies of arctic and non-arctic desert soil, where both show that repeated short-term events are particularly important for the microbial biomass (Schimel *et al.*, 2007). In the Arctic, a major drop in active microorganisms upon repeated freezing has been reported (Schimel and Clein, 1996; Jefferies *et al.*, 2010). The importance of extreme events on arctic soil microorganisms contrasts with the paucity of studies of these events and the likelihood that extreme events will become more frequent, particularly in the Arctic.

Concluding Remarks

Arctic soils are different from soils in other ecosystems, and knowledge derived from temperate and tropical soil cannot easily be transferred to an arctic setting. Active layer soil microorganisms experience large seasonal fluctuations not only in temperature, but also in soil water availability, and their temperature response may not follow a classic Q_{10} response due to soil water limitation at sub-zero temperatures and death of microorganisms during freezing. An increase in soil microbial biomass during spring represents a sink for nutrients and therefore intensified competition for nutrients with plants. In contrast, future warmer arctic winters will enhance microbial organic matter decomposition, which may increase nutrient availability for microorganisms and plants at the onset of the growing season (Semenchuk *et al.*, 2015). On the other hand, warmer winters may increase the risk of leaching losses at snow melt. Such interactions highlight the complexity of forecasting the integrated biological effects of climate change in the Arctic.

Appendix 1

Reference	Ecosystem	Location	Region
Haugwitz *et al.* (2011)	Fellfield	Abisko, Sweden	Subarctic
Jonasson *et al.* (1996)	Shrub heath, fellfield	Abisko, Sweden	Subarctic
Jonasson *et al.* (1999)	Shrub heath, fellfield	Abisko, Sweden	Subarctic
Rinnan *et al.* (2007a)	Shrub heath	Abisko, Sweden	Subarctic
Rinnan *et al.* (2007b)	Shrub heath	Abisko, Sweden	Subarctic
Stark and Grellmann (2002)	Fellfield	Finnmarksvidda, Norway	Subarctic
Allen *et al.* (2010)	Drained lake basin	Barrow, Alaska	Low arctic
Buckeridge and Grogan (2008)	Birch hummock tundra	Daring Lake, Canada	Low arctic
Cheng and Virginia (1993)	Dwarf-shrub tundra, inter-tussock tundra, lichen heath, stone stripe, riparian area, wet tussock tundra, water track	Toolik Lake, Alaska	Low arctic
Chu and Grogan (2010)	Birch hummock tundra, dry heath, tall birch tundra, wet sedge meadow	Daring Lake, Canada	Low arctic
Edwards and Jeffries (2013)	Dry and wet sedge meadows	Churchill, Manitoba, Canada	Low arctic
Sistla and Schimel (2013)	Tussock-forming sedge meadow	Toolik Lake, Alaska	Low arctic
Bekku *et al.* (2004)	Glacier toe ecosystems	Ny-Ålesund, Svalbard, Norway	High arctic
Capek *et al.* (2015)	Shrub and moss tundra	Taymir peninsula, Russia	High arctic
Christiansen *et al.* (2012)	Arid lowland open heath	Zackenberg, Greenland	High arctic
Illeris *et al.* (2003)	Arid lowland open heath	Zackenberg, Greenland	High arctic
Melle *et al.* (2015)	Tussock soil	Northern Alaska	High arctic
Rinnan *et al.* (2005)	Arid lowland open heath	Zackenberg, Greenland	High arctic

References

Aanderud, Z. T., Jones, S. E., Schoolmaster Jr., D. R., Fierer, N. and Lennon, J. T. (2013). Sensitivity of soil respiration and microbial communities to altered snowfall. *Soil Biol. Biochem.* 57, 217–227.

Aerts, R. (2006). The freezer defrosting: global warming and litter decomposition rates in cold biomes. *J. Ecol.* 94, 713–724.

Aerts, R., Cornelissen, J. H. C. and Dorrepaal, E. (2006). Plant performance in a warmer world: general responses of plants from cold, northern biomes and the importance of winter and spring events. *Plant Ecol.* 182, 65–77.

Albert, M. R. and Perron, F. E. (2000). Ice layer and surface crust permeability in a seasonal snow pack. *Hydrol. Process.* 14, 3204–3214.

Allen, B., Willner, D., Oechel, W. C. and Lipson, D. (2010). Top-down control of microbial activity and biomass in an Arctic soil ecosystem. *Environ. Microbiol.* 12, 642–648.

Ayala-Del-Rio, H. L., Chain, P. S., Grzymski, J. J., Ponder, M. A., Ivanova, N., Bergholz, P. W., *et al.* (2010). The genome sequence of *P. arcticus* 273-4, a psychroactive Siberian permafrost bacterium, reveals mechanisms for adaptation to low-temperature growth. *Appl. Environ. Microbiol.* 76, 2304–2312.

Bekku, Y. S., Nakatsubo, T., Kume, A., Adachi, M. and Koizumi, H. (2003). Effect of warming on the temperature dependence of soil, respiration rate in arctic, temperate and tropical soils. *Appl. Soil Ecol.* 22, 205–210.

Bekku, Y. S., Nakatsubo, T., Kume, A. and Koizumi, H. (2004). Soil microbial biomass, respiration rate, and temperature dependence on a successional glacier foreland in Ny-Ålesund, Svalbard. *Arc. Antarc. Alpine Res.* 36, 395–399.

Björkman, M. P., Morgner, E., Björk, R. G., Cooper, E. J., Elberling, B. and Klemedtsson, L. (2010). A comparison of annual and seasonal carbon dioxide effluxes between sub-Arctic Sweden and High-Arctic Svalbard. *Pol. Res.* 29, 75–84.

Blagodatskaya, E. and Kuzyakov, Y. (2013). Active microorganisms in soil: critical review of estimation criteria and approaches. *Soil Biol. Biochem.* 67, 192–211.

Blok, D., Elberling, B. and Michelsen, A. (2015). Initial stages of tundra shrub litter decomposition may be accelerated by deeper winter snow but slowed down by spring warming. *Ecosystems.* doi: 10.1007/s10021-015-9924-3.

Bokhorst, S., Bjerke, J. W., Street, L. E., Callaghan, T. V. and Phoenix, G. K. (2011). Impacts of multiple extreme winter warming events on sub-Arctic heathland: phenology, reproduction, growth, and CO_2 flux responses. *Glob. Change Biol.* 17, 2817–2830.

Bokhorst, S., Bjerke, J. W., Tømmervik, H., Callaghan, T. V. and Phoenix, G. K. (2009). Winter warming events damage sub-Arctic vegetation: consistent evidence from an experimental manipulation and a natural event. *J. Ecol.* 97, 1408–1415.

Bradford, M. A. (2013). Thermal adaptation of decomposer communities in warming soils. *Front. Microbiol.* 4, article 333.

Bradford, M. A., Davies, C. A., Frey, S. D., *et al.* (2008). Thermal adaptation of soil microbial respiration to elevated temperature. *Ecol. Lett.* 11, 1316–1327.

Broeckling, C. D., Broz, A. K., Bergelson, J., Manter, D. K. and Vivanco, J. M. (2008). Root exudates regulate soil fungal community composition and diversity. *Appl. Environ. Microbiol.* 74, 738–744.

Buckeridge, K. M. and Grogan, P. (2008). Deepened snow alters soil microbial nutrient limitations in arctic birch hummock tundra. *Appl. Soil Ecol.* 39, 210–222.

Buckeridge, K. and Grogan, P. (2010). Deepened snow increases late thaw biogeochemical pulses in mesic low arctic tundra. *Biogeochemistry* 101, 105–121.

Capek, P., Diáková, K., Dickopp, J.-E., *et al.* (2015). The effect of warming on the vulnerability of subducted organic carbon in arctic soils. *Soil Biol. Biochem.* 90, 19–29.

Chattopadhyay, M. (2006). Mechanism of bacterial adaptation to low temperature. *J. Biosci.* 31, 157–165.

Cheng, W. and Virginia, R. A. (1993). Measurement of microbial biomass in arctic tundra soils using fumigation–extraction and substrate-induced respiration procedures. *Soil Biol. Biochem.* 25, 135–141.

Christiansen, C. T., Svendsen, S. H., Schmidt, N. M. and Michelsen, A. (2012). High arctic heath soil respiration and biogeochemical dynamics during summer and autumn freeze-in—effects of long-term enhanced water and nutrient supply. *Glob. Change Biol.* 18, 3224–3236.

Chu, H., Fierer, N., Lauber, C. L., Caporaso, J., Knight, R. and Grogan, P. (2010). Soil bacterial diversity in the Arctic is not fundamentally different from that found in other biomes. *Environ. Microbiol.* 12, 2998–3006.

Chu, H. and Grogan, P. (2010). Soil microbial biomass, nutrient availability and nitrogen mineralization potential among vegetation-types in a low arctic tundra landscape. *Plant Soil* 329, 411–420.

Clemmensen, K. E., Michelsen, A., Jonasson, S. and Shaver, G. R. (2006). Increased ectomycorrhizal fungal abundance after long-term fertilization and warming of two arctic tundra ecosystems. *New Phytol.* 171, 391–404.

Coolen, M. J. L. and Orsi, W. D. (2015). The transcriptional response of microbial communities in thawing Alaskan permafrost soils. *Front. Microbiol.* 6, article 197.

Davidson, E. A. and Janssens, I. A. (2006). Temperature sensitivity of soil carbon decomposition and feedbacks to climate change. *Nature* 440, 165–173.

D'Amico, S., Collins, T., Marx, J.-C., Feller, G. and Gerday, C. (2006). Psychrophilic microorganisms: challenges for life. *EMBO Rep.* 7, 385–389.

de Boer, W., Folman, L. B., Summerbell, R. C. and Boddy, L. (2005). Living in a fungal world: impact of fungi on soil bacterial niche development. *FEMS Microbiol. Rev.* 29, 795–811.

de Maayer, P., Anderson, D., Cary, C. and Cowan, D. A. (2014). Some like it cold: understanding the survival strategies of psychrophiles. *EMBO Rep.* embr.201338170.

Deslippe, J. R., Hartmann, M., Mohn, W. W. and Simard, S. W. (2011). Long-term experimental manipulation of climate alters the ectomycorrhizal community of betula nana in arctic tundra. *Glob. Change Biol.* 17, 1625–1636.

Deslippe, J. R., Hartmann, M., Simard, S. W. and Mohn, W. W. (2012). Long-term warming alters the composition of arctic soil microbial communities. *FEMS Microbiol. Ecol.* 82, 303–315.

Deslippe, J. R. and Simard, S. W. (2011). Below-ground carbon transfer among Betula nana may increase with warming in arctic tundra. *New Phytol.* 192, 689–698.

Dorrepaal, E., Toet, S., van Logtestijn, R. S. P., Swart, E., van de Weg, M. J., Callaghan, T. V., *et al.* (2009). Carbon respiration from subsurface peat accelerated by climate warming in the subarctic. *Nature* 460, 616–619.

Dungait, J. A. J., Hopkins, D. W., Gregory, A. S. and Whitmore, A. P. (2012). Soil organic matter turnover is governed by accessibility not recalcitrance. *Glob. Change Biol.* 18, 1781–1796.

Dutta, K., Schuur, E. A. G., Neff, J. C. and Zomov S. A. (2006). Potential carbon release from permafrost soils of Northeastern Siberia. *Glob. Change Biol.* 12, 2336–2351.

Edwards, K. A. and Jefferies, R. L. (2013). Inter-annual and seasonal dynamics of soil microbial biomass and nutrients in wet and dry low-arctic sedge meadows. *Soil Biol. Biochem.* 57, 83–90.

Edwards, K. A., McCulloch, J., Kershaw, P. G. and Jefferies, R. L. (2006). Soil microbial and nutrient dynamics in a wet Arctic sedge meadow in late winter and early spring. *Soil Biol. Biochem.* 38, 2843–2851.

Edwards, A. C., Scalenghe, R. and Freppaz, M. (2007). Changes in the seasonal snow cover of alpine regions and its effect on soil processes: a review. *Quaternary Int.* 162–163, 172–181.

Elberling, B. (2003). Seasonal trends of soil CO_2 dynamics in a soil subject to freezing. *J. Hydrol.* 276, 159–175.

Elberling, B. (2005). Temperature and oxygen control on pyrite oxidation in frozen mine tailings. *Cold Regions Sci. Technol.* 41, 121–133.

Elberling, B. and Brandt, K. K. (2003). Uncoupling of microbial CO_2 production and CO_2 release in frozen soil and its implications for field studies of arctic C cycling. *Soil Biol. Biochem.* 35, 263–272.

Elberling, B., Christiansen, H. H. and Hansen, B. U. (2010). High nitrous oxide production from thawing permafrost. *Nat. Geosci.* 3, 332–335.

Elberling, B., Michelsen, A., Schädel, C., Schuur, E. A. G., Christiansen, H. H., Berg, L., *et al.* (2013). Long-term CO_2 production following permafrost thaw. *Nat. Clim. Change* 3, 890–894.

Elmendorf, S. C., Henry, G. H. R., Hollister, R. D., Björk, R. G., Boulanger-Lapointe, N., Cooper, E. J., *et al.* (2012). Plot-scale evidence of tundra vegetation change and links to recent summer warming. *Nat. Clim. Change* 2, 453–457.

Ewert, M. and Deming, J. W. (2014). Bacterial responses to fluctuations and extremes in temperature and brine salinity at the surface of Arctic winter sea ice. *FEMS Microbiol. Ecol.* 89, 476–489.

Fahnestock, J. T., Jones, M. H., Brooks, P. D., Walker, D. A. and Welker, J. M. (1998). Winter and early spring CO_2 efflux from tundra communities of northern Alaska. *J. Geophys. Res. Atmos.* 103, 29023–29027.

Fahnestock, J. T., Jones, M. H. and Welker, J. M. (1999). Wintertime CO_2 efflux from Arctic soils: Implications for annual carbon budgets. *Glob. Biogeochem. Cycles* 13, 775–779.

Fang, C. and Moncrieff, J. B. (2001). The dependence of soil CO_2 efflux on temperature. *Soil Biol. Biochem.* 33, 155–165.

Feng, S., Powell, S. M., Wilson, R. and Bowman, J. P. (2014). Extensive gene transfer in the extremely psychrophilic bacterial species *Psychroflexus torquis* and the link to sea-ice ecosystem specialism. *Genome Biol. Evol.* 6, 133–148.

Fierer, N. and Jackson, R. B. (2006). The diversity and biogeography of soil bacterial communities. *Proc. Natl. Acad. Sci. USA* 103, 626–631.

Frey, S. D., Lee, J., Melillo, J. M. and Six, J. (2013). The temperature response of soil microbial efficiency and its feedback to climate. *Nat. Clim. Change* 3, 395–398.

Geml, J., Morgado, L. N., Semenova, T. A., Welker, J. M., Walker, M. D. and Smets, E. (2015). Long-term warming alters richness and composition of taxonomic and functional groups of arctic fungi. *FEMS Microbiol. Ecol.* 91, article fiv095.

Gilichinsky, D. A. (2002). Astrobiology: the quest for the conditions of life. In Horneck, G. and Baumstark-Khan, C. (eds.), *Permafrost Model of Extraterrestrial Habitats*, Chapter 8. Springer, Berlin, pp. 125–142.

Gittel, A., Bárta, J., Kohoutová, I., Mikutta, R., Owens, S., Gilbert, J., *et al.* (2014b). Distinct microbial communities associated with buried soils in the Siberian tundra. *ISME J.* 8, 841–853.

Gittel, A., Bárta, J., Kohoutová, I., Schnecker, J., Wild, B., Capek, P., *et al.* (2014a). Site- and horizon-specific patterns of microbial community structure and enzyme activities in permafrost-affected soils of Greenland. *Front. Microbiol.* 5, article 541.

Grogan, P., Michelsen, A., Ambus, P. and Jonasson, S. (2004). Freeze–thaw regime effects on carbon and nitrogen dynamics in sub-arctic heath tundra mesocosms. *Soil Biol. Biochem.* 36, 641–654.

Haei, M., Rousk, J., Ilstedt, U., Öquist, M., Bååth, E. and Laudon, H. (2011). Effects of soil frost on growth, composition and respiration of the soil microbial decomposer community. *Soil Biol. Biochem.* 43, 2069–2077.

Hagerty, S. B., Van Groenigen, K. J., Allison, S. D., Hungate, B. A., Schwartz, E., Koch, G. W., *et al.* (2014). Accelerated microbial turnover but constant growth efficiency with warming in soil. *Nat. Clim. Change* 4, 903–906.

Hall, E. K., Singer, G. A., Kainz, M. J. and Lennon, J. T. (2010). Evidence for a temperature acclimation mechanism in bacteria: an empirical test of a membrane-mediated trade-off. *Funct. Ecol.* 24, 898–908.

Hansen, J., Sato, M. and Ruedy, R. (2012). Perception of climate change. Proc. Natl Acad. Sci. USA 109, E2415–E2423.

Haugwitz, M., Schmidt, I. K. and Michelsen, A. (2011). Long-term microbial control of nutrient availability and plant biomass in a subarctic-alpine heath after addition of carbon, fertilizer and fungicide. *Soil Biol. Biochem.* 43, 179–187.

Henry, G. H. R. and Molau, U. (1997). Tundra plants and climate change: the international tundra experiment (itex). *Glob. Change Biol.* 3, 1–9.

Hobbie, J. E. and Hobbie, E. A. (2006). N-15 in symbiotic fungi and plants estimates nitrogen and carbon flux rates in arctic tundra. *Ecology* 87, 816–822.

Hultman, J., Waldrop, M. P., Mackelprang, R., David, M. M., McFarland, J., Blazewicz, S. J., *et al.* (2015). Multi-omics of permafrost, active layer and thermokarst bog soil microbiomes. *Nature* 521, 208–212.

Illeris, L., Michelsen, A. and Jonasson, S. (2003). Soil plus root respiration and microbial biomass following water, nitrogen, and phosphorus application at a high arctic semi desert. *Biogeochemistry* 65, 15–19.

Intergovernmental Panel on Climate Change (IPCC) (2013). Climate Change 2013: The Physical Science Basis. Contribution of Working Group I to the Fifth Assessment Report of the Intergovernmental Panel on Climate Change. In Stocker, T. F., Qin, D., Plattner, G.-K., Tignor, M., Allen, S. K. and Boschung, J. (eds.), *Summary for Policymakers*. Cambridge University Press, Cambridge, UK, pp. 3–29.

Jansson, J. K. and Taş, N. (2014). The microbial ecology of permafrost. *Nat. Rev. Microbiol.* 12, 414–425.

Jefferies, R. L., Walker, N. A., Edwards, K. A. and Dainty, J. (2010). Is the decline of soil microbial biomass in late winter coupled to changes in the physical state of cold soils? *Soil Biol. Biochem.* 42, 129–135.

Jonasson, S., Michelsen, A., Schmidt, I. K. and Nielsen E. V. (1999). Responses in microbes and plants to changed temperature, nutrient, and light regimes in the arctic. *Ecology* 80, 1828–1843.

Jonasson, S., Michelsen, A., Schmidt, I. K., Nielsen, E. V. and Callaghan, T. V. (1996). Microbial biomass C, N and P in two arctic soils and responses to addition of NPK fertilizer and sugar: implications for plant nutrient uptake. *Oecologia* 106, 507–515.

Jones, M. H., Fahnestock, J. T., Walker, D. A., Walker, M. D. and Welker, J. M. (1998). Carbon dioxide fluxes in moist and dry arctic tundra during the snow-free season: responses to increases in summer temperature and winter snow accumulation. *Arc. Alpine Res.* 30, 373–380.

Kandror, O., Bretschneider, N., Kreydin, E., Cavalieri, D. and Goldberg, A. L. (2004). Yeast adapt to near-freezing temperatures by STRE/Msn2,4-dependent induction of trehalose synthesis and certain molecular chaperones. *Mol. Cell* 13, 771–781.

Karhu, K., Auffret, M. D., Dungait, J. A. J., Hopkins, D. W., Prosser, J. I., Singh, B. K., *et al.* (2014). Temperature sensitivity of soil respiration rates enhanced by microbial community response. *Nature* 513, 81–84.

Karhu, K., Fritze, H., Hamalainen, K., Vanhala, P., Jungner, H., Oinonen, M., *et al.* (2010). Temperature sensitivity of soil carbon fractions in boreal forest soil. *Ecology* 91, 370–376.

Keuper, F., van Bodegom, P. M., Dorrepaal, E., Weedon, J. T., van Hal, J., van Logtestijn, R. S. P., *et al.* (2012). A frozen feast: thawing permafrost increases plant-available nitrogen in subarctic peatlands. *Glob. Change Biol.* 18, 1998–2007.

Knoblauch, C., Beer, C., Sosnin, A., Wagner, D. and Pfeiffer, E.-M. (2013). Predicting long-term carbon mineralization and trace gas production from thawing permafrost of Northeast Siberia. *Glob. Change Biol.* 19, 1160–1172.

Larsen, K. S., Grogan, P., Jonasson, S. and Michelsen, A. (2007). Respiration and microbial dynamics in two subarctic ecosystems during winter and spring thaw: effects of increased snow depth. *Arc. Antarc. Alpine Res.* 39, 268–276.

Lee, H., Schuur, E. A. G. and Vogel, J. G. (2010). Soil CO_2 production in upland tundra where permafrost is thawing. *J. Geophys. Res.* 115, G01009.

Lipson, D. A., Schmidt, S. K. and Monson, R. K. (2000). Carbon availability and temperature control the post-snowmelt decline in alpine soil microbial biomass. *Soil Biol. Biochem.* 32, 441–448.

Lupascu, M., Welker, J. M., Xu, X. and Czimczik, C. I. (2014). Rates and radiocarbon content of summer ecosystem respiration in response to long-term deeper snow in the High Arctic of NW Greenland. *J. Geophys. Res. Biogeosci.* 119, 1180–1194.

Mackelprang, R., Waldrop, M. P., De Angelis, K. M., David, M. M., Chavarria, K. L., Blazewicz, S. J., *et al.* (2011). Metagenomic analysis of a permafrost microbial community reveals a rapid response to thaw. *Nature* 480, 368–371.

Männistö, M. K., Tiirola, M. and Häggblom, M. M. (2009). Effect of freeze–thaw cycles on bacterial communities of Arctic tundra soil. *Microb. Ecol.* 58, 621–631.

Manzoni, S., Schimel, J. P. and Porporato, A. (2011). Responses of soil microbial communities to water stress: results from a meta-analysis. *Ecology* 93, 930–938.

Melle, C., Wallenstein, M., Darrouzet-Nardi, A. and Weintraub, M. N. (2015). Microbial activity is not always limited by nitrogen in Arctic tundra soils. *Soil Biol. Biochem.* 90, 51–61.

Michelsen, A., Andersson, M., Jensen, M., Kjøller, A. and Gashew, M. (2004). Carbon stocks, soil respiration and microbial biomass in fire-prone tropical grassland, woodland and forest ecosystems. *Soil Biol. Biochem.* 36, 1707–1717.

Mikan, C. J., Schimel, J. P. and Doyle, A. P. (2002). Temperature controls of microbial respiration in arctic tundra soils above and below freezing. *Soil Biol. Biochem.* 34, 1785–1795.

Molau, U. and Mølgaard, P. (Eds.) (1996). *International Tundra Experiment, Itex Manual.* Danish Polar Centre, Copenhagen, P. 85 .

Moni, C., Lerch, T. Z., de Zarruk, K. K., Strand, L. T., Forte, C., Certini, G., *et al.* (2015). Temperature response of soil organic matter mineralisation in arctic soil profiles. *Soil Biol. Biochem.* 88, 236–246.

Morgado, L. N., Semenova, T. A., Welker, J. M., Walker, M. D., Smets, E. and Geml, J. (2015). Summer temperature increase has distinct effects on the ectomycorrhizal fungal communities of moist tussock and dry tundra in arctic Alaska. *Glob. Change Biol.* 21, 959–972.

Morgner, E., Elberling, B., Strebel, D. and Cooper, E. J. (2010). The importance of winter in annual ecosystem respiration in the High Arctic: effects of snow depth in two vegetation types. *Pol. Res.* 29, 58–74.

Myers-Smith, I. H., Forbes, B. C., Wilmking, M., Hallinger, M., Lantz, T, Blok, D., *et al.* (2011). Shrub expansion in tundra ecosystems: dynamics, impacts and research priorities. *Environ. Res. Lett.* 6, 045509.

Mykytczuk, N. C. S., Foote, S. J., Omelon, C. R., Southam, G., Greer, C. W. and Whyte, L. G. (2013). Bacterial growth at -15°C: molecular insights from the permafrost bacterium *Planococcus halocryophilus* Or1. *ISME J.* 7, 1211–1226.

Nadelhoffer, K. J., Giblin, A. E., Shaver, G. and Linkins, A. E. (1992). Arctic ecosystems in a changing climate: an ecophysiological perspective. Chapin, F. S. I., Jefferies, R., Reynolds, J. F., Shaver, G. and Svoboda, J. (eds.), *Microbial Processes and Plant Nutrient Availability in Arctic Soils*. Academic Press, San Diego, pp. 281–300.

Natali, S. M., Schuur, E. A. G. and Rubin, R. L. (2012). Increased plant productivity in Alaskan tundra as a result of experimental warming of soil and permafrost. *J. Ecol.* 100, 488–498.

Natali, S. M., Schuur, E. A. G., Trucco, C., Pries, C. E. H., Crummer, K. G. and Lopez, A. F. B. (2011). Effects of experimental warming of air, soil and permafrost on carbon balance in Alaskan tundra. *Glob. Change Biol.* 17, 1394–1407.

Natali, S. M., Schuur, E. A. G., Webb, E. E., Pries, C. E. H. and Crummer, K. G. (2014). Permafrost degradation stimulates carbon loss from experimentally warmed tundra. *Ecology* 95, 602–608.

Nedwell, D. B. (1999). Effect of low temperature on microbial growth: lowered affinity for substrates limits growth at low temperature. *FEMS Microbiol. Ecol.* 30, 101–111.

Nobrega, S. and Grogan, P. (2007). Deeper snow enhances winter respiration from both plant-associated and bulk soil carbon pools in birch hummock tundra. *Ecosystems* 10, 419–431.

Nunn, B. L., Slattery, K. V., Cameron, K. A., Timmins-Schiffman E. and Junge, K. (2015). Proteomics of *Colwellia psychrerythraea* at subzero temperatures—a life with limited movement, flexible membranes and vital DNA repair. *Environ. Microbiol.* 17, 2319–2335.

Olefeldt, D., Turetsky, M. R., Crill, P. M. and McGuire, A. D. (2013). Environmental and physical controls on northern terrestrial methane emissions across permafrost zones. *Glob. Change Biol.* 19, 589–603.

Olofsson, J., Ericson, L., Torp, M., Stark, S. and Baxter, R. (2011). Carbon balance of Arctic tundra under increased snow cover mediated by a plant pathogen. *Nat. Clim. Change* 1, 220–223.

Onipchenko, V. G., Makarov, M. I., Van Logtestijn, R. S. P., Ivanov, V. B., Akhmetzhanova, A. A., Tekeev, *et al.* (2009). New nitrogen uptake strategy: specialized snow roots. *Ecol. Lett.* 12, 758–764.

Öquist, M. G., Sparrman, T., Klemedtsson, L., Drotz, S. H., Grip, H., Schleucher, J., *et al.* (2009). Water availability controls microbial temperature responses in frozen soil CO_2 production. *Glob. Change Biol.* 15, 2715–2722.

Ostroumov, V. E. and Siegert, C. (1996). Exobiological aspects of mass transfer in microzones of permafrost deposits. *Adv. Space Res.* 18, 79–86.

Panikov, N. S., Flanagan, P. W., Oechel, W. C., Mastepanov, M. A. and Christensen, T. R. (2006). Microbial activity in soils frozen to below -39°C. *Soil Biol. Biochem.* 38, 785–794.

Panikov, N. S. and Sizova, M. V. (2007). Growth kinetics of microorganisms isolated from Alaskan soil and permafrost in solid media frozen down to -35°C. *FEMS Microb. Ecol.* 59, 500–512.

Phoenix, G. K. and Lee, J. A. (2004). Predicting impacts of Arctic climate change: past lessons and future challenges. *Ecol. Res.* 19, 65–74.

Ponder, M., Thomasshow, M. F. and Tiedje, J. M. (2008). Metabolic activity of Siberian permafrost isolates, *Psychrobacter arcticus* and *Exiguobacterium sibiricum*, at low water activities. *Extremophiles* 12, 481–490.

Pries, C. E. H., Schuur, E. A. G., Vogel, J. G. and Natali, S. M. (2013). Moisture drives surface decomposition in thawing tundra. *J. Geophys. Res. Biogeosci.* 118, 1133–1143.

Rinnan, R., Keinänen, M. M., Kasurinen, A., Asikainen, J., Kekki, T. K., Holopainen, T., *et al.* (2005). Ambient ultraviolet radiation in the Arctic reduces root biomass and alters microbial community composition but has no effects on microbial biomass. *Glob. Change Biol.* 4, 564–574.

Rinnan, R., Michelsen, A. and Baath, E. (2013). Fungi benefit from two decades of increased nutrient availability in tundra heath soil. *PLoS One* 8, article e56532.

Rinnan, R., Michelsen, A., Baath, E. and Jonasson S. (2007a). Fifteen years of climate change manipulations alter soil microbial communities in a subarctic heath ecosystem. *Glob. Change Biol.* 13, 28–39.

Rinnan, R., Michelsen, A., Baath, E. and Jonasson, S. (2007b). Mineralization and carbon turnover in subarctic heath soil as affected by warming and additional litter. *Soil Biol. Biochem.* 39, 3014–3023.

Rivkina, E. M., Gilichinsky, D., Wagener, S., Tiedje, J. and McGrath, J. (1998). Biogeochemical activity of anaerobic microorganisms from buried permafrost sediments. *Geomicrobiology* 15, 187–193.

Robinson, C. H., Wookey, P. A., Parsons, A. N., Potter, J. A., Callaghan, T. V., Lee, J. A., *et al.* (1995). Responses of plant litter decomposition and nitrogen mineralisation to simulated environmental change in a high arctic polar semi-desert and a subarctic dwarf shrub heath. *Oikos* 74, 503–512.

Rogers, M. C., Sullivan, P. F. and Welker, J. M. (2011). Evidence of nonlinearity in the response of net ecosystem CO_2 exchange to increasing levels of winter snow depth in the High Arctic of Northwest Greenland. *Arc. Antarc. Alpine Res.* 43, 95–106.

Romanovsky, V. E. and Osterkamp, T. E. (2000). Effects of unfrozen water on heat and mass transport processes in the active layer and permafrost. *Permafrost Periglac. Proc.* 11, 219–239.

Rustad, L. E., Campbell, J. L., Marion, G. M., Norby, R. J., Mitchell, M. J., Hartley, A. E., *et al.* (2001). A meta-analysis of the response of soil respiration, net nitrogen mineralization, and aboveground plant growth to experimental ecosystem warming. *Oecologia* 126, 543–562.

Schädel, C., Schuur, E. A. G., Bracho, R., Elberling, B., Knoblauch, C., Lee, H., *et al.* (2014). Circumpolar assessment of permafrost C quality and its vulnerability over time using long-term incubation data. *Glob. Change Biol.* 20, 641–652.

Schadt, C. W., Martin, A. P., Lipson, D. A. and Schmidt, S. K. (2003). Seasonal dynamics of previously unknown fungal lineages in tundra soils. *Science* 301, 1359–1361.

Schimel, J. P., Balser, T. C. and Wallenstein, M. (2007). Microbial stress-response physiology and its implications for ecosystem function. *Ecology* 88, 1386–1394.

Schimel, J. P., Bilbrough, C. and Welker, J. M. (2004). Increased snow depth affects microbial activity and nitrogen mineralization in two arctic tundra communities. *Soil Biol. Biochem.* 36, 217–227.

Schimel, J. P. and Clein, J. S. (1996). Microbial response to freeze–thaw cycles in tundra and taiga soils. *Soil Biol. Biochem.* 28, 1061–1066.

Schimel, J. P. and Mikan, C. (2005). Changing microbial substrate use in Arctic tundra soils through a freeze–thaw cycle. *Soil Biol. Biochem.* 37, 1411–1418.

Schimel, J. P. and Schaeffer, S. M. (2012). Microbial control over carbon cycling in soil. *Front. Microbiol.* 3, article 348.

Schmidt, M. V. I., Torn, M. S., Abiven, S., Dittmar, T., Guggenberger, G., Janssens, I. A., *et al.* (2011). Persistence of soil organic matter as an ecosystem property. *Nature* 478, 49–56.

Schostag, M., Stibal, M., Jacobsen, C. S., Bælum, J., Taş, N., Elberling, B., *et al.* (2015). Distinct summer and winter bacterial communities in the active layer of Svalbard permafrost revealed by DNA- and RNA-based analyses. *Front. Microbiol.* 6, article 399.

Schuur, E. A. G., McGuire, A. D., Schädel, C., Grosse, G., Harden, J. W., Hayes, D. J., *et al.* (2015). Climate change and the permafrost carbon feedback. *Nature* 520, 171–179.

Schuur, E. A. G., Vogel, J. G., Crummer, K. G., Lee, H., Sickman, J. O. and Osterkamp, T. E. (2009). The effect of permafrost thaw on old carbon release and net carbon exchange from tundra. *Nature* 459, 556–559.

Semenchuk, P. R., Elberling, B., Amtorp, C., Winkler, J., Rumpf, S., Michelsen, A., *et al.* (2015). Deeper snow alters soil nutrient availability and leaf nutrient status in high Arctic tundra. *Biogeochemistry* 124, 81–94.

Semenchuk, P. R., Elberling, B. and Cooper, E. J. (2013). Snow cover and extreme winter warming events control flower abundance of some, but not all species in high arctic Svalbard. *Ecol. Evol.* 3, 2585–2599.

Semenova, T. A., Morgado, L. N., Welker, J. M., Walker, M. D., Smets, E. and Geml, J. (2015). Long-term experimental warming alters community composition of ascomycetes in alaskan moist and dry arctic tundra. *Mol. Ecol.* 24, 424–437.

Shabbar, A. and Bonsal, B. (2003). An assessment of changes in winter cold and warm spells over Canada. Nat. Hazards 29, 173–188.

Sistla, S. A., Moore, J. C., Simpson, R. T., Gough, L., Shaver, G. R. and Schimel J. P. (2013). Long-term warming restructures arctic tundra without changing net soil carbon storage. *Nature* 497, 615–618.

Sistla, S. A. and Schimel J. P. (2013). Seasonal patterns of microbial extracellular enzyme activities in an arctic tundra soil: identifying direct and indirect effects of long-term summer warming. *Soil Biol. Biochem.* 66, 119–129.

Sitch, S., McGuire, A. D., Kimball, J., Gedney, N., Gamon, J., Engstrom, R., *et al.* (2007). Assessing the carbon balance of circumpolar arctic tundra using remote sensing and process modelling. *Ecol. Appl.* 17, 213–234.

Sorensen, P., Lett, S. and Michelsen, A. (2012). Moss-specific changes in nitrogen fixation following two decades of warming, shading, and fertilizer addition. *Plant Ecol.* 213, 695–706.

Stark, S. and Grellmann, D. (2002). Soil microbial responses to herbivory in an arctic tundra heath at two levels of nutrient availability. *Ecology* 83, 2736–2744.

Steven, B., Pollard, W. H., Greer, C. W. and Whyte, L. G. (2008). Microbial diversity and activity through a permafrost/ground ice core profile from the Canadian high Arctic. *Environ. Microbiol.* 10, 3388–3403.

Stres, B., Philippot, L., Faganeli, J. and Tiedje, J. M. (2010). Frequent freeze–thaw cycles yield diminished yet resistant and responsive microbial communities in two temperate soils: a laboratory experiment. *FEMS Microbiol. Ecol.* 74, 323–335.

Tarnocai, C., Canadell, J. G., Schuur, E. A. G., Kuhry, P., Mazhitova, G. and Zimov, S. (2009). Soil organic carbon pools in the northern circumpolar permafrost region. *Glob. Biogeochem. Cycl.* 23, article GB2023.

Tilston, E. L., Sparrman, T. and Öquist, M. G. (2010). Unfrozen water content moderates temperature dependence of sub-zero microbial respiration. *Soil. Biol. Biochem.* 42, 1396–1407.

Timling, I. and Taylor, D. L. (2012). Peeking through a frosty window: molecular insights into the ecology of Arctic soil fungi. *Fung. Ecol.* 5, 419–429.

Timling, I., Walker, D. A., Nusbaum, C., Lennon, N. J. and Taylor, D. L. (2014). Rich and cold: diversity, distribution and drivers of fungal communities in patterned-ground ecosystems of the North American Arctic. *Mol. Ecol.* 23, 3258–3272.

Treat, C., Natali, S., Ernakovich, J., Iversen, C., Lupascu, M., McGuire, A. D., *et al.* (2015). Controls on CH_4 and CO_2 production from permafrost soil carbon under saturated conditions. *Glob. Change Biol.* 21, 2787–2803.

Tuorto, S. J., Darias, P., McGuinness, L. R., Panikov, N., Zhang, T., Häggblom, M. M., *et al.* (2014). Bacterial genome replication at subzero temperatures in permafrost. *ISME J.* 8, 139–149.

Tveit, A., Schwacke, R., Svenning, M. M. and Urich, T. (2013). Organic carbon transformations in high-Arctic peat soils: key functions and microorganisms. *ISME J.* 7, 299–311.

Vishnivetskaya, T. A., Petrova, M. A., Urbance, J., Ponder, M., Moyer, C. L., Gilichinsky, D. A., *et al.* (2006). Bacterial community in ancient Siberian permafrost as characterized by culture and culture-independent methods. *Astrobiology* 6, 400–414.

Waldrop, M. P., Wickland, K. P., White, R., Berhe, A. A., Harden, J. W. and Romanovsky, V. E. (2010). Molecular investigations into a globally important carbon pool: permafrost-protected carbon in Alaskan soils. *Glob. Change Biol.* 16, 2543–2554.

Walker, V. K., Palmer, G. R. and Voordouw, G. (2006). Freeze–thaw tolerance and clues to the winter survival of a soil community. *Appl. Environ. Microbiol.* 72, 1784–1792.

Wallenstein, M. D., McMahon, S. and Schimel, J. (2007). Bacterial and fungal community structure in Arctic tundra tussock and shrub soils. *FEMS Microbiol. Ecol.* 59, 428–435.

Wardle, D. A. (1992). A comparative assessment of factors which influence microbial biomass carbon and nitrogen levels in soil. *Biol. Rev.* 67, 321–358.

Welker, J. M., Fahnestock, J. T. and Jones, M. H. (2000). Annual CO_2 flux in dry and moist arctic tundra: field responses to increases in summer temperatures and winter snow depth. *Clim. Change* 44, 139–150.

Wieder, W. R., Bonan, G. B. and Allison, S. D. (2013). Global soil carbon projections are improved by modelling microbial processes. *Nat. Clim. Change* 3, 909–912.

Wipf, S. and Rixen, C. (2010). A review of snow manipulation experiments in Arctic and alpine tundra ecosystems. *Pol. Res.* 29, 95–109.

Wipf, S., Sommerkorn, M., Stutter, M., Wubs, E. R. J. and Van der Wal, R. (2015). Snow cover, freeze–thaw, and the retention of nutrients in an oceanic mountain ecosystem. *Ecosphere* 6, art207.

Wookey, P. A., Aerts, R., Bardgett, R. D., Baptist, F., Bråthen, K. A., Cornelissen, J. H. C., *et al.* (2009). Ecosystem feedbacks and cascade processes: understanding their role in the responses of Arctic and alpine ecosystems to environmental change. *Glob. Change Biol.* 15, 1153–1172.

Yergeau, E., Hogues, H., Whyte, L. G. and Greer, C. W. (2010). The functional potential of high Arctic permafrost revealed by metagenomic sequencing, qPCR and microarray analyses. *ISME J.* 4, 1206–1214.

Yuste, J. C., Fernandez-Gonzalez, A. J., Fernandez-Lopez, M., Ogaya, R., Penuelas, J., Sardans, J., *et al.* (2014). Strong functional stability of soil microbial communities under semiarid Mediterranean conditions and subjected to long-term shifts in baseline precipitation. *Soil Biol. Biochem.* 69, 223–233.

Zhang, T., Barry, R., Knowles, K., Heginbottom, J. and Brown, J. (2008). Statistics and characteristics of permafrost and ground-ice distribution in the Northern Hemisphere. *Pol. Geogr.* 31, 47–68.

Zhang, T., Osterkamp, T. E. and Stamnes, K. (1996). Some characteristics of the climate in northern Alaska, U.S.A. *Arc. Alpine Res.* 28, 509–518.

Zimov, S. A., Zimova, G. M., Daviodov, S. P., Daviodova, A. I., Voropaev, Y. V., Voropaeva, Z. V., *et al.* (1993). Winter biotic activity and production of CO_2 in Siberian soils: a factor in the greenhouse effect. *J. Geophys. Res. Atmos.* 98, 5017–5023.

Index

A
AA mineralisation, 191
Abanilla and Santomera semi-arid soils, 175
abiotic–biotic interactions, 242
above-ground biodiversity, 88
acclimation, 143
acetate, 168
Acetoclastic methanogens, 229, 246
Acetyglucosamine, 53
  chitin, 53
Acid hydrolysis, 194
acidic, 110
Acidobacteria, 254
acid soil, 9–11, 17, 19, 73
actinomycetes, 45, 153, 167
activation energy, 154
active layer, 276, 278–279, 288–289, 292, 294–295
active microorganisms, 168
adaptations, 235, 245
added N interaction, 58
adenosine 5′ diphosphate (ADP), 22–23, 26
adenosine 5′ monophosphate (*see also* AMP), 22–23, 26
Adenosine 5′ Triphosphate (*see also* ATP), 16, 19, 22–27, 103
adenylate energy change (AEC), 26–27, 105
aeration, 105
aerenchymatous tissue, 235
aerobic, 25
Aerobic CH_4 oxidisers, 230
aerobic methanotrophic bacteria, 229
aggregate, 157
agricultural inputs, 189
agricultural land uses, 86
agricultural practices, 86–87
Agricultural reference soil, 176
agricultural systems, 85, 87
agrochemicals, 238
agro-ecosystems, 67, 69–70, 77, 79, 88–89, 91

air-drying, 5–6, 8, 17
Al^{3+}, 74
algae, 167
algal blooms, 238
algal growth, 106
altered precipitation, 236
Amanita muscaria, 131
ambient CH_4 concentrations, 177
amendments, 110, 189
amino acid (*see also* AA), 55–56,183, 185–186, 191, 194, 196–197
Amino Acid Stable Isotope Probing (AA SIP), 183
ammonia oxidation, 256
ammonium, 56
anaerobic, 25, 27, 103
anaerobic microbes, 249
anaerobic respiration, 250
anammox, 230–231, 253
Andisols, 103, 111, 172, 181–182
aniline blue, 9
ANIMO, 83
anoxia, 233
anoxic conditions, 228–229, 257
anoxic environments, 253
antibiotic-resistant bacteria, 238
antimicrobial compounds, 130
ants, 119
AOA, 259
AOB, 259
applied substrate, 186
appropriate application rates, 171
arable land, 89
arable soils, 180
arable systems, 74
arbuscular mycorrhizal fungi, 125
archaea, 142, 153, 229
arctic, 275–285, 287, 291–296
arctic wetlands, 236
Aristotle, 120
Armillaria luteobubalina, 131
Arrhenius function, 154
assimilation, 169, 182, 184–187
atmospheric CH_4, 181
atmospheric CH_4 oxidation, 180
atmospheric CH_4 oxidation capacity, 182
ATPases, 22
Australia, 76, 90, 131
autoclaving, 5–6, 8
autolysis, 104

B

bacteria, 2, 8, 12, 44–45, 51–52, 119, 127, 132, 142, 148, 150, 153, 158, 167, 173–175
Bacterial, 53, 175
bacterial cells, 78
bacterial energy channel, 125–127
bacterial membrane lipid C, 182
Basal respiration, 75
below-ground biomass, 177
beneficial microbes, 218
Biochar, 108
biochemical components, 184
biochemical functions, 79
biochemical pathways, 189
biochemical pools, 189
biochemical processes, 69, 77
biochemistry, 185
biodiversity, 76, 227
biodiversity–function relationships, 121, 124
biogeochemical cycles, 82, 141, 242
biogeochemical functioning, 228
biogeochemical processes, 70

biogeography, 230, 255
Bioinformatics, 79
biological invasions, 131–132
biological N fixation, 42
biomarkers, 167, 177, 183
biomass C, 176, 178
biomass C pools, 169
biomass nitrogen, 17, 29
biomass of CH_4-oxidising bacteria, 178
biomass phosphorous, 18
biomass protein, 189
biomass sulphur, 18
biomes, 71–72
biomolecular components, 182
biopolymers, 189
biosolids, 30
biosynthetic pathways, 185, 187
biosynthetic proximity, 185
bioturbation, 256
biovolume, 9
black box, 71, 76, 79, 239
body size, 129
boreal forest, 125
Boreal pine forest soil, 173
Broadbalk Continuous Wheat Experiment, 74
Bromus tectorum, 132
Bronydd-Mawr (Wales), 178–179
Brookes, Phil, 2, 16
bulk density, 181
bulk IRMS, 182
bulk N isotope, 186
bulk soil, 198
bulk soil C, 182
bulk soil C pool, 184
bulk stable isotope approaches, 169
Bulk $\delta^{15}N$ values, 188

C
^{12}C, 77–78
^{13}C, 77
^{13}C- and ^{15}N-SIP, 171
^{13}C incorporation, 180
^{13}C-PLFA SIP, 169
^{13}C, ^{15}N, 171
^{13}C-,^{15}N-litter, 175
^{13}C-acetate, 169
^{13}C-alanine and glutamic acid, 175
^{13}C-alkyl halides, 169
^{13}C-aromatics, 169
  PAHs, 169
  phenolic, 169
^{13}C-aspartic acid, 174
^{13}C-aspen litter, 175
^{13}C-barley residues, 176
^{13}C-butyrate, 169
^{13}C-Cu-N-cyclo-hexyldiazenium dioxide, 174
^{13}C-*E.coli*, 176
^{13}C-eucalyptus leaf litter, 176
^{13}C-Glu, 191
^{13}C-glucose, 174–175
$^{13}CH_3OH$, 169
$^{13}CH_4$, 169, 172, 177, 183–184, 191
$^{13}CH_4$ incubation, 190
$^{13}CH_4$ time-series ^{13}C-PLFA labelling, 177
^{13}C-label, 168, 183–184, 187, 198
^{13}C-labelled and unlabelled DNA, 169
^{13}C-labelled hopanoids, 177
^{13}C-labelled microbial cultures, 177
^{13}C-labelled plant litter, 177
^{13}C-labelled plants, 169
^{13}C-labelled PLFA, 178
^{13}C-labelled root exudate, 177

^{13}C-labelled substrate, 170–171, 177, 182
^{13}C-labelling, 169, 187
^{13}C-lignin and ^{13}C-cellulose, 175
^{13}C-methanotrophs, 176
^{13}C-mono- and polysaccharides, 169
^{13}C mobilisation, 175
$^{13}CO_2$, 169, 172–174
$^{13}CO_2$ pulse-chase experiments, 177
^{13}C or ^{15}N-labelled tracers, 168
^{13}C-phenanthrene, 174, 177
^{13}C-PLFA, 168–169, 177, 179
^{13}C-PLFA labelling, 181
^{13}C-PLFA SIP, 168–170
^{13}C-propionate, 169
^{13}C-pyruvate, 169
^{13}C-SIP, 169
^{13}C-SIP experiments, 171
^{14}C-labelled, 4–5, 9
δ^{13}C, 178, 185
δ^{13}C value determinations, 170
δ^{13}C values, 199
C_4 plant litters, 171
CH_4, 182
CH_4 cycling, 253
CH_4 emissions, 180–181, 233–234, 249
CH_4 flux, 181, 232
CH_4 oxidation, 180–181, 229, 232, 234
CH_4 oxidation capacity, 180
CH_4 oxidation in soils, 179
CH_4 oxidation rate, 180–181, 190
CH_4-oxidising bacteria, 251
CH_4 production, 232
CH_4 sink capacity, 178, 180
CH_4 sources and sinks, 181
CO_2 emission, 285, 294
CO_2 flux, 287, 293–294
CaPK, 179
C-to-N or lignin-to-N, 54
C to N ratio, 56, 60–61
    ratios of C:N:P, 48
C turnover, 174
C uptake, 182
cadmium, 30
calibration, 170
carbohydrates, 182, 189
carbon, 1
carbon dioxide (*see also* CO_2), 42, 229, 288
carbon dioxide exchange, 60
carbon mineralisation
    mineralisation of soil C, 58
carbon sequestration, 236
carbon sink, 228
carbon, soil, 146
carbon substrates, 168
C content, 181
C dynamics, 181, 183
Cell size, 153
cellular biochemical, 168, 185, 189
cellular biochemical products, 187
cellular biomass, 168
cellulose, 51–52
cheaters, 130
chemoautotrophs, 42
chemotaxis, 251
chemotaxonomic assessments, 178
China, 89
Chip-SIP, 79
chloroform, 2, 5–6
chloroform fumigation, 104–105
chloropicrin, 4
Cl-, 179–180
climate change, 60, 123, 235–236, 242, 275, 279–280, 285, 292, 294–295

elevated CO_2, 60–61
greenhouse gases, 61
Climate models, 231
climatic feedback, 233
clone libraries, 169
C:N ratio, 49
cold-shock proteins, 245
Collembola, 173
comammox, 230
community, 175
community composition, 88, 243, 245
community diversity, 73
community shifts, 242
compound-specific ^{15}N SIP, 187
compound-specific AA ^{15}N-SIP, 186–187
compound-specific approaches, 171
compound-specific stable C isotope analyses, 189
conservation agriculture, 88
Contaminated soil, 174
continuous incubation experiments, 171
control, 11, 13–14
controlled-release N fertiliser, 107
Conventionally tilled Luvisol, 176
copiotrophic, 243
copiotrophic bacteria, 127
copiotroph–oligotroph, 155
copper, 31
cropping systems, 88
crop production, 70
Cross-feeding, 178
cryoprotectants, 245
cryoturbation, 294
cultivation-independent technique, 168
cultured microbes, 169
culture-independent, 207–208, 210
culture-independent microbiological methods, 180
cyanobacteria, 31, 42

D

DAISY, 83
damage/change of vegetation, 231
data analysis protocols, 170
decay, 182–183
decay coefficient, 145–146
decision-making on-farm, 90
decomposition, 120, 143–144, 148, 155–156, 158
denitrification, 107, 230, 232, 259
denitrifiers, 107, 142, 238, 249
de novo AA biosynthesis, 188
de novo biosynthesis, 185
density, 168
derivatisation, 170, 187, 189
Desert soil, 173
desiccation, 247–248
desiccation-rewetting, 234
Desulfofrigus spp, 168
Desulfosporosinus, 255
Desulfotomaculum acetoxidans, 168
developing countries, 88
diazotrophs, 231
Differential utilisation of substrates, 172, 174
Differing microbial communities, 173
Dimethyldisulphide
geometries, 192
Dimethyl disulphide (DMDS) derivatives, 192
direct counting, 2–3
direct infusion IRMS, 189
discrimination, 186
dispersal ability, 127
dissimilation, 182

dissimilatory nitrate reduction to ammonia, 259
dissolved organic matter (DOM), 57
  dissolved organic C and N, 46
disturbance, 231, 239, 244, 255
diversity–function relationships, 124
DNA, 2, 76, 168, 173, 176, 182
DNA/RNA, 169, 187
DNA/RNA-SIP, 168–169
DNRA, 259
dormancy, 243
double-bond positions, 192
Douglas-fir forest soil, 172, 174
Drained boreal organic peat soil, 173
drought, 227, 233–234, 243, 247
dual ^{13}C, 169, 191
dynamic $^{13}CH_4$ ^{13}C-PLFA-SIP, 180
dynamic $^{13}CH_4$ SIP, 181, 190
dynamic ^{13}C-SIP, 169, 172
Dynamic Compound-Specific Stable Isotope Probing, 167
dynamic-SIP, 169, 171, 187, 190
  Dynamic $^{13}CH_4$ SIP, 177
  small-molecule dynamic-SIP, 169
dynamics of N cycling, 186

E

early warning, 35
Earth Microbiome Project, 76
Earth System Models (ESM), 242
earthworms, 45, 119–120, 131
ecological function, 87
ecological research, 170
ecosystem function, 69–71, 76, 81
ecosystem functioning, 67, 231, 256, 275, 279, 282
ecosystem models, 120
ecosystem processes, 68, 71, 80
ecosystem services, 142, 228, 235, 260
ectomycorrhizal, 54
electron acceptor, 249, 251
elemental analysis-isotope ratio mass spectrometry (EA-IRMS), 196
elemental cycles, 228
Elemental imbalances, 233
elevated atmospheric CO_2 conditions, 106
elevated CO_2, 62, 156, 173
emissions, 288
enchytraieids, 119
endophytic bacteria, 74
endosphere, 249
England, 76
enumeration of high-affinity methanotroph, 177
environmental drivers, 259
environmental risks, 112
environmental sensing, 251
environmental variables, 177–178
enzyme, 8, 50–53, 55–56
Escherichia coli, 132
ethanol, 11
Eukaryotes, 176, 182
Eutric cambisol, 172
eutrophication, 238, 256–257
evolution
  Experimental, 143
exophosphatases, 253
exopolysaccharides, 247
exotic plant species, 132
experimental design, 171
Experimental Methods, 170, 190
experimental warming, 281
Explicitly modelled microbial traits, 243

extracellular enzyme, 144–145, 148–149, 152, 154–155, 157
extracellular enzymes, 80
extraction, 187
Extraction and derivatisation of carbohydrates, 193
Extraction, isolation and derivatisation of hydrolysable amino acids, 193
Extreme weather events, 295
eye of the needle, 69

F

farming practices, 85
farmyard manure, 30, 180
fatty acids, 50, 168
fauna, 43, 45
faunal, 44
feedback, 232, 244, 260
fermentative metabolism, 252
fertiliser, 4, 180
fertiliser additions, 180
fertiliser-N, 185
fertiliser recommendation system, 84–85
fertiliser-use efficiency, 89
FI, 2, 8–11, 16
fingerprinting, 169
flooded water, 106
flooding, 227
flow-through chamber, 177, 180, 184
flow-through incubation chamber, 190
fluctuating redox environments, 251
flush of decomposition, 6, 9, 12, 14
fluxes, 185, 187
flux measurement, 180
food production, 85
food security, 88
food web, 43, 123, 126
foodweb-based models, 243
Fore-dune soil, 173
Forest soil, 175
formaldehyde, 4, 28
fractionation, 186
France, 72
free atmospheric, 60
free fatty acids, 191
free radical, 20
freeze–thaw, 276, 283–285, 291–293
freezing, 276–277, 282–283, 293–295
from ocean, 169
frost events, 282
fumigants, 4
fumigation, 4–6, 10, 28
fumigation–extraction (*see also* FE), 2, 15–16
fumigation–extraction method, 74, 82
fumigation–incubation, 2
fumigation–incubation method, 71
functional dissimilarity, 124
functional diversity, 86
functional genes, 79
functional redundancy, 239–241
functional traits, 124, 127
functioning, 279
fungal, 44, 175
fungal diseases, 28
fungal energy channel, 126
Fungal-to-bacterial PLFA-C, 173
fungi, 2, 8, 12, 45, 51–53, 119, 127, 142, 148, 150, 153, 158, 167, 173, 175
Fungal lignin, 53

G

gas chromatography-combustion-IRMS (*see also* GC-C-IRMS), 168, 170–171, 177–178, 182–183, 185–186, 189, 192, 195–196

GC, 178, 192, 194
GC columns, 170
GC-MS, 178, 192, 195
gene-based framework, 261
gene probing, 180
general salt effects, 179
genomics, 142, 261
geochemistry, 260
global C-cycling budgets, 80
global challenges, 88
global change, 141
global change scenarios, 248
Global Circulation Models (GCM), 242
Global environmental change, 260
global scale, 71
global warming, 291
globins, 251
glomalin, 126
glutamate, 174, 185–187
glycine, 185–187
glycolipids, 191
GM plant, 173
government legislation, 90
gradients, 228
gradient ultracentrifugation, 169
Gram-negative, 182
grassland, 132
grassland soil, 185
graze, 176
grazing, 235, 249, 260
greenhouse gas, 67, 69, 71, 79, 89, 111
greenhouse gas budgets, 228, 239
greenhouse gas emissions, 180
greenhouse gases, 68, 77, 110, 249
gross N immobilisation, 82
gross nitrification, 75, 82
gross N mineralisation, 75, 81–82

H

$^2H/^1H$, 189
habitable space, 73
haemocytometer, 3
halophiles, 248
Haplic Chernozem, 174
Hapludand, 181
Heat-shock proteins, 245
heavy metal contamination, 106
heavy metal presence, 245
heavy metals, 29–30, 237, 257
helophytes, 235
hemerythrins, 251
hemicellulose, 51
herbivores, 129
heterogeneity, 112
heterotrophic bacteria, 42
high-affinity methanotrophic bacteria, 179–181
high-affinity methanotrophic populations, 178
high-affinity methanotrophs, 180–181
highly enriched stable isotopically labelled substrates, 171
high-molecular-weight lipids, 189
high-performance liquid chromatography-IRMS, 189
high sensitivity, 170
high specificity, 178
high-temperature GC-C-IRMS, 189
high-throughput screening, 261
hopanoids, 191
horizontal gene transfer, 255
human gut, 169
Human-made changes, 235
Humid and semi-arid Cambisol and Haplic Kastanozem, 176
hydrogenotrophic methanogenesis, 246

hydrogenotrophic methanogens, 229, 251
hydrological cycles, 234
Hydrologic models, 233
hydrology, 231, 239
hydrolysable AA pool, 188
Hydrolysates, 194
hydrolysis, 189
hydroxylamine, 256
Hyphae, 9
hyphal networks, 127

I

identify, 178
imidazole, 23
immobilisation, 54, 56
incorporation, 183
incorporation of ^{13}C-label, 178
incorporation, turnover and decay, 172
Incubation experiments, 191
inhibition, 180
inorganic fertilisation, 172
inorganic fertilisers, 89, 180–181
inorganic fertiliser treatments, 181
inorganic N, 187
inorganic N-fertilisers, 182
Instrumental analyses, 194
instrumentation, 170
Integrated biogeochemistry, 262
interactions, 260
Intergovernmental Panel on Climate Change, 242
internal phosphorus loading, 257
International Rice Research Institute, 104
International Tundra Experiment, 280
invasive bamboo, 132
invasive plants, 131
invasive plant species, 123
invasive species, 235, 237, 249, 260
iron oxides, 249
iron oxidisers, 249
iron-oxidising bacteria, 251
irradiation, 5–6, 8, 10
isotope ratio mass spectrometry (IRMS), 168
isotopes, 77
isotope studies
    3C-labelled DOM, 57
    ^{13}C- or ^{15}N-labelled, 58
isotopically labelled substrates, 168, 171
Isotopic corrections, 199
isotopic tracer, 77, 79

J

Jena Biodiversity Experiment, 174
Jenkinson, David, 1–6, 22
Jones-Mollison, 3, 8–9
    agar film, 3, 9

K

K+, 179
K_2SO_4, 179
$K^{15}NO_3$, 191
kC, 6, 8, 10, 16
KCl, 179
kinetics, 175, 232
Konosu long-term paddy experiment, 105

L

labelled substrate, 183
laboratory and field experiments, 171
laboratory mesocosms, 185
landfill cap soil, 172, 184
land use, 242
land-use change, 32, 110, 228, 260

lateral gene transfer, 241
leaching, 107
lignin, 42, 49, 51–52, 54
Lipid extraction and fractionation, 193
lipids, 50, 182
liquid fertiliser, 110
litter, 57, 147–148
litter decomposition, 80
living organisms, 176
Loamy luvisol, 175
long-term, 105
long-term C allocation, 173
long-term inorganic fertiliser applications, 178
Long-Term Rice–Wheat Rotation, 108
low affinity methanotrophs, 181–182, 190
low concentration, 178
low-input fertilised soil, 179
low-input fertiliser inhibition, 181

M

macrofauna, 47
management practices, 86
manures, 75
Manuring, 180
Mass balance calculations, 182
mathematical modelling, 83
McGrath, Steve, 30
mechanisms, 175
mechanistic models, 260
mesophyles, 245
meta-analysis, 74
metabolic activity, 79
metabolic function, 168
metabolic pathways, 79
metabolic quotients, 112
metabolites, 184
metabolomics, 79, 157, 261
metagenomics, 169, 215
metagonomic approaches, 125
metatranscriptomics, 216, 219
methane, 110, 142, 229
methane fluxes, 232
methane monooxygenase, 256
Methane oxidising bacteria, 256
methane production, 103
methanogenesis, 229, 244
Methanogenic, 110
methanogens, 107, 246, 249, 251, 261
Methanoregula, 254
Methanosaeta, 254
methanotrophic bacteria, 181–182, 254
methanotrophic biomass, 180
methanotrophic biomass determinations, 178
methanotrophic communities, 177
methanotrophic population, 177–178, 181
methanotrophs, 107, 177, 182–183, 231, 255, 261
    methanotroph population characteristics, 172
Methanotrophs affected by N fertilisation, 172
methanotrophy, 183
methyl bromide, 4–5, 28
Michaelis-Menten kinetics, 144, 148–149, 152, 159
micorrhizae
    mycorrhizal, 57
microarrays, 169
microarthropods, 119
microbial, 175
microbial activity, 238
microbial adaptations, 253

microbial biochemical assimilation pathways, 186
microbial biomass, 175, 178, 180
microbial biomass K, 107
microbial biomass N extraction, 187
Microbial cells, 176
microbial community, 105, 107, 168–169, 184, 247
microbial community composition, 239
microbial community structure, 75, 81
microbial diversity, 239
microbial ecology, 240, 260
microbial-enzyme, 154
microbial-explicit, 145–147, 154, 156–160
microbial functioning, 239
microbial genomes, 241
microbial growth efficiency, 147, 151–152, 154, 157
microbial guilds, 246
microbial phylogeny, 79
microbial pools, 178
microbial population, 180, 183
microbial population structure and function, 182
microbial processing, 186
Microbial residues, 154
microbial trait, 240, 243, 250, 258
microbial turnover, 159
microbiome, 142, 229
microhabitats, 76
microorganisms, 53, 55
mineralisation, 54, 57–58, 185
mineralisation-immobilisation turnover (MIT), 56, 62, 82
    gross N-mineralisation, 54
    immobilises, 49
    mineralises, 49
mineral soils, 181
models, 83
    agent-based, 156–157
    biogeochemical, 155
    cellular automaton, 153
    CENTURY, 143
    CORPSE, 156
    Earth system, 145, 160
    emulation, 157
    first-order, 145
    linear, 144
    mathematical, 142
    microbial-explicit, 144, 152
    MIMICS, 154
    RothC, 143
    trait-based, 160
    vegetation, 147
model, biogeochemical, 157
model, emulation, 158
model, first-order, 146, 154
model, Guild Decomposition, 147
modelling frameworks, 261
model, microbial-explicit, 153
Model predictions, 243–244
model, trait-based, 148, 155
moisture, 112
molecular-based SOM cycling models, 189
molecular characterisation, 169
molecular level, 183, 185
molecular techniques, 76
moles, 119
monosaccharides, 191, 193
most probable number techniques, 180
Mountain meadow dystric cambisol, 174
multi-functionality, 125
multi-trophic interactions, 121

mycorrhizae, 54
mycorrhizal, 55, 174
mycorrhizal fungi, 68, 74, 120, 129
mycorrhizal networks, 130

N

$^{15}NO_3^-$, 192
^{14}N, 77–78
^{15}N, 77–78, 82, 169, 191
^{15}N,^{13}C-labelled glutamic acid, 191
$^{15}N/^{14}N$, 189, 197
^{15}N AA SIP, 188
^{15}N-DNA/RNA SIP, 169
^{15}N-Glu, 192
$^{15}NH_4^+$, 187, 192
^{15}N isotope pool dilution technique, 82
^{15}N isotopic pool dilution, 82
^{15}N-labelled ammonium chloride ($^{15}NH_4Cl$), 191
^{15}N-labelled fertilisers, 188
^{15}N-labelled potassium nitrate, 191
^{15}N-labelled substrates, 186–187
^{15}N-labelled substrates/amendments, 186
$^{15}NO_3^-$, 187
^{15}N-SIP, 169
^{15}N-SIP incubation experiments, 184
^{15}N-SIP of AA, 196
^{15}N tracer, 104
$(NH_4)2SO_4$, 179
$\delta^{15}N$, 185
$\delta^{15}N$ determinations, 170
$\delta^{15}N$ value(s), 187, 196–198
N_2O, 110–111, 232, 243
N_2O emissions, 82
Na_2SO_4, 179
N-acetyl, *O*-isopropyl derivatives, 194
N additions, 173
n-alkanes, 191
n-alkanols, 191
N and C cycles, 185
nanoscale secondary ion mass spectrometry (*see also* nanoSIMS), 77–79, 240
native grasses, 110
natural abundance-labelled substrates, 171
NCaPK, 179
N-containing substrates, 186
N cycle, 81, 169
N cycling, 57, 187–188
nematodes, 45, 47, 76, 119
net ecosystem production, 44
NEP, 46
net ecosystem productivity
NPP, 43–44, 46, 48, 62
net primary productivity, 41, 71, 81
NPP, 49, 54, 57–61
net secondary production, 44
NSP, 48–49, 53–54, 57, 61–62
newly synthesised soil protein, 197
New Zealand, 131
next-generation sequencing, 240
N fertiliser, 89, 180, 187
NH_4^+, 179–180
NH_4Cl, 179
NH_4^+ oxidation, 179
nifH, 253
Nil-graze soils, 179
ninhydrin, 17
nirK gene, 250
nirS gene, 250
nitrate ammonifiers, 230
nitrification, 56, 230, 243, 256, 259
nitrifiers, 142, 261
nitrifying bacteria, 180
nitrogen, 3, 238, 257

nitrogenase enzyme, 253
nitrogen cycle, 259
nitrogen deposition, 237
nitrogen enrichment, 123
nitrogen fixation, 231, 244, 253, 258
nitrogen-fixation genes, 253
nitrogen isotope pool dilution, 54
  ^{15}N-labelled N, 58
  ^{15}N pool dilution, 56
nitrogen mineralisation
  ^{15}N-labelling, 55
  gross N mineralisation, 56
  mineralisable N, 53
  N and P mineralisation, 54
  N mineralisation, 54, 56–57, 61
nitrogen (N) mineralisation, 42, 50, 54–55
nitrogen-specific inhibitory mechanism, 179
nitrosative stress, 251, 254
Nitrosomonas europaea, 6
nitrous oxide, 229
NO_3^-, 179
non-living SOM, 176
non-methanotrophic consumers, 182
non-methanotrophs, 180
NPK, 175
N turnover time, 106
nucleic acids, 169
nutrient availability, 239
nutrient cycling, 41–42, 48, 53–54, 58, 60–61, 67, 77, 82, 86
  N cycling, 55, 62
  N deposition, 61
  nutrient, 53
  nutrients recycled, 59
nutrient load, 235
nutrient loading, 237
nutrient starvation, 253
nutrient stress, 253
nutrient supply, 81

O

16:1ω5, 178
18:1ω7c, 178
$^{18}O/^{16}O$, 189
Odcombe (Somerset, UK), 182
oil, 237
oligotrophic, 243
on-farm use, 87
open-top chambers, 281
optimising experimental designs, 171
  changes in microbial populations with time, 171
  cross-feeding between microbes, 171
  equilibrium conditions, 171
  estimates for nutrient fluxes and pool sizes, 171
  optimal incubation times, 171
organic acids, 90
organic amendments, 89
organic farming, 88
organic matter, 229
organic N, 187
organic N cycling, 184
osmoprotectant, 248
osmotic shock, 245
osmotic stress, 248
oven-drying, 5
Overall plant residue, 176
oxalate, 174
oxic–anoxic interfaces, 228–229
oxic/anoxic zones, 254
oxygen, 230, 243
oxygen, uptake, 4, 8, 10

P

Paddy fields, 103
paddy soil ecosystems, 112
Paddy soil (Hydragric Anthrosol), 173
paleosol, 111
paraquat, 22–23
parental cultivar, 173
partial sterilisation, 1, 6
participatory action research, 90
particulate organic C, 59–60
pasture, 110
pathogens, 130–131, 249
Paul, Eldor, 13, 22
pectins, 51
peptides, 55
perfusion, 21
permafrost, 232, 276, 278, 288–291, 294
Permanent grassland, 175
perturbing the activities of microbial biomass, 171
pesticides, 89, 237
P-eutrophication, 233
P fertilisers, 90
pH, 72, 74–75, 112
pharmaceuticals, 237
phenol, 174
phosgene, 11
phosphate rock, 89
phospholipid fatty acids, 107
phospholipids, 50, 168
phosphorus, 42, 238, 257
    P, 46, 48, 52, 54, 59
photoautotrophs, 42
Photosynthate, 173
photosynthesis, 42–43
    photosynthetic, 49
phototrophic, 107
phylogenetic guilds, 243
phylogenetic signal, 241
physicochemical properties, 72
Phytophthora ramorum, 131
phytoplankton, 48, 148, 155, 158
plant community dynamics, 129
plant defence, 124, 131
plant disease suppression, 130
plant diversity, 129–130
plant health, 130
plant inputs, 174
Plant litter, 175
plant–microbe interactions, 208–210
plant production, 82
plant productivity, 71
plate counts, 2
    plate counting, 2
PLFA(s), 169, 172–176, 182, 184, 189, 191–192, 199
PLFA analysis, 192
PLFAs detected, 177
PLFA-SIP, 168, 187
P-limited, 254
pollutants, 235, 255
pollution, 231, 237, 242, 260
polyhydroxybutyrate, 254
polymerase chain reaction amplification, 169
polyphosphates, 254, 257–258
polysaccharides, 44, 51, 53
pore connectivity, 72
Pore size, 80
Position-specific incorporation of ^{13}C atoms, 175
positive bacteria, 182
Powlson, David, 1–2, 6
predictions, 72, 243
Predictive tools, 260
primary consumers (methanotrophs), 184
primary metabolic processes, 185

priming, 156
priming effect, 42, 58, 61–62
propionate, 168
proteases, 56
protein pool, 186
proteins, 49, 55, 182, 189
proteomics, 261
protozoa, 45, 47, 167, 176, 255
Prunus serotina, 132
Pseudomonas, 12
Psychrophyles, 244
Psychrotolerant, 245
pulse-chase, 171
Pythium cinnamomi, 131

Q

Q10, 145, 232, 291, 293, 295
quantify individual PLFAs, 178
Quantitative assessments, 178
quantitative estimates, 187
quantitative insights, 186

R

radiative forcing, 236
r and K selection, 47
Rape oil, 174
rate constants, 185
rates, 187
rates of uptake, 189
rate-yield trade-off, 154
reaction amplification, 169
reactive oxygen species, 247
reallocation of C, 182
recycling, 111, 178, 235
Redfield Ratio, 48
redistribution of ^{13}C-label in C pools other, 182
redox conditions, 248
redox potential, 60
redox reactions, 235
redox sensitive enzymes, 251
redox stress, 250
redundancy, 254
Regulatory Gate, 18, 21
Remnant native woodland soil, 176
resilience, 128, 234, 248
resistance, 248
resistance and resilience, 126
resource stoichiometry, 80
respiration
  basal, 148, 150, 152
  heterotrophic, 156–157
  metabolic, 152
  soil, 143, 145, 158
respiration, metabolic, 153
respiration, microbial, 154
respiration, soil, 4
response, 154
response to temperature, 176
reverse methanogenesis, 253
rhizobacteria, 78
Rhizobium, 31
rhizoplane, 74, 249
rhizosphere, 53, 55, 58, 74, 130, 132, 156, 187, 230, 249
rhizosphere microbiome, 131
ribosomal copy number, 247
rice, 103, 106, 110
Rice field/forest soils, 172
rice paddies, 245
rice roots, 249, 251
rice soils, 251
rising CO_2 concentrations, 80
RNA, 76, 168, 173–174, 176, 182
rock phosphate, 90
root-associated pathogens, 129
root exudates, 81, 84, 130
root-feeding fauna, 120

root herbivores, 131
roots, 235, 249
ROS-scavenging molecules, 251
Rothamsted, 1–3, 180
Rothamsted Classical Experiment, 172
Rothamsted Experimental Station, 105, 111
RothC model, 34
r-strategists, 256

S

salination, 247
saline soils, 110
salinisation, 234
salinity, 234
Sand–clay growth medium, 175
Sandy grassland soil, 174
Sandy loam agricultural soil, 173
Sandy peat agricultural soil, 173
saprotrophic fungi, 174
secondary consumers, 184
seed bank community, 247
selectivity, 170, 189
selectivity GC-C-IRMS, 187
semi-arid regions, 89
sensitivity, 170, 180, 187
sewage sludge, 30
siderophores, 90
signal transduction, 251
sink for tropospheric CH_4, 180
sinks of greenhouse gases, 235
SIS CH_4 incubation, 190
SIS experiment, 183
SIS technique, 187
small-molecule SIP, 170
snow addition, 280, 284
snow fences, 285, 287
snow manipulation, 285–287, 294
SO_42^-, 179–180
soil
    carbon, 142, 144, 147, 154, 156, 158–159
    functioning, 141
    moisture, 141, 145
    organic matter, 141, 144–145, 147, 154, 156, 158
    processes, 141
    state factor, 156
    texture, 144–145
    warming, 147
soil aggregates, 156
soil aggregate stability, 125
soil animals, 126
soil biodiversity, 54, 119–126, 129–131, 133
soil biomass dynamic, 169
soil biota, 69, 71, 189
soil-borne diseases, 76
soil-borne pathogens, 130–131
soil-borne pests, 28
soil C, 45, 57, 59, 61
soil fertility, 70, 90
soil food web, 124, 126–127, 132, 176, 182
soil formation, 125
soil habitat, 42, 69–70
    Habitat, 44–45
soil health, 90
soil matrix, 78
soil methanotrophs, 167
soil microbes, 177
soil microbial biomass, 1, 5, 41, 43–46, 49, 51–54, 56, 61, 120, 167, 187, 190
    microbial biomass, 43, 45–49, 51, 54–56, 58–59, 62
    soil biomass, 47–48, 53
    soil microbial, 47

soil microbial community, 177
soil microbial ecology, 207–208, 220
soil microbiomes, 208–210, 218, 220
soil N, 45
soil N supply, 84
soil organic C allocation, 182
soil organic C (SOC), 111
soil organic matter, 3, 44, 234
    organic matter, 42, 50
    SOM, 43, 45, 48–49, 54, 57–58
soil pH, 71, 73
soil physical properties, 125
soil–plant interface, 77
soil protein, 184, 187
soil protein pool, 187–188
soil quality, 68, 86–87, 108
soil/root respiration, 112
soil sealing, 123
soil SIP, 169
soil, state factor, 160
soil texture, 71–72
soil water availability, 282
soil water potential, 277
Soil–wood–sawdust mixture, 174
solarisation, 246
soluble organic C, 110
SOM, 46, 57
SPANC balance, 241, 254
spatial distribution, 111
spatial variability, 81
species concept, 241
specific compounds, 171
spores, 27
stabilise soil organic matter, 80
stability, 126
stable isotope label, 168, 182, 189
stable isotope probing (*see also* SIP), 167, 189, 240
stable isotope switching (*see also* SIS), 177, 182, 189, 198
stable isotopically labelled substrates, 167–168, 170
standardisation, 170
starch, 50–51
starvation, 254
Statistical analyses, 198–199
Stem injection of substrates, 174
steppe soil, 110
sterols, 191
stoichiometry, 153
strategies, 234
straw, 34
straw incorporation, 33
stress, 233
stressors, 257
submerse macrophytes, 235
sub-oxic, 251
substrate-induced respiration, 275, 278
substrates, 169
substrate uptake, 152–153
sudden oak death, 131
sugars, 50, 183
sulphur
    S, 46
sulphate reducers, 231, 249
sulphate-reducing bacteria (SRB), 168
sulphate reduction, 246
summer warming, 279–280, 282
SUNDIAL, 83–85
sustainable agricultural systems, 68
symbiosis, 57
symbiotic, 53
symbiotic mutualists, 131
synthetic air, 177, 190
synthetic fertilisers, 67

T
taxonomic, 169
taxonomic assessments, 177
taxonomic fingerprint, 168
taxonomic groups, 187
taxonomic tools, 187
taxonomy, 241
temperature, 244
Temperature change, 231
temperature programmes, 170
temperature sensitivity, 151, 157
temporal and quantitative insights, 177
tertiary consumers, 184
Thaw, 232, 282–285, 287–291, 294
theory
game, 143
social evolution, 143
tillage, 178
tilled soils, 180
time-course, 187
time-course experiments, 171
time-course incubation experiments, 187
tipping point, 238
TIP reagent, 23
toxic intermediates, 256
trade-offs, 241, 254, 258
trait(s), 230, 240, 243, 246, 251
community-averaged, 157
databases, 158
effect, 152
functional, 147
life-history, 153
response, 152
trait-based, 148, 158, 240, 244, 261
Trait-based approaches, 260
Trait combinations, 241–242, 246, 254
trait, life-history, 155
trait packages, 254–255, 261
transcriptomics, 261
transfer coefficient, 154
transmission electron microscopy, 78
Tree-rhizosphere, 174
trichloroacetic acid (*see also* TCA), 22–23, 27
TCA, 22
trichloro(nitro)methane, 4
trigger molecules, 27–28
Trimethylsilylation, 193
trophic interactions, 169
tropical forest soils, 112
tundra, 275, 279–285, 287, 294, 296
turnover, 111, 182–183, 189
Enzyme, 151
microbial, 154
turnover and fate of OM, 176
turnover, microbial, 151, 153
turnover rates, 169
turnover, soil carbon, 156
turnover times, 74, 75
type II methanotrophs, 180
type I methanotrophs, 180

U
U-^{13}C,^{15}N-glutamate, 185–186
U-^{13}C,^{15}N-glycine, 185–186
UK, 74, 81, 87
UK soil CH_4 sink, 181
Ultic Fulvudand soils, 181
ultrasonic dispersion, 22
unculturable high-affinity methanotrophic bacteria, 177
unculturable soil microorganisms, 167
Upland grassland, 172–173
urbanisation, 236
USA, 87

V
Vance, Eric, 16
Various cellular biochemicals, 172
volatile organic compounds, 130
volcanic, 103
volcanic ash, 181
Volcanic soils, 111, 172
volcano, 110
Voroney, Paul, 13

W
Waksman, Selman, 2, 141–142, 147
Wales, 76
warming of wetlands, 233
water availability, 277, 283, 293, 295
water-holding capacity, 181
water-limited yield potential, 76
waterlogged, 105
waterlogged soil, 103
water status, 173
wetland biogeochemical cycling, 239
wetland biogeochemistry, 233, 237
wetland ecosystem functioning, 259
wetland ecosystems, 260
wetland functioning, 239
Wetland microbes, 229, 247
wetland microbiomes, 248
wetland plant roots, 253
wetlands, 103, 227–228, 231, 234, 249
wetland vegetation, 234, 237, 248
wheat rhizosphere, 78
Wheat yield, 75
Woburn Grassland and Broadbalk soils, 175

Y
yeasts, 8

Z
zinc, 31